Graduate Texts in Mathematics 5

Springer
New York
Berlin
Heidelberg
Barcelona
Hong Kong
London
Milan
Paris
Singapore
Tokyo

Graduate Texts in Mathematics

(continued after index)

Saunders Mac Lane

Categories for the Working Mathematician

Second Edition

Springer

Saunders Mac Lane
Professor Emeritus
Department of Mathematics
University of Chicago
Chicago, IL 60637-1514
USA

Mathematics Subject Classification (2000): 18-01

Library of Congress Cataloging-in-Publication Data
Mac Lane, Saunders, 1909–
Categories for the working mathematician/Saunders Mac Lane. — 2nd ed.
p. cm. — (Graduate texts in mathematics; 5)
Includes bibliographical references and index.
ISBN 978-1-4419-3123-8
1. Categories (Mathematics). I. Title. II. Series.
QA169.M33 1998
512′.55—dc21 97-45229

Printed on acid-free paper.

Printed in the United States of America.

9 8 7 6 5 4 3 2

Springer-Verlag New York Berlin Heidelberg
A member of BertelsmannSpringer Science+Business Media GmbH

Preface to the Second Edition

This second edition of "Categories Work" adds two new chapters on topics of active interest. One is on symmetric monoidal categories and braided monoidal categories and the coherence theorems for them—items of interest in their own right and also in view of their use in string theory in quantum field theory. The second new chapter describes 2-categories and the higher-dimensional categories that have recently come into prominence. In addition, the bibliography has been expanded to cover some of the many other recent advances concerning categories.

The earlier 10 chapters have been lightly revised, clarifying a number of points, in many cases due to helpful suggestions from George Janelidze. In Chapter III, I have added a description of the colimits of representable functors, while Chapter IV now includes a brief description of characteristic functions of subsets and of the elementary topoi.

Dune Acres, March 27, 1997 — Saunders Mac Lane

Preface to the First Edition

Category theory has developed rapidly. This book aims to present those ideas and methods that can now be effectively used by mathematicians working in a variety of other fields of mathematical research. This occurs at several levels. On the first level, categories provide a convenient conceptual language, based on the notions of category, functor, natural transformation, contravariance, and functor category. These notions are presented, with appropriate examples, in Chapters I and II. Next comes the fundamental idea of an adjoint pair of functors. This appears in many substantially equivalent forms: that of universal construction, that of direct and inverse limit, and that of pairs of functors with a natural isomorphism between corresponding sets of arrows. All of these forms, with their interrelations, are examined in Chapters III to V. The slogan is "Adjoint functors arise everywhere."

Alternatively, the fundamental notion of category theory is that of a monoid—a set with a binary operation of multiplication that is associative and that has a unit; a category itself can be regarded as a sort of generalized monoid. Chapters VI and VII explore this notion and its generalizations. Its close connection to pairs of adjoint functors illuminates the ideas of universal algebra and culminates in Beck's theorem characterizing categories of algebras; on the other hand, categories with a monoidal structure (given by a tensor product) lead inter alia to the study of more convenient categories of topological spaces.

Since a category consists of arrows, our subject could also be described as learning how to live without elements, using arrows instead. This line of thought, present from the start, comes to a focus in Chapter VIII, which covers the elementary theory of abelian categories and the means to prove all of the diagram lemmas without ever chasing an element around a diagram.

Finally, the basic notions of category theory are assembled in the last two chapters: more exigent properties of limits, especially of filtered limits; a calculus of "ends"; and the notion of Kan extensions. This is the deeper form of the basic constructions of adjoints. We end with the observations that all concepts of category theory are Kan extensions (§7 of Chapter X).

I have had many opportunities to lecture on the materials of these chapters: at Chicago; at Boulder, in a series of colloquium lectures to the American Mathematical Society; at St. Andrews, thanks to the Edinburgh Mathematical Society; at Zurich, thanks to Beno Eckmann and the Forschungsinstitut für Mathematik; at London, thanks to A. Fröhlich and Kings and Queens Colleges; at Heidelberg, thanks to H. Seifert and Albrecht Dold; at Canberra, thanks to Neumann, Neumann, and a Fulbright grant; at Bowdoin, thanks to Dan Christie and the National Science Foundation; at Tulane, thanks to Paul Mostert and the Ford Foundation; and again at Chicago, thanks ultimately to Robert Maynard Hutchins and Marshall Harvey Stone.

Many colleagues have helped my studies. I have profited much from a succession of visitors to Chicago (made possible by effective support from the Air Force Office of Scientific Research, the Office of Naval Research, and the National Science Foundation): M. André, J. Bénabou, E. Dubuc, F.W. Lawvere, and F.E.J. Linton. I have had good counsel from Michael Barr, John Gray, Myles Tierney, and Fritz Ulmer, and sage advice from Brian Abrahamson, Ronald Brown, W.H. Cockcroft, and Paul Halmos. Daniel Feigin and Geoffrey Phillips both managed to bring some of my lectures into effective written form. My old friend, A.H. Clifford, and others at Tulane were of great assistance. John MacDonald and Ross Street gave pertinent advice on several chapters; Spencer Dickson, S.A. Huq, and Miguel La Plaza gave a critical reading of other material. Peter May's trenchant advice vitally improved the emphasis and arrangement, and Max Kelly's eagle eye caught many soft spots in the final manuscript. I am grateful to Dorothy Mac Lane and Tere Shuman for typing, to Dorothy Mac Lane for preparing the index, and to M.K. Kwong for careful proofreading—but the errors that remain, and the choice of emphasis and arrangement, are mine.

Dune Acres, March 27, 1971 Saunders Mac Lane

Contents

Introduction

Category theory starts with the observation that many properties of mathematical systems can be unified and simplified by a presentation with diagrams of arrows. Each arrow $f: X \to Y$ represents a function; that is, a set X, a set Y, and a rule $x \mapsto fx$ which assigns to each element $x \in X$ an element $fx \in Y$; whenever possible we write fx and not $f(x)$, omitting unnecessary parentheses. A typical diagram of sets and functions is

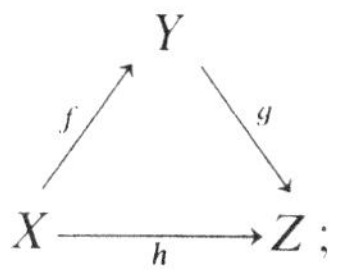

it is commutative when h is $h = g \circ f$, where $g \circ f$ is the usual composite function $g \circ f: X \to Z$, defined by $x \mapsto g(fx)$. The same diagrams apply in other mathematical contexts; thus in the "category" of all topological spaces, the letters X, Y, and Z represent topological spaces while f, g, and h stand for continuous maps. Again, in the "category" of all groups, X, Y, and Z stand for groups, f, g, and h for homomorphisms.

Many properties of mathematical constructions may be represented by universal properties of diagrams. Consider the cartesian product $X \times Y$ of two sets, consisting as usual of all ordered pairs $\langle x, y \rangle$ of elements $x \in X$ and $y \in Y$. The projections $\langle x, y \rangle \mapsto x$, $\langle x, y \rangle \mapsto y$ of the product on its "axes" X and Y are functions $p: X \times Y \to X$, $q: X \times Y \to Y$. Any function $h: W \to X \times Y$ from a third set W is uniquely determined by its composites $p \circ h$ and $q \circ h$. Conversely, given W and two functions f and g as in the diagram below, there is a unique function h which makes the diagram commute; namely, $hw = \langle fw, gw \rangle$ for each w in W:

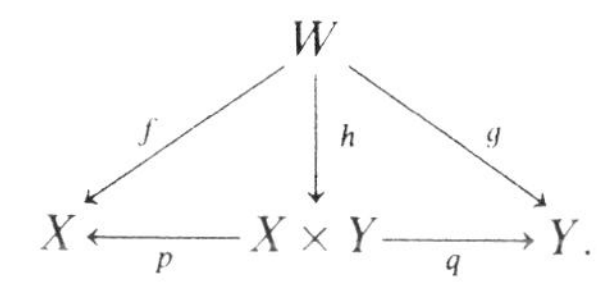

Thus, given X and Y, $\langle p, q\rangle$ is "universal" among pairs of functions from some set to X and Y, because any other such pair $\langle f, g\rangle$ factors uniquely (via h) through the pair $\langle p, q\rangle$. This property describes the cartesian product $X \times Y$ uniquely (up to a bijection); the same diagram, read in the category of topological spaces or of groups, describes uniquely the cartesian product of spaces or the direct product of groups.

Adjointness is another expression for these universal properties. If we write $\hom(W, X)$ for the set of all functions $f: W \to X$ and $\hom(\langle U, V\rangle, \langle X, Y\rangle)$ for the set of all pairs of functions $f: U \to X$, $g: V \to Y$, the correspondence $h \mapsto \langle ph, qh\rangle = \langle f, g\rangle$ indicated in the diagram above is a bijection

$$\hom(W, X \times Y) \cong \hom(\langle W, W\rangle, \langle X, Y\rangle).$$

This bijection is "natural" in the sense (to be made more precise later) that it is defined in "the same way" for all sets W and for all pairs of sets $\langle X, Y\rangle$ (and it is likewise "natural" when interpreted for topological spaces or for groups). This natural bijection involves two constructions on sets: The construction $W \mapsto W, W$ which sends each set to the diagonal pair $\Delta W = \langle W, W\rangle$, and the construction $\langle X, Y\rangle \mapsto X \times Y$ which sends each pair of sets to its cartesian product. Given the bijection above, we say that the construction $X \times Y$ is a *right adjoint* to the construction Δ, and that Δ is left adjoint to the product. Adjoints, as we shall see, occur throughout mathematics.

The construction "cartesian product" is called a "functor" because it applies suitably to sets *and* to the functions between them; two functions $k: X \to X'$ and $t: Y \to Y'$ have a function $k \times t$ as their cartesian product:

$$k \times t: X \times Y \to X' \times Y', \qquad \langle x, y\rangle \mapsto \langle kx, ty\rangle.$$

Observe also that the one-point set $1 = \{0\}$ serves as an identity under the operation "cartesian product", in view of the bijections

$$1 \times X \xrightarrow{\lambda} X \xleftarrow{\varrho} X \times 1 \tag{1}$$

given by $\lambda\langle 0, x\rangle = x$, $\varrho\langle x, 0\rangle = x$.

The notion of a monoid (a semigroup with identity) plays a central role in category theory. A monoid M may be described as a set M together with two functions

$$\mu: M \times M \to M, \qquad \eta: 1 \to M \tag{2}$$

such that the following two diagrams in μ and η commute:

$$\begin{array}{ccc} M \times M \times M & \xrightarrow{1 \times \mu} & M \times M \\ \downarrow{\scriptstyle \mu \times 1} & & \downarrow{\scriptstyle \mu} \\ M \times M & \xrightarrow{\mu} & M, \end{array} \qquad \begin{array}{ccccc} 1 \times M & \xrightarrow{\eta \times 1} & M \times M & \xleftarrow{1 \times \eta} & M \times 1 \\ \downarrow{\scriptstyle \lambda} & & \downarrow{\scriptstyle \mu} & & \downarrow{\scriptstyle \varrho} \\ M & = & M & = & M; \end{array} \tag{3}$$

here 1 in $1 \times \mu$ is the identity function $M \to M$, and 1 in $1 \times M$ is the one-point set $1 = \{0\}$, while λ and ϱ are the bijections of (1) above. To say that these diagrams commute means that the following composites are equal:

$$\mu \circ (1 \times \mu) = \mu \circ (\mu \times 1)\,, \quad \mu \circ (\eta \times 1) = \lambda\,, \quad \mu \circ (1 \times \eta) = \varrho\,.$$

These diagrams may be rewritten with elements, writing the function μ (say) as a product $\mu(x, y) = xy$ for $x, y \in M$ and replacing the function η on the one-point set $1 = \{0\}$ by its (only) value, an element $\eta(0) = u \in M$. The diagrams above then become

$$\begin{array}{ccc} \langle x, y, z\rangle & \longmapsto & \langle x, yz\rangle \\ \downarrow & & \downarrow \\ \langle xy, z\rangle & \longmapsto & (xy)z = x(yz)\,, \end{array} \qquad \begin{array}{ccc} \langle 0, x\rangle & \longmapsto & \langle u, x\rangle \\ \downarrow & & \downarrow \\ x & = & ux\,, \end{array} \qquad \begin{array}{ccc} \langle x, u\rangle & \longleftarrow & \langle x, 0\rangle \\ \downarrow & & \downarrow \\ xu & = & x\,. \end{array}$$

They are exactly the familiar axioms on a monoid, that the multiplication be associative and have an element u as left and right identity. This indicates, conversely, how algebraic identities may be expressed by commutative diagrams. The same process applies to other identities; for example, one may describe a group as a monoid M equipped with a function $\zeta : M \to M$ (of course, the function $x \mapsto x^{-1}$) such that the following diagram commutes:

$$\begin{array}{ccccc} M & \xrightarrow{\delta} & M \times M & \xrightarrow{1 \times \zeta} & M \times M \\ \downarrow & & & & \downarrow{\scriptstyle \mu} \\ 1 & \longrightarrow & & & M \end{array} \qquad \begin{array}{ccccc} x & \longmapsto & \langle x, x\rangle & \longmapsto & \langle x, x^{-1}\rangle \\ \downarrow & & & & \downarrow \\ 0 & \longmapsto & u & = & xx^{-1}\,, \end{array} \tag{4}$$

here $\delta : M \to M \times M$ is the diagonal function $x \mapsto \langle x, x\rangle$ for $x \in M$, while the unnamed vertical arrow $M \to 1 = \{0\}$ is the evident (and unique) function from M to the one-point set. As indicated just to the right, this diagram does state that ζ assigns to each element $x \in M$ an element x^{-1} which is a right inverse to x.

This definition of a group by arrows μ, η, and ζ in such commutative diagrams makes no explicit mention of group elements, so applies to other circumstances. If the letter M stands for a topological space (not just a set) and the arrows are continuous maps (not just functions), then the conditions (3) and (4) define a topological group – for they specify that M is a topological space with a binary operation μ of multiplication which is continuous (simultaneously in its arguments) and which has a continuous right inverse, all satisfying the usual group axioms. Again, if the letter M stands for a differentiable manifold (of class C^∞) while 1 is the one-point manifold and the arrows μ, η, and ζ are smooth mappings of manifolds, then the diagrams (3) and (4) become the definition of a Lie group. Thus groups, topological groups, and Lie groups can all be described as "diagrammatic" groups in the respective categories of sets, of topological spaces, and of differentiable manifolds.

This definition of a group in a category depended (for the inverse in (4)) on the diagonal map $\delta: M \to M \times M$ to the cartesian square $M \times M$. The definition of a monoid is more general, because the cartesian product $\times$ in $M \times M$ may be replaced by any other operation $\Box$ on two objects which is associative and which has a unit 1 in the sense prescribed by the isomorphisms (1). We can then speak of a monoid in the system $(C, \Box, 1)$, where C is the category, $\Box$ is such an operation, and 1 is its unit. Consider, for example, a monoid M in (**Ab**, $\otimes$, **Z**), where **Ab** is the category of abelian groups, $\times$ is replaced by the usual tensor product of abelian groups, and 1 is replaced by **Z**, the usual additive group of integers; then (1) is replaced by the familiar isomorphism

$$\mathbf{Z} \otimes X \cong X \cong X \otimes \mathbf{Z}, \qquad X \text{ an abelian group}.$$

Then a monoid M in (**Ab**, $\otimes$, **Z**) is, we claim, simply a ring. For the given morphism $\mu: M \otimes M \to M$ is, by the definition of $\otimes$, just a function $M \times M \to M$, call it multiplication, which is bilinear; i.e., distributive over addition on the left and on the right, while the morphism $\eta: \mathbf{Z} \to M$ of abelian groups is completely determined by picking out one element of M; namely, the image u of the generator 1 of **Z**. The commutative diagrams (3) now assert that the multiplication μ in the abelian group M is associative and has u as left and right unit – in other words, that M is indeed a ring (with identity = unit).

The (homo)-morphisms of an algebraic system can also be described by diagrams. If $\langle M, \mu, \eta \rangle$ and $\langle M', \mu', \eta' \rangle$ are two monoids, each described by diagrams as above, then a morphism of the first to the second may be defined as a function $f: M \to M'$ such that the following diagrams commute:

$$\begin{array}{ccccccc} M & & M \times M & \xrightarrow{\mu} & M & 1 \xrightarrow{\eta} & M \\ \downarrow f & & \downarrow f \times f & & \downarrow f & \| & \downarrow f \\ M', & & M' \times M' & \xrightarrow{\mu'} & M', & 1 \xrightarrow{\eta'} & M'. \end{array} \qquad (5)$$

In terms of elements, this asserts that $f(xy) = (fx)(fy)$ and $fu = u'$, with u and u' the unit elements; thus a homomorphism is, as usual, just a function preserving composite and units. If M and M' are monoids in (**Ab**, $\otimes$, **Z**), that is, rings, then a homomorphism f as here defined is just a morphism of rings (preserving the units).

Finally, an *action* of a monoid $\langle M, \mu, \eta \rangle$ on a set S is defined to be a function $\nu : M \times S \to S$ such that the following two diagrams commute:

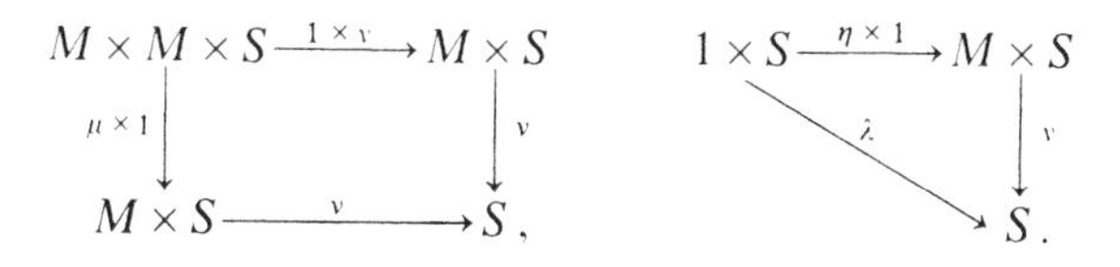

If we write $\nu(x, s) = x \cdot s$ to denote the result of the action of the monoid element x on the element $s \in S$, these diagrams state just that

$$x \cdot (y \cdot s) = (xy) \cdot s, \quad u \cdot s = s$$

for all $x, y \in M$ and all $s \in S$. These are the usual conditions for the action of a monoid on a set, familiar especially in the case of a group acting on a set as a group of transformations. If we shift from the category of sets to the category of topological spaces, we get the usual continuous action of a topological monoid M on a topological space S. If we take $\langle M, \mu, \eta \rangle$ to be a monoid in $(\mathbf{Ab}, \otimes, \mathbf{Z})$, then an action of M on an object S of **Ab** is just a left module S over the ring M.

I. Categories, Functors, and Natural Transformations

1. Axioms for Categories

First we describe categories directly by means of axioms, without using any set theory, and call them "metacategories". Actually, we begin with a simpler notion, a (meta)graph.

A *metagraph* consists of *objects* $a, b, c, \ldots$, *arrows* $f, g, h, \ldots$, and two operations, as follows:

Domain, which assigns to each arrow f an object $a = \operatorname{dom} f$;
Codomain, which assigns to each arrow f an object $b = \operatorname{cod} f$.

These operations on f are best indicated by displaying f as an actual arrow starting at its domain (or "source") and ending at its codomain (or "target"):

$$f : a \to b \quad \text{or} \quad a \xrightarrow{f} b .$$

A finite graph may be readily exhibited: Thus $\cdot \to \cdot \to \cdot$ or $\cdot \rightrightarrows \cdot$.

A *metacategory* is a metagraph with two additional operations:
Identity, which assigns to each object a an arrow $\mathrm{id}_a = 1_a : a \to a$;
Composition, which assigns to each pair $\langle g, f \rangle$ of arrows with $\operatorname{dom} g = \operatorname{cod} f$ an arrow $g \circ f$, called their *composite*, with $g \circ f : \operatorname{dom} f \to \operatorname{cod} g$. This operation may be pictured by the diagram

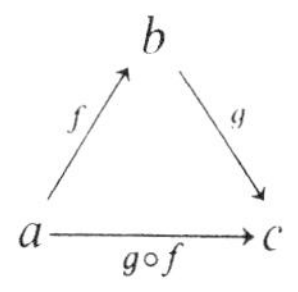

which exhibits all domains and codomains involved. These operations in a metacategory are subject to the two following axioms:

Associativity. For given objects and arrows in the configuration

$$a \xrightarrow{f} b \xrightarrow{g} c \xrightarrow{k} d$$

one always has the equality

$$k \circ (g \circ f) = (k \circ g) \circ f . \tag{1}$$

This axiom asserts that the associative law holds for the operation of composition whenever it makes sense (i.e., whenever the composites on either side of (1) are defined). This equation is represented pictorially by the statement that the following diagram is commutative:

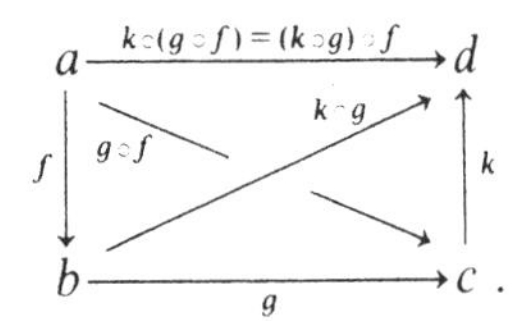

Unit law. For all arrows $f: a \to b$ and $g: b \to c$ composition with the identity arrow 1_b gives

$$1_b \circ f = f \quad \text{and} \quad g \circ 1_b = g\,. \tag{2}$$

This axiom asserts that the identity arrow 1_b of each object b acts as an identity for the operation of composition, whenever this makes sense. The Eqs. (2) may be represented pictorially by the statement that the following diagram is commutative:

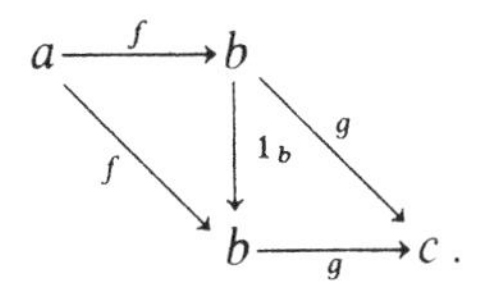

We use many such diagrams consisting of vertices (labelled by objects of a category) and edges (labelled by arrows of the same category). Such a diagram is *commutative* when, for each pair of vertices c and c', any two paths formed from directed edges leading from c to c' yield, by composition of labels, equal arrows from c to c'. A considerable part of the effectiveness of categorical methods rests on the fact that such diagrams in each situation vividly represent the actions of the arrows at hand.

If b is any object of a metacategory C, the corresponding identity arrow 1_b is uniquely determined by the properties (2). For this reason, it is sometimes convenient to identify the identity arrow 1_b with the object b itself, writing $b: b \to b$. Thus $1_b = b = \mathrm{id}_b$, as may be convenient.

A metacategory is to be any interpretation which satisfies all these axioms. An example is the *metacategory of sets*, which has objects all sets and arrows all functions, with the usual identity functions and the usual composition of functions. Here "function" means a function with specified domain and specified codomain. Thus a function $f: X \to Y$ consists of a set X, its domain, a set Y, its codomain, and a rule $x \mapsto fx$ (i.e., a suitable set of ordered pairs $\langle x, fx \rangle$) which assigns, to each element $x \in X$, an element $fx \in Y$. These values will be written as fx, f_x, or $f(x)$, as may be convenient. For example, for any set S, the assignment $s \mapsto s$ for all $s \in S$ describes the *identity function* $1_S : S \to S$; if S is a subset of Y, the assignment $s \mapsto s$ also describes the *inclusion* or *insertion function* $S \to Y$; these functions are *different* unless $S = Y$. Given functions $f : X \to Y$ and $g : Y \to Z$, the *composite* function $g \circ f : X \to Z$ is defined by $(g \circ f)x = g(fx)$ for all $x \in X$. Observe that $g \circ f$ will mean first apply f, then g – in keeping with the practice of writing each function f to the left of its argument. Note, however, that many authors use the opposite convention.

To summarize, the metacategory of all sets has as objects, all sets, as arrows, all functions with the usual composition. The metacategory of all groups is described similarly: Objects are all groups G, H, K; arrows are all those functions f from the set G to the set H for which $f : G \to H$ is a homomorphism of groups. There are many other metacategories: All topological spaces with continuous functions as arrows; all compact Hausdorff spaces with the same arrows; all ringed spaces with their morphisms, etc. The arrows of any metacategory are often called its *morphisms*.

Since the objects of a metacategory correspond exactly to its identity arrows, it is technically possible to dispense altogether with the objects and deal only with arrows. The data for an *arrows-only metacategory* C consist of arrows, certain ordered pairs $\langle g, f \rangle$, called the composable pairs of arrows, and an operation assigning to each composable pair $\langle g, f \rangle$ an arrow $g \circ f$, called their composite. We say "$g \circ f$ is defined" for "$\langle g, f \rangle$ is a composable pair".

With these data one *defines* an identity of C to be an arrow u such that $f \circ u = f$ whenever the composite $f \circ u$ is defined *and* $u \circ g = g$ whenever $u \circ g$ is defined. The data are then required to satisfy the following three axioms:

(i) The composite $(k \circ g) \circ f$ is defined if and only if the composite $k \circ (g \circ f)$ is defined. When either is defined, they are equal (and this *triple composite* is written as kgf).

(ii) The triple composite kgf is defined whenever both composites kg and gf are defined.

(iii) For each arrow g of C there exist identity arrows u and u' of C such that $u' \circ g$ and $g \circ u$ are defined.

In view of the explicit definition given above for identity arrows, the last axiom is a quite powerful one; it implies that u' and u are unique in (iii), and it gives for each arrow g a codomain u' and a domain u. These axioms are equivalent to the preceding ones. More explicitly, given a metacategory of objects and arrows, its arrows, with the given composition, satisfy the "arrows-only" axioms; conversely, an arrows-only metacategory satisfies the objects-and-arrows axioms when the identity arrows, defined as above, are taken as the objects (Proof as exercise).

2. Categories

A category (as distinguished from a metacategory) will mean any interpretation of the category axioms within set theory. Here are the details. A *directed graph* (also called a "diagram scheme") is a set O of objects, a set A of arrows, and two functions

$$A \underset{\text{cod}}{\overset{\text{dom}}{\rightrightarrows}} O. \tag{1}$$

In this graph, the set of composable pairs of arrows is the set

$$A \times_O A = \{\langle g, f\rangle \mid g, f \in A \quad \text{and} \quad \operatorname{dom} g = \operatorname{cod} f\},$$

called the "product over O".

A *category* is a graph with two additional functions

$$\begin{array}{ll} O \xrightarrow{\text{id}} A, & A \times_O A \longrightarrow A, \\ c \longmapsto \text{id}_c, & \langle g, f\rangle \longmapsto g \circ f, \end{array} \tag{2}$$

called identity and composition also written as gf, such that

$$\operatorname{dom}(\operatorname{id} a) = a = \operatorname{cod}(\operatorname{id} a), \quad \operatorname{dom}(g \circ f) = \operatorname{dom} f, \quad \operatorname{cod}(g \circ f) = \operatorname{cod} g \tag{3}$$

for all objects $a \in O$ and all composable pairs of arrows $\langle g, f\rangle \in A \times_O A$, and such that the associativity and unit axioms (1.1) and (1.2) hold. In treating a category C, we usually drop the letters A and O, and write

$$c \in C \qquad f \text{ in } C \tag{4}$$

for "c is an object of C" and "f is an arrow of C", respectively. We also write

$$\hom(b, c) = \{f \mid f \text{ in } C,\ \operatorname{dom} f = b,\ \operatorname{cod} f = c\} \tag{5}$$

for the set of arrows from b to c. Categories can be defined directly in terms of composition acting on these "hom-sets" (§ 8 below); we do not follow this custom because we put the emphasis not on sets (a rather special category), but on axioms, arrows, and diagrams of arrows. We will later observe that our definition of a category amounts to saying that a category is a monoid for the product $\times_O$, in the general sense described in the introduction. For the moment, we consider examples.

0 is the empty category (no objects, no arrows);
1 is the category $\mathfrak{D}$ with one object and one (identity) arrow;
2 is the category $\mathfrak{D} \to \mathfrak{D}$ with two objects a, b, and just one arrow $a \to b$ not the identity;

3 is the category with three objects whose non-identity arrows are arranged as in the triangle $\cdot\!\!\to\!\!\cdot$ (with apex $\cdot$);

$\downdownarrows$ is the category with two objects a, b and just two arrows $a \rightrightarrows b$ not the identity arrows. We call two such arrows *parallel arrows*.

In each of the cases above there is only one possible definition of composition.

Discrete Categories. A category is *discrete* when every arrow is an identity. Every set X is the set of objects of a discrete category (just add one identity arrow $x \to x$ for each $x \in X$), and every discrete category is so determined by its set of objects. Thus, discrete categories are sets.

Monoids. A monoid is a category with one object. Each monoid is thus determined by the set of all its arrows, by the identity arrow, and by the rule for the composition of arrows. Since any two arrows have a composite, a monoid may then be described as a set M with a binary operation $M \times M \to M$ which is associative and has an identity ($=$ unit). Thus a monoid is exactly a semigroup with identity element. For any category C and any object $a \in C$, the set $\hom(a, a)$ of all arrows $a \to a$ is a monoid.

Groups. A group is a category with one object in which every arrow has a (two-sided) inverse under composition.

Matrices. For each commutative ring K, the set $\mathbf{Matr}_K$ of all rectangular matrices with entries in K is a category; the objects are all positive integers $m, n, \ldots$, and each $m \times n$ matrix A is regarded as an arrow $A : n \to m$, with composition the usual matrix product.

Sets. If V is any set of sets, we take $\mathbf{Ens}_V$ to be the category with objects all sets $X \in V$, arrows *all* functions $f : X \to Y$, with the usual composition of functions. By **Ens** we mean any one of these categories.

Preorders. By a preorder we mean a category P in which, given objects p and p', there is at most one arrow $p \to p'$. In any preorder P, define a binary relation $\leqq$ on the objects of P with $p \leqq p'$ if and only if there is an arrow $p \to p'$ in P. This binary relation is reflexive (because there is an identity arrow $p \to p$ for each p) and transitive (because arrows can be composed). Hence a preorder is a set (of objects) equipped with a reflexive and transitive binary relation. Conversely, any set P with such a relation determines a preorder, in which the arrows $p \to p'$ are exactly those ordered pairs $\langle p, p' \rangle$ for which $p \leqq p'$. Since the relation is transitive, there is a unique way of composing these arrows; since it is reflexive, there are the necessary identity arrows.

Preorders include *partial orders* (preorders with the added axiom that $p \leqq p'$ and $p' \leqq p$ imply $p = p'$) and *linear orders* (partial orders such that, given p and p', either $p \leqq p'$ or $p' \leqq p$).

Ordinal Numbers. We regard each ordinal number n as the linearly ordered set of all the preceding ordinals $n = \{0, 1, \ldots, n-1\}$; in particular,

0 is the empty set, while the first infinite ordinal is $\omega = \{0, 1, 2, \ldots\}$. Each ordinal n is linearly ordered, and hence is a category (a preorder). For example, the categories **1**, **2**, and **3** listed above are the preorders belonging to the (linearly ordered) ordinal numbers 1, 2, and 3. Another example is the linear order ω. As a category, it consists of the arrows

$$0 \to 1 \to 2 \to 3 \to \cdots,$$

all their composites, and the identity arrows for each object.

Δ is the category with objects all finite ordinals and arrows $f: m \to n$ all order-preserving functions ($i \leqq j$ in m implies $f_i \leqq f_j$ in n). This category Δ, sometimes called the *simplicial category*, plays a central role (Chapter VII).

Finord = $\mathbf{Set}_\omega$ is the category with objects all finite ordinals n and arrows $f: m \to n$ all functions from m to n. This is essentially the category of all finite sets, using just one finite set n for each finite cardinal number n.

Large Categories. In addition to the metacategory of all sets – which is not a set – we want an actual category **Set**, the category of all *small* sets. We shall assume that there is a big enough set U, the "universe", then describe a set x as "small" if it is a member of the universe, and take **Set** to be the category whose set U of objects is the set of all small sets, with arrows all functions from one small set to another. With this device (details in § 7 below) we construct other familiar large categories, as follows:

Set: Objects, all small sets; arrows, all functions between them.

$\mathbf{Set}_*$: Pointed sets: Objects, small sets each with a selected base point; arrows, base-point-preserving functions.

Ens: Category of all sets and functions within a (variable) set V.

Cat: Objects, all small categories; arrows, all functors (§ 3).

Mon: Objects, all small monoids; arrows, all morphisms of monoids.

Grp: Objects, all small groups; arrows, all morphisms of groups.

Ab: Objects, all small (additive) abelian groups, with morphisms of such.

Rng: All small rings, with the ring morphisms (preserving units) between them.

CRng: All small commutative rings and their morphisms.

R-**Mod**: All small left modules over the ring R, with linear maps.

Mod-R: Small right R-modules.

K-**Mod**: Small modules over the commutative ring K.

Top: Small topological spaces and continuous maps.

Toph: Topological spaces, with arrows homotopy classes of maps.

$\mathbf{Top}_*$: Spaces with selected base point, base point-preserving maps.

Particular categories (like these) will always appear in bold-face type. Script capitals are used by many authors to denote categories.

3. Functors

A *functor* is a morphism of categories. In detail, for categories C and B a functor $T: C \to B$ with domain C and codomain B consists of two suitably related functions: The *object function* T, which assigns to each object c of C an object Tc of B and the *arrow function* (also written T) which assigns to each arrow $f: c \to c'$ of C an arrow $Tf: Tc \to Tc'$ of B, in such a way that

$$T(1_c) = 1_{Tc}, \qquad T(g \circ f) = Tg \circ Tf, \tag{1}$$

the latter whenever the composite $g \circ f$ is defined in C. A functor, like a category, can be described in the "arrows-only" fashion: It is a function T from arrows f of C to arrows Tf of B, carrying each identity of C to an identity of B and each composable pair $\langle g, f \rangle$ in C to a composable pair $\langle Tg, Tf \rangle$ in B, with $Tg \circ Tf = T(g \circ f)$.

A simple example is the power set functor $\mathscr{P}$: **Set** $\to$ **Set**. Its object function assigns to each set X the usual power set $\mathscr{P}X$, with elements all subsets $S \subset X$; its arrow function assigns to each $f: X \to Y$ that map $\mathscr{P}f: \mathscr{P}X \to \mathscr{P}Y$ which sends each $S \subset X$ to its image $fS \subset Y$. Since both $\mathscr{P}(1_X) = 1_{\mathscr{P}X}$ and $\mathscr{P}(g \circ f) = \mathscr{P}g \circ \mathscr{P}f$, this clearly defines a functor $\mathscr{P}$: **Set** $\to$ **Set**.

Functors were first explicitly recognized in algebraic topology, where they arise naturally when geometric properties are described by means of algebraic invariants. For example, singular homology in a given dimension n (n a natural number) assigns to each topological space X an abelian group $H_n(X)$, the n-th homology group of X, and also to each continuous map $f: X \to Y$ of spaces a corresponding homomorphism $H_n(f): H_n(X) \to H_n(Y)$ of groups, and this in such a way that H_n becomes a functor **Top** $\to$ **Ab**. For example, if $X = Y = S^1$ is the circle, $H_1(S^1) = \mathbf{Z}$, so the group homomorphism $H_1(f): \mathbf{Z} \to \mathbf{Z}$ is determined by an integer d (the image of 1); this integer is the usual "degree" of the continuous map $f: S^1 \to S^1$. In this case and in general, homotopic maps $f, g: X \to Y$ yield the same homomorphism $H_n(X) \to H_n(Y)$, so H_n can actually be regarded as a functor **Toph** $\to$ **Grp**, defined on the homotopy category. The Eilenberg-Steenrod axioms for homology start with the axioms that H_n, for each natural number n, is a functor on **Toph**, and continue with certain additional properties of these functors. The more recently developed extraordinary homology and cohomology theories are also functors on **Toph**. The homotopy groups $\pi_n(X)$ of a space X can also be regarded as functors; since they depend on the choice of a base point in X, they are functors $\mathbf{Top}_* \to$ **Grp**. The leading idea in the use of functors in topology is that H_n or π_n gives an algebraic picture or image not just of the topological spaces, but also of all the continuous maps between them.

Functors arise naturally in algebra. To any commutative ring K the set of all non-singular $n \times n$ matrices with entries in K is the usual general linear group $\mathrm{GL}_n(K)$; moreover, each homomorphism $f: K \to K'$ of rings produces in the evident way a homomorphism $\mathrm{GL}_n f: \mathrm{GL}_n(K) \to \mathrm{GL}_n(K')$ of groups. These data define for each natural number n a functor $\mathrm{GL}_n: \mathbf{CRng} \to \mathbf{Grp}$. For any group G the set of all products of commutators $xyx^{-1}y^{-1}(x, y \in G)$ is a normal subgroup $[G, G]$ of G, called the *commutator* subgroup. Since any homomorphism $G \to H$ of groups carries commutators to commutators, the assignment $G \mapsto [G, G]$ defines an evident functor $\mathbf{Grp} \to \mathbf{Grp}$, while $G \mapsto G/[G, G]$ defines a functor $\mathbf{Grp} \to \mathbf{Ab}$, the factor-commutator functor. Observe, however, that the center $Z(G)$ of G (all $a \in G$ with $ax = xa$ for all x) does not naturally define a functor $\mathbf{Grp} \to \mathbf{Grp}$, because a homomorphism $G \to H$ may carry an element in the center of G to one not in the center of H.

A functor which simply "forgets" some or all of the structure of an algebraic object is commonly called a *forgetful* functor (or, an *underlying* functor). Thus the forgetful functor $U: \mathbf{Grp} \to \mathbf{Set}$ assigns to each group G the set UG of its elements ("forgetting" the multiplication and hence the group structure), and assigns to each morphism $f: G \to G'$ of groups the same function f, regarded just as a function between sets. The forgetful functor $U: \mathbf{Rng} \to \mathbf{Ab}$ assigns to each ring R the additive abelian group of R and to each morphism $f: R \to R'$ of rings the same function, regarded just as a morphism of addition.

Functors may be composed. Explicitly, given functors

$$C \xrightarrow{T} B \xrightarrow{S} A$$

between categories A, B, and C, the composite functions

$$c \mapsto S(Tc) \qquad f \mapsto S(Tf)$$

on objects c and arrows f of C define a functor $S \circ T: C \to A$, called the *composite* (in that order) of S with T. This composition is associative. For each category B there is an identity functor $I_B: B \to B$, which acts as an identity for this composition. Thus we may consider the metacategory of all categories: its objects are all categories, its arrows are all functors with the composition above. Similarly, we may form the category **Cat** of all small categories – but not the category of all categories.

An *isomorphism* $T: C \to B$ of categories is a functor T from C to B which is a bijection, both on objects and on arrows. Alternatively, but equivalently, a functor $T: C \to B$ is an isomorphism if and only if there is a functor $S: B \to C$ for which both composites $S \circ T$ and $T \circ S$ are identity functors; then S is the *two-sided inverse* $S = T^{-1}$.

Certain properties much weaker than isomorphism will be useful.

A functor $T: C \to B$ is *full* when to every pair c, c' of objects of C and to every arrow $g: Tc \to Tc'$ of B, there is an arrow $f: c \to c'$ of C with $g = Tf$. Clearly the composite of two full functors is a full functor.

A functor $T : C \to B$ is *faithful* (or an embedding) when to every pair c, c' of objects of C and to every pair $f_1, f_2 : c \to c'$ of parallel arrows of C the equality $Tf_1 = Tf_2 : Tc \to Tc'$ implies $f_1 = f_2$. Again, composites of faithful functors are faithful. For example, the forgetful functor **Grp**$\to$**Set** is faithful but not full and not a bijection on objects.

These two properties may be visualized in terms of hom-sets (see (2.5)). Given a pair of objects $c, c' \in C$, the arrow function of $T: C \to B$ assigns to each $f: c \to c'$ an arrow $Tf: Tc \to Tc'$ and so defines a function

$$T_{c,c'} : \hom(c, c') \to \hom(Tc, Tc'), \qquad f \mapsto Tf.$$

Then T is full when every such function is surjective, and faithful when every such function is injective. For a functor which is both full and faithful (i.e., "fully faithful"), every such function is a bijection, but this need not mean that the functor itself is an isomorphism of categories, for there may be objects of B not in the image of T.

A *subcategory* S of a category C is a collection of some of the objects and some of the arrows of C, which includes with each arrow f both the object $\operatorname{dom} f$ and the object $\operatorname{cod} f$, with each object s its identity arrow 1_s and with each pair of composable arrows $s \to s' \to s''$ their composite. These conditions ensure that these collections of objects and arrows themselves constitute a category S. Moreover, the injection (inclusion) map $S \to C$ which sends each object and each arrow of S to itself (in C) is a functor, the *inclusion functor*. This inclusion functor is automatically faithful. We say that S is a *full subcategory* of C when the inclusion functor $S \to C$ is full. A full subcategory, given C, is thus determined by giving just the set of its objects, since the arrows between any two of these objects s, s' are all morphisms $s \to s'$ in C. For example, the category $\mathbf{Set}_f$ of all finite sets is a full subcategory of the category **Set**.

Exercises

1. Show how each of the following constructions can be regarded as a functor: The field of quotients of an integral domain; the Lie algebra of a Lie group.
2. Show that functors $\mathbf{1} \to C$, $\mathbf{2} \to C$, and $\mathbf{3} \to C$ correspond respectively to objects, arrows, and composable pairs of arrows in C.
3. Interpret "functor" in the following special types of categories: (a) A functor between two preorders is a function T which is monotonic (i.e., $p \leqq p'$ implies $Tp \leqq Tp'$). (b) A functor between two groups (one-object categories) is a morphism of groups. (c) If G is a group, a functor $G \to$ **Set** is a permutation representation of G, while $G \to \mathbf{Matr}_K$ is a matrix representation of G.
4. Prove that there is no functor **Grp**$\to$**Ab** sending each group G to its center (consider $S_2 \to S_3 \to S_2$, the symmetric groups).
5. Find two different functors T: **Grp**$\to$**Grp** with object function $T(G) = G$ the identity for every group G.

4. Natural Transformations

Given two functors $S, T: C \to B$, a *natural transformation* $\tau : S \dot{\to} T$ is a function which assigns to each object c of C an arrow $\tau_c = \tau c : Sc \to Tc$ of B in such a way that every arrow $f: c \to c'$ in C yields a diagram

$$\begin{array}{ccccc} c & \quad & Sc & \xrightarrow{\tau c} & Tc \\ \downarrow{\scriptstyle f} & & {\scriptstyle Sf}\downarrow & & \downarrow{\scriptstyle Tf} \\ c', & & Sc' & \xrightarrow{\tau c'} & Tc' \end{array} \tag{1}$$

which is commutative. When this holds, we also say that $\tau_c : Sc \to Tc$ is *natural* in c. If we think of the functor S as giving a picture in B of (all the objects and arrows of) C, then a natural transformation τ is a set of arrows mapping (or, translating) the picture S to the picture T, with all squares (and parallelograms!) like that above commutative:

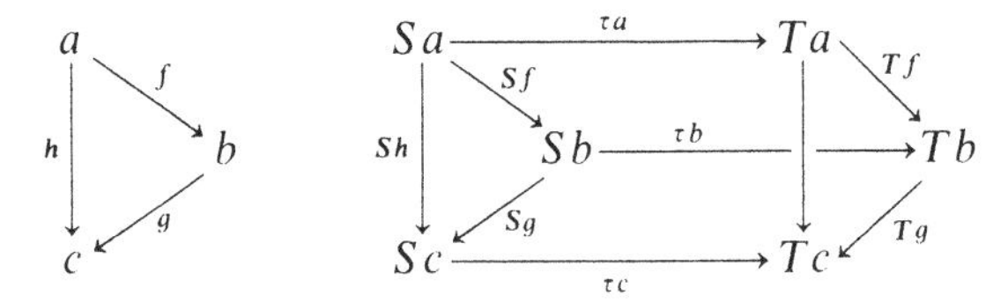

We call $\tau a, \tau b, \tau c, \ldots,$ the *components* of the natural transformation τ.

A natural transformation is often called a *morphism of functors*; a natural transformation τ with every component τc invertible in B is called a *natural equivalence* or better a *natural isomorphism*; in symbols, $\tau : S \cong T$. In this case, the inverses $(\tau c)^{-1}$ in B are the components of a natural isomorphism $\tau^{-1} : T \dot{\to} S$.

The determinant is a natural transformation. To be explicit, let $\det_K M$ be the determinant of the $n \times n$ matrix M with entries in the commutative ring K, while K^* denotes the group of units (invertible elements) of K. Thus M is non-singular when $\det_K M$ is a unit, and $\det_K$ is a morphism $\mathrm{GL}_n K \to K^*$ of groups (an arrow in **Grp**). Because the determinant is defined by the same formula for all rings K, each morphism $f: K \to K'$ of commutative rings leads to a commutative diagram

$$\begin{array}{ccc} \mathrm{GL}_n K & \xrightarrow{\det_K} & K^* \\ {\scriptstyle \mathrm{GL}_n f}\downarrow & & \downarrow{\scriptstyle f^*} \\ \mathrm{GL}_n K' & \xrightarrow{\det_{K'}} & K'^*. \end{array} \tag{2}$$

This states that the transformation $\det : \mathrm{GL}_n \to (\ )^*$ is natural between two functors **CRng** $\to$ **Grp**.

For each group G the projection $p_G : G \to G/[G, G]$ to the factor-commutator group defines a transformation p from the identity functor on **Grp** to the factor-commutator functor **Grp**$\to$**Ab**$\to$**Grp**. Moreover, p is natural, because each group homomorphism $f: G\to H$ defines the evident homomorphism f' for which the following diagram commutes:

$$\begin{array}{ccc} G & \xrightarrow{pG} & G/[G,G] \\ {\scriptstyle f}\downarrow & & \downarrow{\scriptstyle f'} \\ H & \xrightarrow{pH} & H/[H,H]\,. \end{array} \tag{3}$$

The double character group yields a suggestive example in the category **Ab** of all abelian groups G. Let $D(G)$ denote the character group of G, so that $DG = \hom(G, \mathbf{R}/\mathbf{Z})$ is the set of all homomorphisms $t: G\to\mathbf{R}/\mathbf{Z}$ with the familiar group structure, where $\mathbf{R}/\mathbf{Z}$ is the additive group of real numbers modulo 1. Each arrow $f: G'\to G$ in **Ab** determines an arrow $Df: DG\to DG'$ (opposite direction!) in **Ab**, with $(Df)t = tf: G'\to\mathbf{R}/\mathbf{Z}$ for each t; for composable arrows, $D(g\circ f) = Df\circ Dg$. Because of this reversal, D is not a functor (it is a "contravariant" functor on **Ab** to **Ab**, see § II.2); however, the twice iterated character group $G\mapsto D(DG)$ and the identity $I(G)=G$ are both functors **Ab**$\to$**Ab**. For each group G there is a homomorphism

$$\tau_G: G\to D(DG)$$

obtained in a familiar way: To each $g\in G$ assign the function $\tau_G g: DG\to\mathbf{R}/\mathbf{Z}$ given for any character $t\in DG$ by $t\mapsto tg$; thus $(\tau_G g)t = t(g)$. One verifies at once that τ is a natural transformation $\tau: I\dot{\to} DD$; this statement is just a precise expression for the elementary observation that the definition of τ depends on no artificial choices of bases, generators, or the like. In case G is finite, τ_G is an isomorphism; thus, if we restrict all functors to the category $\mathbf{Ab}_f$ of finite abelian groups, τ is a natural isomorphism.

On the other hand, for each finite abelian group G there is an isomorphism $\sigma_G: G\cong DG$ of G to its character group, but this isomorphism depends on a representation of G as a direct product of cyclic groups and so cannot be natural. More explicitly, we can make D into a covariant functor $D': \mathbf{Ab}_{f,i}\to\mathbf{Ab}_{f,i}$ on the category $\mathbf{Ab}_{f,i}$ with objects all finite abelian groups and arrows all isomorphisms f between such groups, setting $D'G = DG$ and $D'f = Df^{-1}$. Then $\sigma_G: G\to D'G$ is a map $\sigma: I\to D'$ of functors $\mathbf{Ab}_{f,i}\to\mathbf{Ab}_{f,i}$, but it is not natural in the sense of our definition.

A parallel example is the familiar natural isomorphism of a finite-dimensional vector space to its double dual.

Another example of naturality arises when we compare the category **Finord** of all finite ordinal numbers n with the category $\mathbf{Set}_f$ of all finite sets (in some universe U). Every ordinal $n=\{0, 1, \ldots, n-1\}$ is a finite set, so the inclusion S is a functor $S: \mathbf{Finord} \to \mathbf{Set}_f$. On the other hand, each finite set X determines an ordinal number $n = \# X$, the number of elements in X; we may choose for each X a bijection $\theta_X : X \to \# X$. For any function $f: X \to Y$ between finite sets we may then define a corresponding function $\# f : \# X \to \# Y$ between ordinals by $\# f = \theta_Y f \theta_X^{-1}$; this ensures that the diagram

$$\begin{array}{ccc} X & \xrightarrow{\theta_X} & \# X \\ {\scriptstyle f}\downarrow & & \downarrow{\scriptstyle \# f} \\ Y & \xrightarrow{\theta_Y} & \# Y \end{array}$$

will commute, and makes $\#$ a functor $\# : \mathbf{Set}_f \to \mathbf{Finord}$. If X is itself an ordinal number, we may take θ_X to be the identity. This ensures that the composite functor $\# \circ S$ is the identity functor I' of **Finord**. On the other hand, the composite $S \circ \#$ is not the identity functor $I: \mathbf{Set}_f \to \mathbf{Set}_f$, because it sends each finite set X to a special finite set – the ordinal number n with the same number of elements as X. However, the square diagram above does show that $\theta : I \dot{\to} S\#$ is a natural isomorphism. All told we have $I \cong S \circ \#$, $I' = \# \circ S$.

More generally, an *equivalence* between categories C and D is defined to be a pair of functors $S: C \to D$, $T: D \to C$ together with natural isomorphisms $I_C \cong T \circ S$, $I_D \cong S \circ T$. This example shows that this notion (to be examined in § IV.4) allows us to compare categories which are "alike" but of very different "sizes".

We shall use many other examples of naturality. As Eilenberg-Mac Lane first observed, "category" has been defined in order to be able to define "functor" and "functor" has been defined in order to be able to define "natural transformation".

Exercises

1. Let S be a fixed set, and X^S the set of all functions $h: S \to X$. Show that $X \mapsto X^S$ is the object function of a functor $\mathbf{Set} \to \mathbf{Set}$, and that evaluation $e_X : X^S \times S \dot{\to} X$, defined by $e(h, s) = h(s)$, the value of the function h at $s \in S$, is a natural transformation.
2. If H is a fixed group, show that $G \mapsto H \times G$ defines a functor $H \times - : \mathbf{Grp} \to \mathbf{Grp}$, and that each morphism $f: H \to K$ of groups defines a natural transformation $H \times - \dot{\to} K \times -$.
3. If B and C are groups (regarded as categories with one object each) and $S, T: B \to C$ are functors (homomorphisms of groups), show that there is a natural transformation $S \dot{\to} T$ if and only if S and T are conjugate; i.e., if and only if there is an element $h \in C$ with $Tg = h(Sg)h^{-1}$ for all $g \in B$.

4. For functors $S, T : C \to P$ where C is a category and P a preorder, show that there is a natural transformation $S \dot{\to} T$ (which is then unique) if and only if $Sc \leqq Tc$ for every object $c \in C$.
5. Show that every natural transformation $\tau : S \dot{\to} T$ defines a function (also called τ) which sends each arrow $f : c \to c'$ of C to an arrow $\tau f : Sc \to Tc'$ of B in such a way that $Tg \circ \tau f = \tau(g f) = \tau g \circ Sf$ for each composable pair $\langle g, f \rangle$. Conversely, show that every such function τ comes from a unique natural transformation with $\tau_c = \tau(1_c)$. (This gives an "arrows only" description of a natural transformation.)
6. Let F be a field. Show that the category of all finite-dimensional vector spaces over F (with morphisms all linear transformations) is equivalent to the category $\mathbf{Matr}_F$ described in § 2.

5. Monics, Epis, and Zeros

In categorical treatments many properties ordinarily formulated by means of elements (elements of a set or of a group) are instead formulated in terms of arrows. For example, instead of saying that a set X has just one element, one can say that for any other set Y there is exactly one function $Y \to X$. We now formulate a few more instances of such methods of "doing without elements".

An arrow $e : a \to b$ is *invertible* in C if there is an arrow $e' : b \to a$ in C with $e'e = 1_a$ and $ee' = 1_b$. If such an e' exists, it is unique, and is written as $e' = e^{-1}$. By the usual proof, $(e_1 e_2)^{-1} = e_2^{-1} e_1^{-1}$, provided the composite $e_1 e_2$ is defined and both e_1 and e_2 are invertible. Two objects a and b are *isomorphic* in the category C if there is an invertible arrow (an *isomorphism*) $e : a \to b$; we write $a \cong b$. The relation of isomorphism of objects is manifestly reflexive, symmetric, and transitive.

An arrow $m : a \to b$ is *monic* in C when for any two parallel arrows $f_1, f_2 : d \to a$ the equality $m \circ f_1 = m \circ f_2$ implies $f_1 = f_2$; in other words, m is monic if it can always be cancelled on the left (is *left cancellable*). In **Set** and in **Grp** the monic arrows are precisely the injections (monomorphisms) in the usual sense; i.e., the functions which are one-one into.

An arrow $h : a \to b$ is *epi* in C when for any two arrows $g_1, g_2 : b \to c$ the equality $g_1 \circ h = g_2 \circ h$ implies $g_1 = g_2$; in other words, h is epi when it is *right cancellable*. In **Set** the epi arrows are precisely the surjections (epimorphisms) in the usual sense; i.e., the functions onto.

For an arrow $h : a \to b$, a *right* inverse is an arrow $r : b \to a$ with $hr = 1_b$. A right inverse (which is usually not unique) is also called a *section* of h. If h has a right inverse, it is evidently epi; the converse holds in **Set**, but fails in **Grp**. Similarly, a *left inverse* for h is called a *retraction* for h, and any arrow with a left inverse is necessarily monic. If $gh = 1_a$, then g is a *split* epi, h a *split* monic, and the composite $f = hg$ is defined and is an idempotent. Generally, an arrow $f: b \to b$ is called *idempotent* when $f^2 = f$; an idempotent is said to *split* when there exist arrows g and h such that $f = hg$ and $gh = 1$.

An object t is *terminal* in C if to each object a in C there is exactly one arrow $a \to t$. If t is terminal, the only arrow $t \to t$ is the identity, and any two terminal objects of C are isomorphic in C. An object s is *initial* in C if to each object a there is exactly one arrow $s \to a$. For example, in the category **Set**, the empty set is an initial object and any one-point set is a terminal object. In **Grp**, the group with one element is both initial and terminal.

A *null object* z in C is an object which is both initial and terminal. If C has a null object, that object is unique up to isomorphism, while for any two objects a and b of C there is a unique arrow $a \to z \to b$ (the composite through z), called the *zero arrow* from a to b. Any composite with a zero arrow is itself a zero arrow. For example, the categories **Ab** and R-**Mod** have null objects (namely 0!), as does $\mathbf{Set}_*$ (namely the one-point set).

A *groupoid* is a category in which every arrow is invertible. A typical groupoid is the *fundamental groupoid* $\pi(X)$ of a topological space X. An object of $\pi(X)$ is a point x of X, and an arrow $x \to x'$ of $\pi(X)$ is a homotopy class of paths f from x to x'. (Such a path f is a continuous function $I \to X$, I the closed interval $I = [0, 1]$, with $f(0) = x$, $f(1) = x'$, while two paths f, g with the same end-points x and x' are homotopic when there is a continuous function $F: I \times I \to X$ with $F(t, 0) = f(t)$, $F(t, 1) = g(t)$, and $F(0, s) = x$, $F(1, s) = x'$ for all s and t in I.) The *composite* of paths $g: x' \to x''$ and $f: x \to x'$ is the path h which is "f followed by g", given explicitly by

$$
\begin{aligned}
h(t) &= f(2t), & 0 \leqq t \leqq 1/2, \\
&= g(2t-1), & 1/2 \leqq t \leqq 1.
\end{aligned}
\tag{1}
$$

Composition applies also to homotopy classes, and makes $\pi(X)$ a category and a groupoid (the inverse of any path is the same path traced in the opposite direction).

Since each arrow in a groupoid G is invertible, each object x in G determines a group $\hom_G(x, x)$, consisting of all $g: x \to x$. If there is an arrow $f: x \to x'$, the groups $\hom_G(x, x)$ and $\hom_G(x', x')$ are isomorphic, under $g \mapsto fgf^{-1}$ (i.e., under conjugation). A groupoid is said to be *connected* if there is an arrow joining any two of its objects. One may readily show that a connected groupoid is determined up to isomorphism by a group (one of the groups $\hom_G(x, x)$) and by a set (the set of all objects). In this way, the fundamental groupoid $\pi(X)$ of a path-connected space X is determined by the set of points in the space and a group $\hom_{\pi(X)}(x, x)$ – the *fundamental group* of X.

Exercises

1. Find a category with an arrow which is both epi and monic, but not invertible (e.g., dense subset of a topological space).
2. Prove that the composite of monics is monic, and likewise for epis.
3. If a composite $g \circ f$ is monic, so is f. Is this true of g?
4. Show that the inclusion $\mathbf{Z} \to \mathbf{Q}$ is epi in the category **Rng**.
5. In **Grp** prove that every epi is surjective (Hint. If $\varphi : G \to H$ has image M not H, use the factor group H/M if M has index 2. Otherwise, let Perm H be the group of all permutations of the set H, choose three different cosets M, Mu and Mv of M, define $\sigma \in \operatorname{Perm} H$ by $\sigma(xu) = xv$, $\sigma(xv) = xu$ for $x \in M$, and σ otherwise the identity. Let $\psi : H \to \operatorname{Perm} H$ send each h to left multiplication ψ_h by h, while $\psi'_h = \sigma^{-1} \psi_h \sigma$. Then $\psi\varphi = \psi'\varphi$, but $\psi \neq \psi'$).
6. In **Set**, show that all idempotents split.
7. An arrow $f : a \to b$ in a category C is *regular* when there exists an arrow $g : b \to a$ such that $fgf = f$. Show that f is regular if it has either a left or a right inverse, and prove that every arrow in **Set** with $a \neq \varnothing$ is regular.
8. Consider the category with objects $\langle X, e, t \rangle$, where X is a set, $e \in X$, and $t : X \to X$, and with arrows $f : \langle X, e, t \rangle \to \langle X', e', t' \rangle$ the functions f on X to X' with $fe = e'$ and $ft = t'f$. Prove that this category has an initial object in which X is the set of natural numbers, $e = 0$, and t is the successor function.
9. If the functor $T : C \to B$ is faithful and Tf is monic, prove f monic.

6. Foundations

One of the main objectives of category theory is to discuss properties of totalities of Mathematical objects such as the "set" of all groups or the "set" of all homomorphisms between any two groups. Now it is the custom to regard a group as a set with certain added structure, so we are here proposing to consider a set of *all* sets with some given structure. This amounts to applying a comprehension principle: Given a property $\varphi(x)$ of sets x, form the set $\{x \mid \varphi(x)\}$ of *all* sets x with this property. However such a principle cannot be adopted in this generality, since it would lead to some of the famous paradoxical sets, such as the set of all sets not members of themselves.

For this reason, the standard practice in naive set theory, with the usual membership relation $\in$, is to restrict the application of the comprehension principle. One allows the formation from given sets u, v of the set $\{u, v\}$ (the set with exactly u and v as elements), of the ordered pair $\langle u, v \rangle$, of an infinite set (the set $\omega = \{0, 1, 2, \dots\}$ of all finite ordinals), and of

The Cartesian Product	$u \times v = \{\langle x, y \rangle \mid x \in u \text{ and } y \in v\}$,
The Power Set	$\mathscr{P}u = \{v \mid v \subset u\}$,
The Union (of a set x of sets)	$\cup x = \{y \mid y \in z \text{ for some } z \in x\}$.

Finally, given a property $\varphi(x)$ (technically, a property expressed in terms of x, the membership relation, and the usual logical connectives, including "for all sets t" and "there exists a set t") and given a set u one allows

Comprehension for elements of u: $\{x \mid x \in u \text{ and } \varphi(x)\}$.

In words: One allows the set of all those x with a given property φ which are members of an already given set u.

To this practice, we add one more assumption: The existence of a universe. A *universe* is defined to be a set U with the following (somewhat redundant) properties:

(i) $x \in u \in U$ implies $x \in U$,

(ii) $u \in U$ and $v \in U$ imply $\{u, v\}$, $\langle u, v\rangle$, and $u \times v \in U$.

(iii) $x \in U$ implies $\mathscr{P}x \in U$ and $\cup x \in U$,

(iv) $\omega \in U$ (here $\omega = \{0, 1, 2, \ldots\}$ is the set of all finite ordinals),

(v) if $f: a \to b$ is a surjective function with $a \in U$ and $b \subset U$, then $b \in U$.

These closure properties for U ensure that any of the standard operations of set theory applied to elements of U will always produce elements of U; in particular, $\omega \in U$ provides that U also contains all the usual sets of real numbers and related infinite sets. We can then regard "ordinary" Mathematics as carried out exclusively within U (i.e., on elements of U) while U itself and sets formed from U are to be used for the construction of the desired large categories.

Now hold the universe U fixed, and call a set $u \in U$ a *small set*. Thus the universe U is the set of all small sets. Similarly, call a function $f: u \to v$ *small* when u and v are small sets. This implies that f itself can be regarded as a small set – say, as the ordered triple $\langle u, G_f, v\rangle$, with $G_f \subset u \times v$ the usual set of all $\langle x, y\rangle$ with $x \in u$, $y = fx$. The limited comprehension principle thus allows the construction of the set A of all those sets which are small functions, since these functions are all elements of U. We can now define the category **Set** of all small sets to be that category in which U (the set of all small sets) is the set of objects and A (the set of all small functions) is the set of arrows. Henceforth **Set** will always denote this category.

A *small group* is similarly a small set with a group structure; i.e., is an ordered pair $\langle u, m\rangle$, where u is a small set and $m: u \times u \to u$ a function (binary operation on u) satisfying the usual group axioms. Since any small group is an element of U, we may form the set of all small groups and the set of all homomorphisms between two small groups. They constitute the category **Grp** of all small groups.

The same process will construct the category of all small Mathematical objects of other types. For example, a category is small if the set of its arrows and the set of its objects are both small sets; we will soon form the category **Cat** of all small categories. Observe, however, that **Set** is not a small category, because the set U of its objects is not a small set (otherwise $U \in U$, and this is contrary to the axiom of regularity, which asserts that there are no infinite chains $\ldots x_n \in x_{n-1} \in x_{n-2} \in \cdots \in x_0$). Similarly, **Grp** is not small.

This description of the foundations may be put in axiomatic form. We are assuming the standard Zermelo-Fraenkel axioms for set theory, plus the existence of a set U which is a universe. The Zermelo-Fraenkel axioms (on a membership relation $\in$) are: Extensionality (sets with the same elements are equal), existence of the null set, existence of the sets $\{u, v\}$, $\langle u, v\rangle$, $\mathscr{P}u$, and $\cup x$ for all sets u, v, and x, the axiom of infinity, the axiom of choice, the axiom of regularity, and the replacement axiom:

Replacement. Let a be a set and $\varphi(x, y)$ a property which is functional for x in a, in the sense that $\varphi(x, y)$ and $\varphi(x, y')$ for $x \in a$ imply $y = y'$, and that for each $x \in a$ there exists a y with $\varphi(x, y)$. Then there exists a set consisting of all those y such that $\varphi(x, y)$ holds for $x \in a$.

Briefly speaking, the replacement axiom states that the image of a set a under a "function" φ is a set. It can be shown that the replacement axiom implies the comprehension axiom, as stated above. Moreover, our conditions defining a universe U imply that all the sets $x \in U$ (all the small sets) do satisfy the Zermelo-Fraenkel axioms – for example, condition (v) in the definition of a universe corresponds to replacement. We shall see that our assumption of one universe suffices for the usual purposes of category theory.

Some authors assume instead sets and "classes", using, for these concepts, the Gödel-Bernays axioms. To explain this, define a *class* C to be any subset $C \subset U$ of the universe. Since $x \in u \in U$ implies $x \in U$, every element of U is also a subset of U, therefore every small set is also a class; but conversely, some classes (such as U itself) are not small sets. These latter are called the *proper classes*. Together, the small sets and the classes satisfy the standard Gödel-Bernays axioms (see Gödel [1940]).

A *large category* is one in which both the set of objects and the set of arrows are classes (proper or otherwise). Using only *small* sets and all classes one can describe many of the needed categories – in particular, our categories **Set**, **Grp**, etc. are proper classes, hence are large categories in this sense. Initially, category theory was restricted to the study of small and large categories (and based on the Gödel-Bernays axioms). However, we will have many occasions to form categories which are not classes. One such is the category **Cls** of all classes: Its objects are all classes; its arrows all functions $f: C \to C'$ between classes. Then the set of objects of **Cls** is the set $\mathscr{P}(U)$ of all subsets of U; it is *not* a class; in fact, its cardinal number is larger than the cardinal of the universe U. Another useful category is **Cat**$'$, the category of all large categories. It is not a class.

In the sequel we shall drop the notation U for the chosen universe and speak simply of small sets, of classes, and of sets, observing that the "sets" include the small sets and the classes, as well as many other sets such as $\mathscr{P}(U)$, $\mathscr{P}\mathscr{P}(U)$, $\{U\}$, and the like. Note, in particular, that $\{U\}$ is a set which has only one element (namely, the universe U). It is thus intuitively very "small", but it is not a small set in our sense; $\{U\} \in U$ would imply $U \in U$, a contradiction to the axiom of regularity. Thus "small set" for us means a member of the universe, and *not* a set with a small cardinal number.

Our foundation by means of one universe does provide, within set theory, an accurate way of discussing the category of all small sets and all small groups, but it does not provide sets to represent certain metacategories, such as the metacategory of *all* sets or that of *all* groups. Grothendieck uses an alternative device. He assumes that for every set X there is a universe U with $X \in U$. This stronger assumption evidently provides for each universe U a category of all those groups which are members of U. However, this does not provide any category of *all* groups. For this reason, there has been considerable discussion of a foundation for category theory (and for all of Mathematics) not based on set theory. This is why we initially gave the definition of a category C in a set-free form, simply by regarding the axioms as first-order axioms on undefined terms "object of C", "arrow of C", "composite", "identity", "domain", and "codomain". In this style, axioms for the elementary (i.e., first-order) theory of the category of all sets, as an alternative to the usual axioms on membership can be given—as an "elementary topos" (cf. Mac Lane-Moerdijk [1992]).

Exercises

1. Given a universe U and a function $f : I \to b$ with domain $I \in U$ and with every value f_i an element of U, for $i \in I$, prove that the usual cartesian product $\Pi_i f_i$ is an element of U.
2. (a) Given a universe U and a function $f : I \to b$ with domain $I \in U$, show that the usual union $\cup_i f_i$ is a set of U.

   (b) Show that this one closure property of U may replace condition (v) and the condition $x \in U$ implies $\cup x \in U$ in the definition of a universe.

7. Large Categories

In many relevant examples, a category consists of all (small) Mathematical objects with a given structure, with arrows all the functions which preserve that structure. We list useful such examples with their monics.

Ab, the category of all small abelian groups, has objects all small (additive) abelian groups $A, B, \ldots$ and arrows all homomorphisms $f: A \to B$ of abelian groups, with the usual composition. In this category, an arrow is monic if and only if it is a monomorphism (one-one into).

Also, an epimorphism (a homomorphism onto) is clearly epi. Conversely, a homomorphism $f: A \to B$ which is epi as an arrow must be onto as a function. For, otherwise, the quotient group B/fA is nonzero, so there are then two different morphisms $B \to B/fA$, the projection p and the zero morphism 0, which have $pf = 0 = 0f$, a contradiction to the assumption that f is epi. In **Ab**, the zero group is both initial and terminal.

A *small ring* R is a small set with binary operations of addition and multiplication which satisfy the usual axioms for a ring – including the existence of a two-sided identity (= unit) 1 for multiplication. **Rng** will denote the category of all small rings; the objects are the small rings R, the arrows $f: R \to S$ the (homo)morphisms of rings – where a morphism of rings is assumed to carry the unit of R to that of S. In this category the zero ring is terminal, and the ring $\mathbf{Z}$ of integers is initial since $\mathbf{Z} \to R$ is the unique arrow carrying $1 \in \mathbf{Z}$ to the unit of the ring R. The monic arrows are precisely the monomorphisms of rings. Every epimorphism of rings is epi as an arrow, but the inclusion $\mathbf{Z} \to \mathbf{Q}$ of $\mathbf{Z}$ in the field $\mathbf{Q}$ of rational numbers is epi, but not an epimorphism.

If R is any small ring, the category R-**Mod** has objects all small left R-modules $A, B, \ldots$ and arrows $f: A \to B$ all morphisms of R-modules (R-linear maps). In this category monics are monomorphisms, epis are epimorphisms, and the zero module is initial and terminal. If F is a field, the category F-**Mod**, also written $\mathbf{Vct}_F$, is that of all vector spaces (linear spaces) over F. By **Mod**-R we denote the category of all small right R-modules. If R and S are two rings, R-**Mod**-S is the category of all small R-S-bimodules (left R-, right S-modules A with $r(as) = (ra)s$ for all $r \in R$, $a \in A$, and $s \in S$). One may similarly construct categories of small algebraic objects of any given type.

The category **Top** of topological spaces has as objects all small topological spaces $X, Y, \ldots$ and as morphisms all continuous maps $f: X \to Y$. Again, the monics are the injections and the epis the surjections. The one-point space is terminal, and the empty space is initial. Similarly, one may form the category of all small Hausdorff spaces or of all small compact Hausdorff spaces.

The category **Toph** has as objects all small topological spaces $X, Y, \ldots$, while a morphism $\alpha: X \to Y$ is a homotopy class of continuous maps $f: X \to Y$; in other words, two homotopic maps $f \simeq g: X \to Y$ determine the same morphism from X to Y. The composition of morphisms is the usual composition of homotopy classes of maps. In this category, the homotopy class of an injection need not be a monic, as one may see, for example, for the injection of a circle into a disc (as the bounding circle of that disc). This category **Toph**, which arises naturally in homotopy theory, shows that an arrow in a category need not be the same thing as a function. There are a number of other categories which are useful in homotopy theory: For example, the categories of CW-complexes, of simplicial sets, of compactly generated spaces (see § VII.8), and of Kan complexes.

Set$_*$ will denote the category of small pointed sets (often called "based" sets). By a *pointed set* is meant a nonvoid set P with a selected element, written $*$ or $*_P$ and called the "base point" of P. A map $f: P \to Q$ of pointed sets is a function on the set P to the set Q which carries base point to base point; i.e., which satisfies $f(*_P) = *_Q$. The pointed sets with these maps as morphisms constitute the category **Set**$_*$. In this category the set $\{*\}$ with just one point (the base point) is both an initial and a terminal object. A morphism f is monic in **Set**$_*$ if and only if it has a left inverse, epi if and only if it has a right inverse, and invertible if and only if it is both monic and epic.

Similarly, **Top**$_*$ denotes the category of small pointed topological spaces: the objects are spaces X with a designated base point $*$; the morphisms are continuous maps $f: X \to Y$ which send the base point of X to that of Y. Again, **Toph**$_*$ is the category with objects pointed spaces and morphisms homotopy classes of continuous base-point-preserving maps (where also the homotopies are to preserve base points). Both categories arise in homotopy theory, where the choice of a base point is always needed in defining the fundamental group or higher homotopy groups of a space, cf. § 5.

Binary relations can be regarded as the arrows of a category **Rel**. The objects are all small sets $X, Y, \ldots$, and the arrows $R: X \to Y$ are the binary relations on X to Y; that is, the subsets $R \subset X \times Y$. If $S: Y \to Z$ is another such relation, the composite relation $S \circ R: X \to Z$ is defined to be the usual relative product

$$S \circ R = \{\langle x, z\rangle \mid \text{for some } y \in Y,\ \langle x, y\rangle \in R \quad \text{and} \quad \langle y, z\rangle \in S\}.$$

The identity arrow $X \to X$ is the identity relation on X, consisting of all $\langle x, x\rangle$ for $x \in X$. The axioms for a category evidently hold. This category **Rel** contains **Set** as a subcategory on the same objects, where each function $f: X \to Y$ is interpreted as the relation consisting of all pairs $\langle x, fx\rangle$ for $x \in X$. But **Rel** has added structure: For each $R: X \to Y$ there is a converse relation $R^{\#}: Y \to X$ consisting of all pairs $\langle y, x\rangle$ with $\langle x, y\rangle \in R$.

A *concrete category* is a pair $\langle C, U\rangle$ where C is a category and U a faithful functor $U: C \to$ **Set**. Since U is faithful, we may identify each arrow f of C with the function Uf. In these terms, a concrete category may be described as a category C in which each object c comes equipped with an "underlying" set Uc, each arrow $f: b \to c$ *is* an actual function $Ub \to Uc$, and composition of arrows is composition of functions. Many of the explicit large categories described above are concrete categories in this sense, each relative to its evident forgetful functor U, but this is not so for **Toph** or for **Rel**. For the applications, the notion of category is simpler (and more "abstract") than that of concrete category.

8. Hom-Sets

For objects a and b in the category C the hom-set

$$\hom_C(a, b) = \{f \mid f \text{ is an arrow } f: a \to b \text{ in } C\}$$

consists of all arrows of the category with domain a and codomain b. The notation for this set is frequently and variously abbreviated as

$$\hom_C(a, b) = C(a, b) = \hom(a, b) = (a, b) = (a, b)_C .$$

A category may be defined in terms of hom-sets as follows. A small category is given by the following data:

(i) A set of objects $a, b, c, \ldots$;

(ii) A function which assigns to each ordered pair $\langle a, b \rangle$ of objects a set hom (a, b);

(iii) For each ordered triple $\langle a, b, c \rangle$ of objects a function

$$\hom(b, c) \times \hom(a, b) \to \hom(a, c) ,$$

called composition, and written $\langle g, f \rangle \mapsto g \circ f$ for $g \in \hom(b, c)$, $f \in \hom(a, b)$;

(iv) For each object b, an element $1_b \in \hom(b, b)$, called the identity of b.

These data are required to satisfy the familiar associativity and unit axioms (1.1) and (1.2), plus an added "disjointness" axiom:

(v) If $\langle a, b \rangle \neq \langle a', b' \rangle$, then $\hom(a, b) \cap \hom(a', b') = \emptyset$, where $\emptyset$ is the empty set.

In particular, the associativity axiom may be restated as the requirement that the following diagram, with each arrow given in the evident way by composition, be a commutative diagram:

$$\begin{array}{ccc} \hom(c, d) \times \hom(b, c) \times \hom(a, b) & \longrightarrow & \hom(b, d) \times \hom(a, b) \\ \downarrow & & \downarrow \\ \hom(c, d) \times \hom(a, c) & \longrightarrow & \hom(a, d) \end{array} .$$

This definition of a category is equivalent to the original definition of § 2. Axiom (v) above requires that "distinct" hom-sets be disjoint; it is included to ensure that each arrow have a definite domain and a definite codomain. Should this axiom fail in an example, it can be readily reinstated by adjusting the hom-sets so that they do become disjoint. For example, we can replace each original set $\hom(a, b)$ by the set $\{a\} \times \hom(a, b) \times \{b\}$; this amounts to "labelling" each $f \in \hom(a, b)$ with its domain a and codomain b. Some authors omit this axiom (v).

A functor $T: C \to B$ may be described in terms of hom-sets as the (usual) object function T together with a collection of functions

$$T_{c,c'}: C(c, c') \to B(Tc, Tc')$$

(namely, the functions $f \mapsto Tf$, for $f \in C(c, c')$) such that each $T_{c,c} 1_c = 1_{Tc}$ and such that every diagram

$$\begin{array}{ccc} C(c', c'') \times C(c, c') & \longrightarrow & C(c, c'') \\ \downarrow{\scriptstyle T_{c',c''} \times T_{c,c'}} & & \downarrow{\scriptstyle T_{c,c''}} \\ B(Tc', Tc'') \times B(Tc, Tc') & \longrightarrow & B(Tc, Tc''), \end{array}$$

with horizontal arrows the composition in B and C, is commutative.

We leave the reader to describe a natural transformation $\tau: S \dot{\to} T$ in terms of functions $C(c, c') \to B(Sc, Tc')$.

In many relevant examples, the hom-sets of a category themselves have some structure; for instance, in the category of vector spaces $V, W, \ldots$ over a fixed field, each $\hom(V, W)$ is itself a vector space (of all linear transformations $V \to W$). The simplest such case is that in which the hom-sets are abelian groups. Formally, define an *Ab-category* (also called a *preadditive* category) to be a category A in which each hom-set $A(a, b)$ is an additive abelian group and for which composition is bilinear: For arrows $f, f': a \to b$ and $g, g': b \to c$,

$$(g + g') \circ (f + f') = g \circ f + g \circ f' + g' \circ f + g' \circ f'.$$

Thus **Ab**, R-**Mod**, **Mod**-R, and the like are all Ab-categories.

Because the composition $\langle g, f \rangle \mapsto g \circ f$ is *bilinear*,

$$A(b, c) \times A(a, b) \to A(a, c),$$

it can also be written (using the tensor product $\otimes = \otimes_{\mathbf{Z}}$) as a *linear* map

$$A(b, c) \otimes A(a, b) \to A(a, c),$$

and the Ab-category A may be described completely in these terms (without assuming ahead of time that it is a category). Thus an Ab-category is given by the data

(i) A set of objects $a, b, c, \ldots$;

(ii) A function which assigns to each ordered pair of objects $\langle b, c \rangle$ an abelian group $A(b, c)$;

(iii) For each ordered triple of objects $\langle a, b, c \rangle$ a morphism

$$A(b, c) \otimes A(a, b) \to A(a, c)$$

of abelian groups called composition, and written $g \otimes f \mapsto g \circ f$;

(iv) For each object, a morphism $\mathbf{Z} \to A(a, a)$. (Here $\mathbf{Z}$ is the additive abelian group of integers; this morphism is completely determined by the image of $1 \in \mathbf{Z}$, which may be written as 1_a.)

These data are required to satisfy the associative and unit laws for composition, stated as in (1.1) and (1.2), or by diagrams. The definition of Ab-category is just like the definition of category by hom-sets: **Set** is replaced by **Ab**, cartesian product $\times$ of sets by tensor product in **Ab**, and the one-point set $*$ is replaced by $\mathbf{Z}$. There is an evident generalization to categories A which have hom-objects $A(b, c)$ in a category like **Ab** which is equipped with a multiplication like $\otimes$ and a unit like $\mathbf{Z}$ for this multiplication. These are called "enriched categories" (Kelly [1982]).

If A and B are Ab-categories, a functor $T: A \to B$ is said to be *additive* when every function $T: A(a, a') \to B(Ta, Ta')$ is a homomorphism of abelian groups; that is, when $T(f + f') = Tf + Tf'$ for all parallel pairs f and f'. Clearly, the composite of additive functors is additive. **Ab-cat** will denote the category of all small Ab-categories, with arrows additive functors.

Notes.

These notes, like those at the end of later chapters, are informal remarks on the background and prospects of our subject, with references to the bibliography (for example, H. Pétard [1980b] refers to the second article by Pétard listed for the year 1980).

The fundamental idea of representing a function by an arrow first appeared in topology about 1940, probably in papers or lectures by W. Hurewicz on relative homotopy groups; see [1941].

His initiative immediately attracted the attention of R. H. Fox (see Fox [1943]) and N. E. Steenrod, whose [1941] paper used arrows and (implicitly) functors; see also Hurewicz-Steenrod [1941]). The arrow $f: X \to Y$ rapidly displaced the occasional notation $f(X) \subset Y$ for a function. It expressed well a central interest of topology. Thus a notation (the arrow) led to a concept (category).

Commutative diagrams were probably also first used by Hurewicz.

Categories, functors, and natural transformations themselves were discovered by Eilenberg-Mac Lane [1942a] in their study of limits (via natural transformations) for universal coefficient theorems in Čech cohomology. In this paper commutative diagrams appeared in print (probably for the first time). Thus Ext was one of the first functors considered. A direct treatment of categories in their own right appeared in Eilenberg-Mac Lane [1945]. Now the discovery of ideas as general as these is chiefly the willingness to make a brash or speculative abstraction, in this case supported by the pleasure of purloining words from the philosophers: "Category" from Aristotle and Kant, "Functor" from Carnap *(Logische Syntax der Sprache)*, and "natural transformation" from then current informal parlance. Initially, categories were used chiefly as a language, notably and effectively in the Eilenberg-Steenrod axioms for homology and cohomology theories. With recent increasing use, the question of proper foundations has come to the fore. Here experts are still not in agreement; our present assumption of "one universe" is an adequate stopgap, not a forecast of the future.

Category theory asks of every type of Mathematical object: "What are the morphisms?"; it suggests that these morphisms should be described at the same time as the objects. Categorists, however, ordinarily name their large categories by the common name of the objects; thus **Set**, **Cat**. Only Ehresmann [1965] and his school have the courage to name each category by the common name of its arrows: our **Cat** is their category of functors. This emphasis on (homo)morphisms is largely due to Emmy Noether, who emphasized the use of homomorphisms of groups and rings.

II. Constructions on Categories

1. Duality

Categorical duality is the process "Reverse all arrows". An exact description of this process will be made on an axiomatic basis in this section and on a set-theoretical basis in the next section. Hence for this section a category will not be described by sets (of objects and of arrows) and functions (domain, codomain, composition) but by axioms as in § I.1.

The *elementary theory* of an *abstract category* (ETAC) consists of certain statements Σ which involve letters a, b, c, ... for objects and letters f, g, h, ... for arrows. These statements are the ones built up from the atomic *statements* which involve the usual undefined terms of category theory; thus, atomic statements are "a is the domain of f", "b is the codomain of f", "i is the identity arrow of a", and "g can be composed with f and h is the composite", "$a = b$" and "$f = g$". These atomic statements can also be written as equations in the familiar way: "$a = \operatorname{dom} f$", "$h = g \circ f$". A statement Σ is defined to be any phrase (well formed formula) built up from the types of atomic statements listed above in the usual fashion by means of the ordinary propositional connectives (and, or, not, implies, if and only if) and the usual quantifiers ("for all a", "for all f", "there exists an a ...", "there exists an f ..."). Thus "$f: a \to b$" is the abbreviation we have adopted for the statement, "a is the domain of f and b is the codomain of f".

A *sentence* is a statement with all variables quantified (i.e., all variables are "bound", none being "free"). For example, "for all f there exist a and b with $f: a \to b$" is a sentence (one which in fact is an axiom, true in every category). The axioms of ETAC (as given in § I.1) are certain such sentences.

The *dual* of any statement Σ of ETAC is formed by making the following replacements throughout in Σ: "domain" by "codomain", "codomain" by "domain", and "h is the composite of g with f" by "h is the composite of f with g"; arrows and composites are reversed. Logic (and, or, ...) is unchanged. This gives the following table (a more extensive table appears in Exercise IV.3.1).

Statement Σ	Dual statement Σ^*
$f: a \to b$	$f: b \to a$
$a = \operatorname{dom} f$	$a = \operatorname{cod} f$
$i = 1_a$	$i = 1_a$
$h = g \circ f$	$h = f \circ g$
f is monic	f is epi
u is a right inverse of h	u is a left inverse of h
f is invertible	f is invertible
t is a terminal object	t is an initial object.

Note that the dual of the dual is the original statement ($\Sigma^{**} = \Sigma$). If a statement involves a diagram, the dual statement involves that diagram with all arrows reversed.

The dual of each of the axioms for a category is also an axiom. Hence in any proof of a theorem about an arbitrary category from the axioms, replacing each statement by its dual gives a valid proof (of the dual conclusion). This is the *duality principle*: If a statement Σ of the elementary theory of an abstract category is a consequence of the axioms, so is the dual statement Σ^*. For example, we noted the (elementary) theorem that a terminal object of a category, if it exists, is unique up to isomorphism. Therefore we have the dual theorem: An initial object, if it exists, is unique up to isomorphism. For more complicated theorems, the duality principle is a handy way to have (at once) the dual theorem. No proof of the dual theorem need be given. We usually leave even the formulation of the dual theorem to the reader.

The duality principle also applies to statements involving several categories and functors between them. The simplest (and typical) case is the elementary theory of one functor; i.e., of two categories C and B and a functor $T: C \to B$. For this theory, the atomic statements are those listed above for the category C, a corresponding list for the category B, as well as the statements "$Tc = b$" or "$Tf = h$", giving the values of the object and arrow functions of T on objects c and arrows f of C. The axioms include the axioms for a category for C and for B and also the statements $T(gf) = (Tg)(Tf)$ and $T(1_a) = 1_{Ta}$ which assert that T is a functor. The dual of a statement is formed by simultaneously dualizing the atomic parts referring to C and to B (i.e., reversing arrows in C and in B). Since the statement that T is a functor is self-dual, the duality principle above is still true.

We emphasize that duality for a statement involving *several* categories and functors between them reverses the arrows in each category but does not reverse the functors.

2. Contravariance and Opposites

To each category C we also associate the *opposite* category C^{op}. The objects of C^{op} are the objects of C, the arrows of C^{op} are arrows f^{op}, in one-one correspondence $f \mapsto f^{\mathrm{op}}$ with the arrows f of C. For each arrow $f: a \rightarrow b$ of C, the domain and codomain of the corresponding f^{op} are as in $f^{\mathrm{op}}: b \rightarrow a$ (the direction is reversed). The composite $f^{\mathrm{op}} g^{\mathrm{op}} = (g f)^{\mathrm{op}}$ is defined in C^{op} exactly when the composite $g f$ is defined in C. This clearly makes C^{op} a category. Moreover, the domain of f^{op} is the codomain of f, f^{op} is monic if and only if f is epi, and so on. Indeed, this process translates any statement Σ about C into the dual statement Σ^* about C^{op}. In detail, an evident induction on the construction of Σ from atomic statements proves that if Σ is any statement with free variables $f, g, \ldots$ in the elementary theory of an abstract category, then Σ is true for arrows $f, g, \ldots$ of a category C if and only if the dual statement Σ^* is true for the arrows $f^{\mathrm{op}}, g^{\mathrm{op}}, \ldots$ of the opposite category C^{op}. In particular, a sentence Σ is true in C^{op} if and only if the dual sentence Σ^* is true in C. This observation allows us to interpret the dual of a property Σ as the original property applied to the opposite category (some authors call C^{op} the "dual" category, and write it $C^{\mathrm{op}} = C^*$).

If $T: C \rightarrow B$ is a functor, its object function $c \mapsto Tc$ and its mapping function $f \mapsto Tf$, rewritten as $f^{\mathrm{op}} \mapsto (Tf)^{\mathrm{op}}$, together define a functor from C^{op} to B^{op}, which we denote as $T^{\mathrm{op}}: C^{\mathrm{op}} \rightarrow B^{\mathrm{op}}$. The assignments $C \mapsto C^{\mathrm{op}}$ and $T \mapsto T^{\mathrm{op}}$ define a (covariant!) functor **Cat**$\rightarrow$**Cat**.

Consider a functor $S: C^{\mathrm{op}} \rightarrow B$. By the definition of a functor, it assigns to each object $c \in C^{\mathrm{op}}$ an object Sc of B and to each arrow $f^{\mathrm{op}}: b \rightarrow a$ of C^{op} an arrow $Sf^{\mathrm{op}}: Sb \rightarrow Sa$ of B, with $S(f^{\mathrm{op}} g^{\mathrm{op}}) = (S f^{\mathrm{op}})(S g^{\mathrm{op}})$ whenever $f^{\mathrm{op}} g^{\mathrm{op}}$ is defined. The functor S so described may be expressed directly in terms of the original category C if we write $\bar{S} f$ for $S f^{\mathrm{op}}$; then $\bar{S}$ is a *contravariant functor* on C to B, which assigns to each object $c \in C$ an object $\bar{S} c \in B$ and to each arrow $f: a \rightarrow b$ an arrow $\bar{S} f: \bar{S} b \rightarrow \bar{S} a$ (in the *opposite* direction), all in such a way that

$$\bar{S}(1_c) = 1_{\bar{S}c}, \qquad \bar{S}(f g) = (\bar{S} g)(\bar{S} f), \tag{1}$$

the latter whenever the composite $f g$ is defined in C. Note that the arrow function $\bar{S}$ of a contravariant functor inverts the order of composition.

Specific examples of contravariant functors may be conveniently presented in this form; i.e., as functions $\bar{S}$ inverting composition. An example is the contravariant *power-set* functor $\bar{P}$ on **Set** to **Set**: For each set X, $\bar{P} X = \{S \mid S \subset X\}$ is the set of all subsets of X; for each function $f: X \rightarrow Y$, $\bar{P} f: \bar{P} Y \rightarrow \bar{P} X$ sends each subset $T \subset Y$ to its inverse image $f^{-1} T \subset X$. Another example is the familiar process which assigns to each vector space V its dual (conjugate) vector space V^* and to each linear transformation $f: V \rightarrow W$ its dual $f^*: W^* \rightarrow V^*$; these assignments describe a contravariant functor on the category of all vector spaces (over a fixed field) to itself.

To contrast, a functor $T: C \to B$ as previously defined, in § I.3, is called a *covariant* functor on C to B. For general discussions it is much more convenient to represent a contravariant functor $\bar{S}$ on C to B as a covariant functor $S: C^{\mathrm{op}} \to B$, or sometimes as a covariant functor $S^{\mathrm{op}}: C \to B^{\mathrm{op}}$. In this book an arrow between (symbols for) categories will always denote a covariant functor $T: C \to B$ or $S: C^{\mathrm{op}} \to B$ between the designated categories.

Hom-sets provide an important example of co- and contravariant functors. Suppose that C is a category with small hom-sets, so that each $\hom(a, b) = \{f \mid f: a \to b \text{ in } C\}$ is a small set, hence an object of the category **Set** of all small sets. Thus we have for each object $a \in C$ the *covariant hom-functor*

$$C(a, -) = \hom(a, -): C \to \mathbf{Set}; \tag{2}$$

its object function sends each object b to the set $\hom(a, b)$; its arrow function sends each arrow $k: b \to b'$ to the function

$$\hom(a, k): \hom(a, b) \to \hom(a, b') \tag{3}$$

defined by the assignment $f \mapsto k \circ f$ for each $f: a \to b$. To simplify the notation, this function $\hom(a, k)$ is sometimes written k_* and called "composition with k on the left", or "the map induced by k".

The *contravariant hom-functor*, for each object $b \in C$, will be written covariantly, as

$$C(-, b) = \hom(-, b): C^{\mathrm{op}} \to \mathbf{Set}; \tag{4}$$

it sends each object a to the set $\hom(a, b)$, and each arrow $g: a \to a'$ of C to the function

$$\hom(g, b): \hom(a', b) \to \hom(a, b) \tag{5}$$

defined by $f \mapsto f \circ g$. Omitting the object b, this function $\hom(g, b)$ is sometimes written simply as g^* and called "composition with g on the right". Thus, for each $f: a' \to b$,

$$k_* f = k \circ f, \qquad g^* f = f \circ g.$$

For two such arrows $g: a \to a'$ and $k: b \to b'$ the diagram

$$\begin{array}{ccc} \hom(a', b) & \xrightarrow{g^*} & \hom(a, b) \\ {\scriptstyle k_*}\downarrow & & \downarrow{\scriptstyle k_*} \\ \hom(a', b') & \xrightarrow{g^*} & \hom(a, b') \end{array} \tag{6}$$

in **Set** is commutative, because both paths send $f \in \hom(a', b)$ to $k f g$.

These hom-functors have been defined only for a category C with small hom-sets. The familiar large categories **Grp**, **Set**, **Top**, etc. do have this property. To include categories without this property, we can proceed as follows: Given a category C, take a set V large enough to include all subsets of the set of arrows of C (for example, V could be the power set of the set of arrows of C). Let $\mathbf{Ens} = \mathbf{Set}_V$ be the category with objects all sets $X \in V$, arrows all functions $f: X \to Y$ between two such sets and composition the usual composition of functions. Then each hom-set $C(a, b) = \hom(a, b)$ is an object of this category **Ens**, so the above procedure defines two hom-functors

$$C(a, -): C \to \mathbf{Ens}, \qquad C(-, b): C^{\mathrm{op}} \to \mathbf{Ens}. \tag{7}$$

In particular, when V is the universe of all small sets, **Ens** = **Set**; in general, **Ens** is a (variable) category of sets which acts as a receiving category for the hom-functors of a category or categories of interest.

There are many other examples of contravariant functors. For X a topological space, the set **Open**(X) of all open subsets U of X, when ordered by inclusion, is a partial order and hence a category; there is an arrow $V \to U$ precisely when $V \subset U$. Let $\bar{C}(U)$ denote the set of all continuous real-valued functions $h: U \to \mathbf{R}$; the assignment $h \mapsto h|V$ restricting each h to the subset V is a function $\bar{C}(U) \to \bar{C}(V)$ for each $V \subset U$. This makes $\bar{C}$ a contravariant functor on **Open**(X) to **Set**. This functor is called the *sheaf* of germs of continuous functions on X. On a smooth manifold, the sheaf of germs of C^{∞}-differentiable functions is constructed in similar fashion (cf. Mac Lane-Moerdijk [1992]).

Mod-R is a contravariant functor from rings R to categories. Specifically, if $\varrho: R \to S$ is any morphism of (small) rings, each right S-module B becomes a right R-module $B\varrho = (\mathrm{Mod}\,\varrho)B$ by "pull-back" along ϱ: Each $r \in R$ acts on $b \in B$ by $b \cdot r = b \cdot (\varrho r)$. Clearly $\mathrm{Mod}\,\varrho$ is a functor **Mod**-$S \to$ **Mod**-R, and $\mathrm{Mod}(\varrho_1 \varrho_2) = (\mathrm{Mod}\,\varrho_2)(\mathrm{Mod}\,\varrho_1)$, so Mod itself can be regarded as a contravariant functor on **Rng** to **Cat**′, the category of all large categories.

One may also form the category **Mod** of all (right) modules over *all* rings. An object of **Mod** is a pair $\langle R, A \rangle$, where R is a small ring and A a small right R-module. A morphism $\langle R, A \rangle \to \langle S, B \rangle$ is a pair $\langle \varrho, f \rangle$, where $\varrho: R \to S$ is a morphism of rings and $f: A \to (\mathrm{Mod}\,\varrho)B$ is a morphism of right R-modules. With the evident composition, this yields a category **Mod**. A projection functor **Mod** $\to$ **Rng** is given by $\langle R, A \rangle \mapsto R$. Further study of the relation of this functor to the previous functor **Rng** $\to$ **Cat**′ leads to the theory of *fibered categories*. (**Mod** is fibered over **Rng**, the fiber over each R being the category **Mod**-R.)

3. Products of Categories

From two given categories B and C we construct a new category $B \times C$, called the *product* of B and C, as follows. An object of $B \times C$ is a pair $\langle b, c \rangle$ of objects b of B and c of C; an arrow $\langle b, c \rangle \to \langle b', c' \rangle$ of $B \times C$ is a pair $\langle f, g \rangle$ of arrows $f: b \to b'$ and $g: c \to c'$, and the composite of two such arrows

$$\langle b, c \rangle \xrightarrow{\langle f, g \rangle} \langle b', c' \rangle \xrightarrow{\langle f', g' \rangle} \langle b'', c'' \rangle$$

is defined in terms of the composites in B and C by

$$\langle f', g' \rangle \circ \langle f, g \rangle = \langle f' \circ f, g' \circ g \rangle . \tag{1}$$

Functors

$$B \xleftarrow{P} B \times C \xrightarrow{Q} C ,$$

called the *projections* of the product, are defined on (objects and) arrows by

$$P \langle f, g \rangle = f, \qquad Q \langle f, g \rangle = g .$$

They have the following property: Given any category D and two functors

$$B \xleftarrow{R} D \xrightarrow{T} C ,$$

there is a unique functor $F: D \to B \times C$ with $PF = R$, $QF = T$; explicitly, these two conditions require that Fh, for any arrow h in D, must be $\langle Rh, Th \rangle$; conversely, this value for Fh does make F a functor with the required properties. The construction of F (dotted arrow) may be visualized by the following commutative diagram of functors:

$$\begin{array}{ccccc} & & D & & \\ & \swarrow^{R} & \downarrow^{F} & \searrow^{T} & \\ B & \xleftarrow{P} & B \times C & \xrightarrow{Q} & C . \end{array} \tag{2}$$

This property of the product category states that the projections P and Q are "universal" among pairs of functors to B and C. It is exactly like a similar property of the projections from the (cartesian) product of two sets, two groups, or two spaces. The general properties of such products in any category will be considered in Chapter III.

Two functors $U: B \to B'$ and $V: C \to C'$ have a product $U \times V: B \times C \to B' \times C'$ which may be defined explicitly on objects and arrows as

$$(U \times V) \langle b, c \rangle = \langle Ub, Vc \rangle \qquad (U \times V) \langle f, g \rangle = \langle Uf, Vg \rangle .$$

Alternatively, this functor $U \times V$ may be described as the unique functor (as in the diagram above) which makes the following diagram commutative:

$$\begin{array}{ccccc} B & \xleftarrow{P} & B \times C & \xrightarrow{Q} & C \\ {\scriptstyle U}\downarrow & & \downarrow{\scriptstyle U \times V} & & \downarrow{\scriptstyle V} \\ B' & \xleftarrow{P'} & B' \times C' & \xrightarrow{Q'} & C' . \end{array} \tag{3}$$

The product $\times$ is thus a pair of functions: To each pair $\langle B, C \rangle$ of categories, a new category $B \times C$; to each pair of functors $\langle U, V \rangle$, a new functor $U \times V$. Moreover, when the composites $U' \circ U$ and $V' \circ V$ are defined, one clearly has $(U' \times V') \circ (U \times V) = U'U \times V'V$. Hence the operation $\times$ itself is a functor; more exactly, on restricting to small categories, it is a functor

$$\times : \mathbf{Cat} \times \mathbf{Cat} \to \mathbf{Cat} .$$

There are similar functors $\mathbf{Grp} \times \mathbf{Grp} \to \mathbf{Grp}$, $\mathbf{Top} \times \mathbf{Top} \to \mathbf{Top}$, etc.

Our definition of product categories has included in (2) the description of functors $F : D \to B \times C$ to a product category. On the other hand, functors $S : B \times C \to D$ from a product category are called *bifunctors* (on B and C) or functors of two variable objects (in B and in C). Such bifunctors occur frequently; for instance, the cartesian product $X \times Y$ of two sets X and Y is (the object function of) a bifunctor $\mathbf{Set} \times \mathbf{Set} \to \mathbf{Set}$. Thus our definition of product category gives an automatic definition of "functor of two variables" – just as the definition of the product $X \times Y$ of two topological spaces gives an automatic definition of "continuous function of two variables".

Fix one argument in a bifunctor S; the result is an ordinary functor of the remaining argument. The whole bifunctor S is determined by these two arrays of one-variable functors in the following elementary way.

Proposition 1. *Let B, C, and D be categories. For all objects $c \in C$ and $b \in B$, let*

$$L_c : B \to D , \qquad M_b : C \to D$$

be functors such that $M_b(c) = L_c(b)$ for all b and c. Then there exists a bifunctor $S : B \times C \to D$ with $S(-, c) = L_c$ for all c and $S(b, -) = M_b$ for all b if and only if for every pair of arrows $f : b \to b'$ and $g : c \to c'$ one has

$$M_{b'} g \circ L_c f = L_{c'} f \circ M_b g . \tag{4}$$

These equal arrows (4) *in D are then the value $S(f, g)$ of the arrow function of S at f and g.*

Proof. If we write b and c for the corresponding identity arrows, the definition (1) of the composite in $B \times C$ shows that

$$\langle b', g \rangle \circ \langle f, c \rangle = \langle b'f, gc \rangle = \langle f, g \rangle = \langle fb, c'g \rangle = \langle f, c' \rangle \circ \langle b, g \rangle .$$

Applying the functor S to this equation gives

$$S(b', g)\, S(f, c) = S(f, c')\, S(b, g)\,;$$

as a commutative diagram this condition is

$$\begin{array}{ccc} S(b, c) & \xrightarrow{S(b,g)} & S(b, c') \\ {\scriptstyle S(f,c)}\downarrow & & \downarrow{\scriptstyle S(f,c')} \\ S(b', c) & \xrightarrow{S(b',g)} & S(b', c')\,. \end{array}$$

This is just condition (4) rewritten, so that condition (4) is necessary. Conversely, given all L_c and M_b, this condition defines $S(f, g)$ for every pair f, g; it may be verified that this definition does yield a bifunctor S with the required properties.

One may also form products of three or more categories, or combine the construction of product categories and opposite categories. There is an evident isomorphism $(B \times C)^{\mathrm{op}} \cong B^{\mathrm{op}} \times C^{\mathrm{op}}$. A functor $B^{\mathrm{op}} \times C \to D$ is often called a bifunctor, contravariant in B and covariant in C, with values in D. For example, if C is a category with small hom-sets, the hom-sets define such a bifunctor

$$\hom : C^{\mathrm{op}} \times C \to \mathbf{Set}\,.$$

Indeed, the commutative diagram (6) of § 2 shows exactly that the co- and contravariant hom-functors

$$\hom(-, c) : C^{\mathrm{op}} \to \mathbf{Set}\,, \qquad \hom(b, -) : C \to \mathbf{Set}$$

do satisfy the condition (4) of the theorem, necessary to make hom a bifunctor.

Next consider natural transformations between bifunctors $S, S' : B \times C \to D$. Let α be a function which assigns to each pair of objects $b \in B$, $c \in C$ an arrow

$$\alpha(b, c) : S(b, c) \to S'(b, c) \tag{5}$$

in D. Call α *natural* in b if for each $c \in C$ the components $\alpha(b, c)$ for all b define

$$\alpha(-, c) : S(-, c) \dot{\to} S'(-, c)\,,$$

a natural transformation of functors $B \to D$. The reader may readily prove the useful result:

Proposition 2. *For bifunctors S, S', the function α displayed in* (5) *is a natural transformation $\alpha : S \dot{\to} S'$ (i.e., of bifunctors) if and only if $\alpha(b, c)$ is natural in b for each $c \in C$ and natural in c for each $b \in B$.*

Such natural transformations appear in the fundamental definition of adjoint functors (Chapter IV). A functor $F : X \to C$ is the *left adjoint* of a functor $G: C \to X$ (opposite direction) when there is a bijection

$$\hom_C(Fx, c) \cong \hom_X(x, Gc)$$

natural in $x \in X$ and $c \in C$. Here $\hom_C(F-, -)$ is a bifunctor, the composite

$$X^{\text{op}} \times C \xrightarrow{F^{\text{op}} \times \text{Id}} C^{\text{op}} \times C \xrightarrow{\hom_C} \mathbf{Set},$$

and $\hom_X(-, G-)$ similarly (at least when X and C have small hom-sets).

The product category can be visualized in the case $C \times \mathbf{2}$, where $\mathbf{2}$ is the category with one non-identity arrow $0 \to 1$; explicitly $C \times \mathbf{2}$ consists of two copies $C \times 0$ and $C \times 1$ of C with arrows joining the first to the second, as in the figure ("diagonal" arrows omitted) for $C = \mathbf{3}$ which is the triangle category of § I.2:

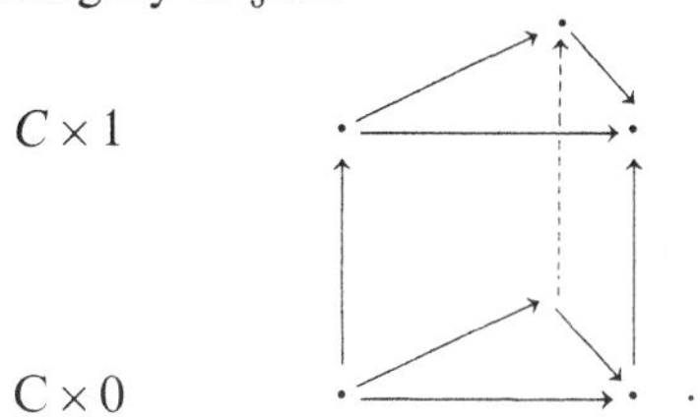

Here the functors $T_0, T_1 : C \to C \times \mathbf{2}$ ("bottom" and "top", respectively) are defined for each arrow f of C by $T_0 f = \langle f, 0 \rangle$ and $T_1 f = \langle f, 1 \rangle$. If $\downarrow$ denotes the unique non-identity arrow $0 \to 1$ of $\mathbf{2}$, then we may define a transformation between $T_0, T_1 : C \to C \times \mathbf{2}$ by

$$\mu : T_0 \dot{\to} T_1, \quad \mu c = \langle c, \downarrow \rangle,$$

for any object c. It maps "bottom" to "top" and is clearly natural. We call μ the *universal natural transformation* from C for the following reason. Given any natural transformation $\tau : S \dot{\to} T$ between $S, T: C \to B$ there is a unique functor $F : C \times \mathbf{2} \to B$ with $F\mu c = \tau c$ for any object c. Specifically, F is, when $f : c \to c'$,

$$F\langle f, 0 \rangle = Sf, \quad F\langle f, 1 \rangle = Tf, \quad F\langle f, \downarrow \rangle = Tf \circ \tau c = \tau c' \circ Sf. \tag{6}$$

It may be readily verified that these assignments do define a bifunctor $F : C \times \mathbf{2} \to B$, and that $F\mu = \tau$.

Exercises

1. Show that the product of categories includes the following known special cases: The product of monoids (categories with one object), of groups, of sets (discrete categories).
2. Show that the product of two preorders is a preorder.

3. If $\{C_i \mid i \in I\}$ is a family of categories indexed by a set I, describe the product $C = \Pi_i C_i$, its projections $P_i : C \to C_i$, and establish the universal property of these projections.
4. Describe the opposite of the category $\mathbf{Matr}_K$ of § I.2.
5. Show that the ring of continuous real-valued functions on a topological space is the object function of a contravariant functor on **Top** to **Rng**.

4. Functor Categories

Given categories C and B, we consider all functors $R, S, T, \ldots : C \to B$. If $\sigma : R \dot{\to} S$ and $\tau : S \dot{\to} T$ are two natural transformations, their components for each $c \in C$ define composite arrows $(\tau \cdot \sigma)c = \tau c \quad \sigma c$ which are the components of a transformation $\tau \cdot \sigma : R \dot{\to} T$. To show $\tau \cdot \sigma$ natural, take any $f : c \to c'$ in C and consider the diagram

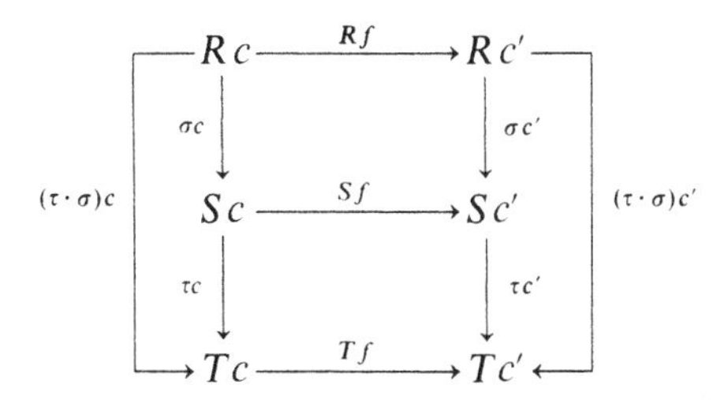

Since σ and τ are natural, both small squares are commutative. Hence the rectangle commutes, so the composite $\tau \cdot \sigma$ is natural.

This composition of transformations is associative; moreover it has for each functor T an identity, the natural transformation $1_T : T \to T$ with components $1_T c = 1_{Tc}$. Hence, given the categories B and C, we may construct formally a *functor category* $B^C = \mathrm{Funct}(C, B)$ with objects the functors $T : C \to B$ and morphisms the natural transformations between two such functors. It is often suggestive to write

$$\mathrm{Nat}(S, T) = B^C(S, T) = \{\tau \mid \tau : S \dot{\to} T \quad \text{natural}\} \tag{1}$$

for the "hom-set" of this category. It need not be a small set.

Functor categories will be used extensively. For example, if B and C are sets (categories with all arrows identities), then B^C is also a set; namely, the familiar "function-set" consisting of all functions $C \to B$. In particular, for $B = \{0, 1\}$ a two-point set, $\{0, 1\}^C$ is (isomorphic to) the set of all subsets of C (the "power set" $\mathscr{P}C$). For any category B, $B^{\mathbf{1}}$ is isomorphic to B, while $B^{\mathbf{2}}$ is called the *category of arrows* of B; its objects are arrows $f : a \to b$ of B, and its arrows $f \to f'$ are those pairs

$\langle h, k \rangle$ of arrows in B for which the square

$$\begin{array}{ccc} a & \xrightarrow{h} & a' \\ {\scriptstyle f}\downarrow & & \downarrow{\scriptstyle f'} \\ b & \xrightarrow{k} & b' \end{array} \tag{2}$$

commutes. If M is a monoid (category with one object) $\mathbf{Set}^M$ is the category with objects the actions of M (on some set) and arrows the morphisms of such actions. An object of the functor category $\mathbf{Grp}^M$ is a group with operators M.

If K is a commutative ring and G a group, then the functor category $(K\text{-}\mathbf{Mod})^G$ is the category of (K-linear) representations of G. Specifically, each functor $T: G \to K\text{-}\mathbf{Mod}$ is determined by a K-module V (the image of the single object of the category G) and a morphism $T: G \to \operatorname{Aut}(V)$ of groups (a representation of G by linear transformations $V \to V$). If T' is a second such representation, a natural transformation $\sigma: T \dot{\to} T'$ is given by a single arrow $\sigma: V \to V'$ (its component at the single object of G) such that the diagram

$$\begin{array}{ccc} V & \xrightarrow{\sigma} & V' \\ {\scriptstyle Tg}\downarrow & & \downarrow{\scriptstyle T'g} \\ V & \xrightarrow{\sigma} & V' \end{array} \tag{3}$$

commutes for every $g \in G$. In representation theory, such a σ is called an *intertwining* operator. Thus $(K\text{-}\mathbf{Mod})^G$ is the category with objects the representations of G and morphisms the intertwining operators.

When the category C is large, the functor category B^C need not be a subset of the universe. For example, if $B = \{0, 1\}$ is the set with just two elements, while C is the set U, then a functor $U \to B$ is just a function on U to a set with two elements. The possible such functions correspond (as characteristic functions) to the possible subsets of U. Therefore the set of objects in $\{0, 1\}^U$ is equivalent to the set $\mathscr{P}(U)$ of all subsets of U, and this set has a larger cardinal number than U.

Exercises

1. For R a ring, describe $R\text{-}\mathbf{Mod}$ as a full subcategory of the functor category $\mathbf{Ab}^R$.
2. Describe B^X, for X a finite set (a finite discrete category).
3. Let $\mathbf{N}$ be the discrete category of natural numbers. Describe the functor category $\mathbf{Ab}^{\mathbf{N}}$ (commonly known as the category of graded abelian groups).
4. If P and Q are preorders, describe the functor category Q^P and show that it is a preorder.

5. If **Fin** is the category of all finite sets and G is a finite group, describe $\mathbf{Fin}^G$ (the category of all permutation representations of G).
6. Let $\mathbf{M}$ be the infinite cyclic monoid (elements $1, m, m^2, \ldots$). In the functor categories $(\mathbf{Matr}_K)^{\mathbf{2}}$ and $(\mathbf{Matr}_K)^{\mathbf{M}}$ show that objects are matrices and isomorphic objects (matrices) are exactly equivalent and similar matrices, respectively, in the usual sense of linear algebra. For **Matr**, see § I.2.
7. Given categories B, C, and the functor category $B^{\mathbf{2}}$, show that each functor $H: C \to B^{\mathbf{2}}$ determines two functors $S, T: C \to B$ and a natural transformation $\tau: S \dot{\to} T$, and show that this assignment $H \mapsto \langle S, T, \tau \rangle$ is a bijection.
8. Relate the functor H of Exercise 7 to the functor F of (3.6).

5. The Category of All Categories

We have defined a "vertical" composite $\tau \cdot \sigma$,

$$C \overset{\downarrow \sigma}{\underset{\downarrow \tau}{\rightrightarrows}} B ,$$

of two natural transformations. There is another "horizontal" composition for natural transformations. Given functors and natural transformations

$$C \underset{T}{\overset{S}{\rightrightarrows}} B \underset{T'}{\overset{S'}{\rightrightarrows}} A \qquad (\downarrow \tau, \ \downarrow \tau') \tag{1}$$

one may form first the composite functors $S' \circ S$ and $T' \ T: C \to A$ and then construct a square

$$\begin{array}{ccc} S'Sc & \xrightarrow{\tau' Sc} & T'Sc \\ {\scriptstyle S'\tau c}\downarrow & & \downarrow{\scriptstyle T'\tau c} \\ S'Tc & \xrightarrow{\tau' Tc} & T'Tc \end{array}$$

which is commutative because of the naturality of τ' for the arrows τc of B. Now define $(\tau' \circ \tau)c$ to be the diagonal of this square;

$$(\tau' \circ \tau)c = T'\tau c \circ \tau' Sc = \tau' Tc \circ S'\tau c . \tag{2}$$

To show $\tau' \circ \tau: S'S \dot{\to} T'T$ natural, form the diagram

$$\begin{array}{ccccccc} S'Sc & \xrightarrow{S'\tau c} & S'Tc & \xrightarrow{\tau' Tc} & T'Tc & & c \\ {\scriptstyle S'Sf}\downarrow & & \downarrow{\scriptstyle S'Tf} & & \downarrow{\scriptstyle T'Tf} & & \downarrow{\scriptstyle f} \\ S'Sb & \xrightarrow[S'\tau b]{} & S'Tb & \xrightarrow[\tau' Tb]{} & T'Tb , & & b \end{array}$$

for any arrow f of C. Horizontally, the composites by definition are $(\tau' \circ \tau)c$ and $(\tau' \circ \tau)b$; the left-hand square commutes because τ is natural and S' is a functor, while the right-hand square commutes because τ' is natural and $Tf: Tc \to Tb$ is an arrow. The commutativity of the outside of the diagram states that $\tau' \circ \tau$ is natural.

This composition $\langle \tau', \tau \rangle \to \tau' \circ \tau$ is readily shown to be associative. It moreover has identities. If $I_B: B \to B$ is the identity functor for the category B and $1_B: I_B \dot{\to} I_B$ the identity natural transformation of that functor to itself, one has $1_B \circ \tau = \tau$ and $\tau' \circ 1_B = \tau'$. Thus 1_B is the identity for the composition $\circ$; it is also the identity for the composition $\cdot$. It is convenient to let the symbol S for a functor also denote the identity transformation $S \dot{\to} S$. With this notation in the situation above we have composite natural transformations

$$S' \circ \tau : S' \circ S \dot{\to} S' \circ T, \qquad \tau' \circ T: S' \circ T \dot{\to} T' \circ T.$$

The definition (2) can then be rewritten, using also the vertical composition, as

$$\tau' \circ \tau = (T' \circ \tau) \cdot (\tau' \circ S) = (\tau' \circ T) \cdot (S' \circ \tau). \tag{3}$$

There is a more general rule. Given three categories and four transformations

$$C \underset{\longrightarrow}{\overset{\longrightarrow}{\xrightarrow{\downarrow\sigma \ \downarrow\tau}}} B \underset{\longrightarrow}{\overset{\longrightarrow}{\xrightarrow{\downarrow\sigma' \ \downarrow\tau'}}} A\,, \tag{4}$$

the "vertical" composites under $\cdot$ and the "horizontal" composites under $\circ$ are related by the identity *(interchange law)*

$$(\tau' \cdot \sigma') \circ (\tau \cdot \sigma) = (\tau' \circ \tau) \cdot (\sigma' \circ \sigma). \tag{5}$$

The reader may enjoy writing down the evident diagrams needed to prove this fact.

These results may be summarized as follows (considering only *small* categories):

Theorem 1. *The collection of all natural transformations is the set of arrows of two different categories under two different operations of composition, $\cdot$ and $\circ$, which satisfy the interchange law (5). Moreover, any arrow (transformation) which is an identity for the composition $\circ$ is also an identity for the composition $\cdot$.*

Note that the objects for the horizontal composition $\circ$ are the categories, for the vertical composition, the functors. In using these compositions, the symbol $\circ$ for the "horizontal" composition is often omitted (as it is usually in writing composition of arrows in a category), while the solid dot designating "vertical" composition is retained. Observe that objects and arrows of C may be written as functors $c: \mathbf{1} \to C$ or $f: \mathbf{2} \to C$; then symbols such as $\sigma \circ c = \sigma c$ have their accepted meaning in a situation such as

$$\mathbf{1} \xrightarrow{c} C \underset{\longrightarrow}{\overset{\longrightarrow}{\downarrow\sigma}} B\,.$$

By a *double category* (Ehresmann) is meant a set which (like the set of all natural transformations) is the set of arrows for two different compositions which together satisfy (5). A 2-*category* (short for two-dimensional category) is a double category in which every identity arrow for the first composition is also an identity for the second composition. For example, the category of all commutative squares in **Set** is a double category (under the evident horizontal and vertical compositions) but not a 2-category. There are also n-categories for higher n, see Chapter XII.

Two (partially defined) binary operations $\cdot$ and $\circ$ are said to satisfy the *interchange law* when (5) holds wherever the composites on either side are defined. Here some other examples. If C is a category and $\cdot : C \times C \to C$ is a functor (for example, a tensor product), while σ, σ', τ and τ' are arrows of C such that the composites $\sigma' \circ \sigma$ and $\tau' \circ \tau$ are defined, then the interchange law (5) holds; indeed, it is precisely the requirement that the functor $\cdot$ preserve composition $\circ$. If σ, σ', τ, and τ' are square matrices such that the usual matrix products $\sigma' \circ \sigma$ and $\tau' \circ \tau$ are defined, while $\tau \cdot \sigma$ denotes the matrix

$$\begin{pmatrix} \tau & 0 \\ 0 & \sigma \end{pmatrix}$$

with blocks τ and σ along the diagonal, zeros elsewhere, then (5) holds.

The functor category B^C is itself a functor of the categories B and C, covariant in B and contravariant in C. Specifically, if we consider only the category **Cat** of all small categories, it is a functor $\mathbf{Cat}^{\mathrm{op}} \times \mathbf{Cat} \to \mathbf{Cat}$; the object function sends a pair of categories $\langle C, B \rangle$ to the functor category B^C, and the arrow function sends a pair of functors $F : B \to B'$ and $G : C' \to C$ to the functor

$$F^G : B^C \to B'^{C'}$$

defined on objects $S \in B^C$ as $F^G S = F \circ S \circ G$ and on arrows $\tau : S \dot{\to} T$ in B^C as $F^G \tau = F \circ \tau \circ G$. Note, for example, that F^C is just "compose with F on the left" while B^G is "compose with G on the right". This functor is an exact analogue to the hom-functor $\mathbf{Set}^{\mathrm{op}} \times \mathbf{Set} \to \mathbf{Set}$.

Exercises

1. For small categories A, B, and C establish a bijection

$$\mathbf{Cat}(A \times B, C) \cong \mathbf{Cat}(A, C^B).$$

and show it natural in A, B, and C. Hence show that $- \times B : \mathbf{Cat} \to \mathbf{Cat}$ has a right adjoint (see Chapter IX).

2. For categories A, B, and C establish natural isomorphisms

$$(A \times B)^C \cong A^C \times B^C, \qquad C^{A \times B} \cong (C^B)^A.$$

Compare the second isomorphism with the bijection of Exercise 1.

3. Use Theorem 1 to show that horizontal composition is a functor

$$\circ : A^B \times B^C \to A^C .$$

4. Let G be a topological group with identity element e, while σ, σ', τ, τ', are continuous paths in G starting and ending at e (thus, if I is the unit interval, $\sigma : I \to G$ is continuous with $\sigma(0) = e = \sigma(1)$). Define $\tau \circ \sigma$ to be the path σ followed by the path τ, as in (I.5.1). Define $\tau \cdot \sigma$ to be the pointwise product of τ and σ, so that $(\tau \cdot \sigma)t = (\tau t)(\sigma t)$ for $0 \leq t \leq 1$. Prove that the interchange law (5) holds.
5. (Hilton-Eckmann). Let S be a set with two (everywhere defined) binary operations $\cdot : S \times S \to S$, $\circ : S \times S \to S$ which both have the same (two-sided) unit element e and which satisfy the interchange identity (5). Prove that $\cdot$ and $\circ$ are equal, and that each is commutative.
6. Combine Exercises 4 and 5 to prove that the fundamental group of a topological group is abelian.
7. If $T: A \to D$ is a functor, show that its arrow functions $T_{a,b} : A(a, b) \to D(Ta, Tb)$ define a natural transformation between functors $A^{\mathrm{op}} \times A \to \mathbf{Set}$.
8. For the identity functor I_C of any category, the natural transformations $\alpha : I_C \dot{\to} I_C$ form a commutative monoid. Find this monoid in the cases $C = \mathbf{Grp}$, **Ab**, and **Set**.

6. Comma Categories

There is another general construction of a category whose objects are certain arrows, as in the following several special cases.

If b is an object of the category C, the category of *objects under* b is the category $(b \downarrow C)$ with objects all pairs $\langle f, c \rangle$, where c is an object of C and $f : b \to c$ an arrow of C, and with arrows $h : \langle f, c \rangle \to \langle f', c' \rangle$ those arrows $h : c \to c'$ of C for which $h \circ f = f'$. Thus an object of $(b \downarrow C)$ is just an arrow in C from b and an arrow of $(b \downarrow C)$ is a commutative triangle with top vertex b. In displayed form:

$$\text{objects } \langle f, c \rangle : \begin{array}{c} b \\ \downarrow f \\ c \end{array}; \quad \text{arrows } \langle f, c \rangle \xrightarrow{h} \langle f', c' \rangle : \begin{array}{ccc} & b & \\ f \swarrow & & \searrow f' \\ c & \xrightarrow{h} & c' \end{array} . \tag{1}$$

The composition of arrows in $(b \downarrow C)$ is then given by the composition in C of the base arrows h of these triangles.

For example, if $*$ denotes any one-point set, while X is any set, each function $* \to X$ is just a selection of a point in the set X; hence $(* \downarrow \mathbf{Set})$ is just the category of pointed sets (§I.7). Similarly, $(\mathbf{Z} \downarrow \mathbf{Ab})$ is the category of abelian groups, each with a selected element.

If a is an object of C, the category $(C \downarrow a)$ of *objects over* a has

$$\text{objects: } \begin{array}{c} c \\ \downarrow f \\ a \end{array}; \quad \text{arrows: } \begin{array}{ccc} c & \xrightarrow{h} & c' \\ f \searrow & & \swarrow f' \\ & a & \end{array} , \tag{2}$$

the triangle commutative. For example, $*$ is terminal in **Set** so there is always a unique $X \to *$; therefore $(\mathbf{Set} \downarrow *)$ is isomorphic to **Set**. Or again, $\mathbf{Z}$ is a ring, and the category $(\mathbf{Rng} \downarrow \mathbf{Z})$ is the category whose objects are rings equipped with a morphism $\varepsilon: R \to \mathbf{Z}$ (called a ring R with an "augmentation" ε) and whose morphisms are morphisms of rings preserving the augmentation.

If b is an object of C and $S: D \to C$ a functor, the category $(b \downarrow S)$ of *objects S-under b* has as objects all pairs $\langle f, d \rangle$ with $d \in \operatorname{Obj} D$ and $f: b \to Sd$ and as arrows $h: \langle f, d \rangle \to \langle f', d' \rangle$ all those arrows $h: d \to d'$ in D for which $f' = Sh \circ f$. In pictures,

$$\text{objects:} \quad \begin{array}{c} b \\ \downarrow f \\ Sd \end{array} ; \qquad \text{arrows } h: \quad \begin{array}{ccc} & b & \\ {}^{f}\swarrow & & \searrow^{f'} \\ Sd & \xrightarrow[Sh]{} & Sd' \end{array} \text{ (commutative)}. \tag{3}$$

Again, composition is given by composition of the arrows h in D. Note especially that equality of arrows in $(b \downarrow S)$ means their equality as arrows of D.

For example, let $U: \mathbf{Grp} \to \mathbf{Set}$ be the forgetful functor. Then for each set x an object of $(x \downarrow U)$ is a function $x \to Ug$ from x into the underlying set of some group g; for example, the function mapping x into the underlying set of the free group generated by the elements of the set x is one such object. This category $(x \downarrow U)$ – and others like it – will be used extensively in the treatment of adjoint functors.

Again, if $a \in C$ and $T: E \to C$ is a functor, one may construct a category $(T \downarrow a)$ of objects T-over a.

Here is the general construction. Given categories and functors

$$E \xrightarrow{T} C \xleftarrow{S} D$$

the *comma category* $(T \downarrow S)$, also written (T, S), has as objects all triples $\langle e, d, f \rangle$, with $d \in \operatorname{Obj} D$, $e \in \operatorname{Obj} E$, and $f: Te \to Sd$, and as arrows $\langle e, d, f \rangle \to \langle e', d', f' \rangle$ all pairs $\langle k, h \rangle$ of arrows $k: e \to e'$, $h: d \to d'$ such that $f' \circ Tk = Sh \circ f$. In pictures,

$$\text{objects } \langle e, d, f \rangle: \quad \begin{array}{c} Te \\ \downarrow f \\ Sd \end{array} ; \qquad \text{arrows } \langle k, h \rangle: \quad \begin{array}{ccc} Te & \xrightarrow{Tk} & Te' \\ \downarrow f & & \downarrow f' \\ Sd & \xrightarrow{Sh} & Sd' \end{array}, \tag{4}$$

with the square commutative. The composite $\langle k', h' \rangle \circ \langle k, h \rangle$ is $\langle k' \circ k, h' \circ h \rangle$, when defined.

This general description of the comma category $(T \downarrow S)$ does include all the cases listed. Indeed, an object b of C may be regarded as a functor

$b: \mathbf{1} \to C$. Taking $T = b$ in this sense, the comma category $(T \downarrow S)$ becomes the category $(b \downarrow S)$ of objects S-under b. If $S = C$ is the identity functor of C, this becomes in particular the category $(b \downarrow C)$ of objects of C under b. Similarly, one may take S to be a functor $\mathbf{1} \to C$; i.e., an object a of C. Again, take $S = T =$ the identity functor of C. Then $(C \downarrow C)$ is exactly the category $C^{\mathbf{2}}$ of all arrows of C. Or take S and T to be objects a and b of C; then $(T \downarrow S) = (b \downarrow a)$ is the category with objects all arrows $f: b \to a$ and morphisms only the identity arrow for each object; in other words $(b \downarrow a)$ is the set (the discrete category) $\hom_C(b, a)$. This case is the reason for the choice of the name "comma category" and the notation (T, S) – a notation which we avoid because the comma is already overworked.

The construction of the comma category $(T \downarrow S)$ may be visualized by the following commutative diagram of categories and functors

$$\begin{array}{c} (T \downarrow S) \\ P \swarrow \quad \downarrow R \quad \searrow Q \\ E \xrightarrow{T} C \xleftarrow{C^{d_0}} C^{\mathbf{2}} \xrightarrow{C^{d_1}} C \xleftarrow{S} D \end{array}, \tag{5}$$

here d_0, d_1 are the two functors $\mathbf{1} \to \mathbf{2}$, the functor category $C^{\mathbf{2}}$ is just the category of arrows f of C, and so the functors C^{d_0}, C^{d_1} (defined as at the end of the last section) are simply the functors which send each arrow f of C to its domain and its codomain, respectively. The functors P and Q (called the *projections* of the comma category) and the functor R are defined (on objects) as suggested in the diagram

$$\begin{array}{c} \langle e, d, f: Te \to Sd \rangle \\ \swarrow \qquad \downarrow \qquad \searrow \\ e \mapsto Te \leftarrow\!\dashv (f: Te \to Sd) \mapsto Sd \leftarrow\!\dashv d . \end{array} \tag{6}$$

Exercises

1. If K is a commutative ring, show that the comma category $(K \downarrow \mathbf{CRng})$ is the (usual) category of all small commutative K-algebras.
2. If t is a terminal object in C, prove that $(C \downarrow t)$ is isomorphic to C.
3. Complete (6) by defining P, Q, and R on arrows.
4. (S.A. Huq). Given functors $T, S: D \to C$, show that a natural transformation $\tau: T \dot\to S$ is the same thing as a functor $\tau: D \to (T \downarrow S)$ such that $P\tau = Q\tau = \mathrm{id}_D$, with P and Q the projections of (5).
5. Given any commutative diagram of categories and functors

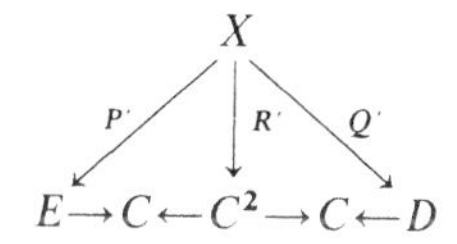

(bottom row as in (5)), prove that there is a unique functor $L: X \to (T \downarrow S)$ for which $P' = PL$, $Q' = QL$, and $R' = RL$. (This describes $(T \downarrow S)$ as a "pull-back", cf. §III.4.)

6. (a) For fixed small C, D, and E, show that $\langle T, S \rangle \mapsto (T \downarrow S)$ is the object function of a functor $(C^E)^{\text{op}} \times (C^D) \to \mathbf{Cat}$.
   (b) Describe a similar functor for variable C, D, and E.

7. Graphs and Free Categories

First, recall the construction of the free monoid FX generated by a set X. It consists of all the finite strings $x_1 x_2 \cdots x_n$ of elements x_i of the set X; the multiplication of these strings is given by juxtaposition, so that the empty string serves as the unit element of FX. The characteristic property of this free monoid may be stated as follows: For any monoid M, let UM denote the set of elements of M. Then any function $f : X \to UM$ extends to a unique morphism of monoids:

$$f : X \to UM \quad \text{extends to a} \quad g : FX \to M .$$

To get the corresponding description of a free category, we replace the starting set X by a directed graph G.

Recall that a (directed) *graph* G (§I.2) is a set O of objects (vertices), and a set A of arrows f (edges), and a pair of functions $A \rightrightarrows O$:

$$A \underset{\partial_1}{\overset{\partial_0}{\rightrightarrows}} O, \qquad \partial_0 f = \text{domain } f, \qquad \partial_1 f = \text{codomain } f.$$

A *morphism* $D: G \to G'$ of graphs is a pair of functions $D_O : O \to O'$ and $D_A : A \to A'$ such that

$$D_O \partial_0 f = \partial_0 D_A f \quad \text{and} \quad D_O \partial_1 f = \partial_1 D_A f$$

for every arrow $f \in A$. These morphisms, with the evident composition, are the arrows of the category **Grph** of all small graphs (a graph is small if both O and A are small sets). Each graph may be pictured by a diagram of vertices (objects) and arrows, just like the diagram for a category except that neither composite arrows nor identity arrows are provided. Hence a graph is often called a *diagram scheme* or a *precategory*.

Every category C determines a graph UC with the same objects and arrows, forgetting which arrows are composites and which are identities. Every functor $F: C \to C'$ is also a morphism $UF: UC \to UC'$ between the corresponding graphs. These observations define the forgetful functor $U: \mathbf{Cat} \to \mathbf{Grph}$ from small categories to small graphs.

Let O be a fixed set. An *O-graph* will be one with O as its set of objects; a morphism D of O-graphs will be one with $D_O : O \to O$ the identity. The simplest O-graph O is $O \rightrightarrows O$, with both functions domain and range the identity. If A and B are (the sets of arrows of) two O-graphs, the product over O is

$$A \times_O B = \{\langle g, f\rangle \mid \partial_0 g = \partial_1 f,\ g \in A,\ f \in B\}; \tag{1}$$

it is the set of "composable pairs" of arrows $\cdot \xrightarrow{f} \cdot \xrightarrow{g} \cdot$. The definitions

$$\partial_0\langle g, f\rangle = \partial_0 f \qquad \partial_1\langle g, f\rangle = \partial_1 g \tag{2}$$

make this set a O-graph. This product operation on O-graphs is associative, since for any three O-graphs A, B, and C there is an evident isomorphism $A \times_O (B \times_O C) \cong (A \times_O B) \times_O C$. For the special O-graph O there is also an isomorphism $A \cong A \times_O O$, given by $f \mapsto \langle f, \partial_0 f\rangle$. Also, $A \cong O \times_O A$.

A category with objects O may be described as an O-graph A equipped with two morphisms $c: A \times_O A \to A$ and $i: O \to A$ of O-graphs (composition and identity) such that the diagrams

$$\begin{array}{ccc} (A \times_O A) \times_O A \cong A \times_O (A \times_O A) & \xrightarrow{1 \times c} & A \times_O A \\ \downarrow{\scriptstyle c \times 1} & & \downarrow{\scriptstyle c} \\ A \times_O A & \xrightarrow{\quad c \quad} & A\,, \end{array} \qquad \begin{array}{ccccc} O \times_O A & \xrightarrow{i \times 1} & A \times_O A & \xleftarrow{1 \times i} & A \times_O O \\ \downarrow{\scriptstyle \cong} & & \downarrow{\scriptstyle c} & & \downarrow{\scriptstyle \cong} \\ A & = & A & = & A \end{array} \tag{3}$$

are commutative, where $1 \times c$ is short for $1 \times_O c$, etc. Indeed, composable arrows $\langle g, f\rangle$ have a composite given by c as $c(g, f)$, each object $b \in O$ has an identity arrow given by $i(b) \in A$, while the first diagram states that composition is associative and the second that each $i(a)$ acts as a left and right identity for composition. In this sense, a category is like a monoid, as described in the introduction: **Set** there is replaced by O-**Grph**, and product of sets by $\times_O$.

Any O-graph G may be used to "generate" a category C on the same set O of objects; the arrows of this category will be the "strings" of composable arrows of G, so that an arrow of C from b to a may be pictured as a path from b to a, consisting of successive edges of G. This category C will be written $C = C(G)$ and called the *free category* generated by the graph G. Its basic properties may be stated as follows.

Theorem 1. *Let $G = \{A \rightrightarrows O\}$ be a small graph. There is a small category $C = C_G$ with O as set of objects and a morphism $P: G \to UC$ of graphs from G to the underlying graph UC of C with the following property. Given any category B and any morphism $D: G \to UB$ of graphs, there is a unique functor $D': C \to B$ with $(UD') \circ P = D$, as in the commutative diagram*

$$\begin{array}{ccccc} C & & G & \xrightarrow{P} & UC \\ \vdots{\scriptstyle D'} & & & \searrow^{D} & \vdots{\scriptstyle UD'} \\ B\,, & & & & UB \end{array} \tag{4}$$

In particular, if B has O as set of objects and D is a morphism of O-graphs, then D' is the identity on objects.

The property of P stated in (4) is equivalent to stating that the arrow $P: G \to UC$ is an initial object in the comma category $(G \downarrow U)$. Hence P is unique up to an isomorphism (of C). Similar properties appear often; we shall say that P is "universal" among morphisms from G to the underlying-graph functor U.

Proof. Take the objects of C to be those of G and the arrows of C to be the finite strings (or "paths")

$$a_1 \xrightarrow{f_1} a_2 \xrightarrow{f_2} a_3 \longrightarrow \cdots \xrightarrow{f_{n-1}} a_n$$

composed of n objects $a_1, \dots, a_n$ of G connected by $n-1$ arrows $f_i: a_i \to a_{i+1}$ of G. Regard each such string as an arrow $\langle a_1, f_1, \dots, f_{n-1}, a_n \rangle : a_1 \to a_n$ in C, and define the composite of two strings by juxtaposition (i.e., by concatenation), identifying the common end. This composition is manifestly associative, and strings $\langle a_1 \rangle$ of length $n=1$ are its identities. Every string of length $n>1$ is a composite of strings of length 2:

$$\langle a_1, f_1, a_2, \dots, a_{n-1}, f_{n-1}, a_n \rangle = \langle a_{n-1}, f_{n-1}, a_n \rangle \circ \cdots \circ \langle a_1, f_1, a_2 \rangle . \tag{5}$$

The desired morphism $P: G \to UC$ of graphs sends each arrow $f: a_1 \to a_2$ of the given graph G to the string $\langle a_1, f, a_2 \rangle$ of length 2.

Now consider any other morphism $D: G \to UB$ of the given graph G to the underlying graph of some category B. If there is a functor $D': C \to B$ with $UD' \circ P = D$, as in the commutative diagram (4), then D' must be $D'\langle a \rangle = Da$ on objects and $D'\langle a_1, f_1, a_2 \rangle = Df_1$ on arrows. Since any string of length $n>1$ is a composite (5) in C, D' must be given by

$$D'\langle a_1, f_1, a_2, \dots, a_{n-1}, f_{n-1}, a_n \rangle = Df_{n-1} \circ \cdots \circ Df_1 .$$

Conversely, this formula does define a functor $D': B \to C$ for which the indicated diagram commutes, q.e.d.

Here are some easy examples. For the graph consisting of a single arrow f with $\partial_0 f = \partial_1 f$, the free category consists of all arrows $1, f, f^2, \dots$. For the graph consisting of a single arrow g with different ends, the free category consists of this arrow plus two identity arrows (one at each end). For the graph $\cdot \to \cdot \to \cdot$ with three different vertices the free category is a commutative triangle (add one composite arrow and three identity arrows).

When O consists of one point, the graph G reduces simply to a set X (the set $X = A$ of arrows) and the theorem provides the familiar construction of a free monoid M generated by X, as follows:

Corollary 2. *To any set X there is a monoid M and a function $p: X \to UM$, where UM is the underlying set of M, with the following* *universal property: For any monoid L and any function* $h: X \to U\,L$ *there is a unique morphism* $h': M \to L$ *of monoids with* $h = U\,h' \circ p$.

The elements of M are the identity and strings $\langle x_1, \ldots, x_{n-1} \rangle$, for $x_i \in X$.

Graphs may be used to describe diagrams. If G is any graph, a *diagram* of the shape G in the category B may be defined to be a morphism $D: G \to U\,B$ of graphs. By the Theorem, these morphisms D correspond exactly to functors $D': C_G \to B$, via the bijection $D' \mapsto D = U\,D' \circ P$. This bijection

$$\mathbf{Cat}(C_G, B) \cong \mathbf{Grph}(G, U\,B) \tag{6}$$

is natural in G and B, so asserts that C : **Grph** $\to$ **Cat** is left adjoint (see Chapter IV) to the forgetful functor U : **Cat** $\to$ **Grph**.

Exercises

1. Define "opposite graph" and "product of two graphs" to agree with the corresponding definitions for categories (i.e., so that the functor U will preserve opposites and products).
2. Show that every finite ordinal number is a free category.
3. Show that each graph G generates a free groupoid F (i.e., one which satisfies Theorem 1 with "category C" replaced by "groupoid F" and "category B" by "groupoid E"). Deduce as a corollary that every set X generates a free group.

8. Quotient Categories

Certain categories may be described by generators and relations, as follows:

Proposition 1. *For a given category* C, *let* R *be a function which assigns to each pair of objects* a, b *of* C *a binary relation* $R_{a,b}$ *on the hom-set* $C(a, b)$. *Then there exist a category* C/R *and a functor* $Q = Q_R : C \to C/R$ *such that* (i) *If* $f R_{a,b} f'$ *in* C, *then* $Qf = Qf'$; (ii) *If* $H : C \to D$ *is any functor from* C *for which* $f R_{a,b} f'$ *implies* $Hf = Hf'$ *for all* f *and* f', *then there is a unique functor* $H' : C/R \to D$ *with* $H' \; Q_R = H$. *Moreover, the functor* Q_R *is a bijection on objects.*

Put briefly: Q is the universal functor on C with $Qf = Qf'$ whenever $f R f'$.

For example, if $C =$ **Top** and $f R f'$ means that f is homotopic to f', then the desired quotient category C/R is just the category **Toph** of § I.7, with objects topological spaces and arrows homotopy classes of continuous maps. This direct construction is possible for **Toph** because the relation of homotopy between maps is an equivalence relation preserved by composition. The general case requires a preliminary construction on the relation R to achieve these properties.

Sketch of proof. Call R a *congruence* on C if (i) for each pair a, b of objects, $R_{a,b}$ is a reflexive, symmetric, and transitive relation on $C(a,b)$; (ii) if $f, f' : a \to b$ have $f R_{a,b} f'$, then for all $g : a' \to a$ and all $h : b \to b'$ one has $(hfg) R_{a',b'} (hf'g)$. Given any R, there is a least congruence R' on C with $R \subset R'$ (proof as exercise). Now take the objects of C/R to be the objects of C, and take each hom-set $(C/R)(a,b)$ to be the quotient $C(a,b)/R'_{a,b}$ of $C(a,b)$ by the equivalence relation R' there. Because the relation is preserved by composition, the composite in C carries over to C/R by the evident projection $Q : C \to C/R$. Now for any functor $H : C \to D$ the sets $S_{a,b} = \{f, f' : a \to b \mid Hf = Hf'\}$ evidently form a congruence on C. Thus, if $S \supset R$ one also has $S \supset R'$, and H factors as $H = H' \circ Q_R$, as required.

In case C is the free category generated by a graph G we call C/R the category with *generators* G and *relations* R. For example, **3** may be described as the category generated by three objects 0, 1, 2, three arrows $f : 0 \to 1$, $g : 1 \to 2$, and $h : 0 \to 2$, and one relation $h = g \circ f$. As a special case (one object), this includes the case of a monoid given by generators and relations.

Exercises

1. Show that the category generated by the graph

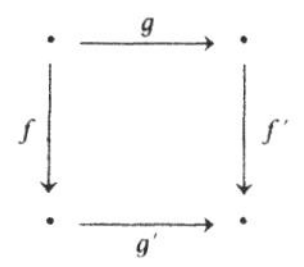

   with the one relation $g'f = f'g$ has four identity arrows and exactly five non-identity arrows f, g, f', g' and $g'f = f'g$.
2. If C is a group G (regarded as a category with one object) show that to each congruence R on C there is a normal subgroup N of G with fRg if and only if $g^{-1}f \in N$.

Notes.

The leading idea of this chapter is to make the simple notion of a functor apply to complex cases by defining suitable complex categories – the opposite category for contravariant functors, the product category for bifunctors, the functor category really as an adjoint to the product, and the comma category to reduce universal arrows to initial objects. The importance of the use of functor categories (sometimes called "categories of diagrams") was emphasized by Grothendieck [1957] and Freyd [1964]. The notion of a comma category, often used in special cases, was introduced in full generality in Lawvere's (unpublished) thesis [1963], in order to give a set-free description of adjoint functors. For a time it was a sort of secret tool in the arsenal of knowledgeable experts.

Duality has a long history. The duality between point and line in geometry, especially projective geometry, led to a sharp description of axiomatic duality in the monumental treatise by Veblen-Young on projective geometry. The explicit description of duality by opposite categories is often preferable, as in the Pontrjagin duality which appears (§ IV.3) as an equivalence between categories, or as an equivalence between a category and an opposite category (see Negrepontis [1971]).

III. Universals and Limits

Universal constructions appear throughout mathematics in various guises – as universal arrows to a given functor, as universal arrows from a given functor, or as universal elements of a set-valued functor. Each universal determines a representation of a corresponding set-valued functor as a hom-functor. Such representations, in turn, are analyzed by the Yoneda Lemma. Limits are an important example of universals – both the inverse limits (= projective limits = limits = left roots) and their duals, the direct limits (= inductive limits = colimits = right roots). In this chapter we define universals and limits and examine a few basic types of limits (products, pullbacks, and equalizers ...). Deeper properties will appear in Chapter IX on special limits, while the relation to adjoints will be treated in Chapter V.

1. Universal Arrows

Given the forgetful functor $U: \mathbf{Cat} \to \mathbf{Grph}$ and a graph G, we have constructed (§ II.7) the free category C on G and the morphism $P: G \to UC$ of graphs which embeds G in C, and we have shown that this arrow P is "universal" from G to U. A similar universality property holds for the morphisms embedding generators into free algebraic systems of other types, such as groups or rings. Here is the general concept.

Definition. *If $S: D \to C$ is a functor and c an object of C, a universal arrow from c to S is a pair $\langle r, u \rangle$ consisting of an object r of D and an arrow $u: c \to Sr$ of C, such that to every pair $\langle d, f \rangle$ with d an object of D and $f: c \to Sd$ an arrow of C, there is a unique arrow $f': r \to d$ of D with $Sf' \circ u = f$. In other words, every arrow f to S factors uniquely through the universal arrow u, as in the commutative diagram*

$$\begin{array}{ccc} c & \xrightarrow{u} & Sr \\ \| & & \downarrow{\scriptstyle Sf'} \\ c & \xrightarrow{f} & Sd, \end{array} \qquad \begin{array}{c} r \\ \downarrow{\scriptstyle f'} \\ d. \end{array} \tag{1}$$

Equivalently, $u: c \to Sr$ is universal from c to S when the pair $\langle r, u\rangle$ is an initial object in the comma category $(c \downarrow S)$, whose objects are the arrows $c \to Sd$. As with any initial object, it follows that $\langle r, u\rangle$ is unique up to isomorphism in $(c \downarrow S)$; in particular, the object r of D is unique up to isomorphism in D. This remark is typical of the use of comma categories.

This notion of a universal arrow has a great variety of examples; we list a few:

Bases of Vector Spaces. Let $\mathbf{Vct}_K$ denote the category of all vector spaces over a fixed field K, with arrows linear transformations, while $U: \mathbf{Vct}_K \to \mathbf{Set}$ is the forgetful functor, sending each vector space V to the set of its elements. For any set X there is a familiar vector space V_X with X as a set of basis vectors; it consists of all formal K-linear combinations of the elements of X. The function which sends each $x \in X$ into the same x regarded as a vector of V_X is an arrow $j: X \to U(V_X)$. For any other vector space W, it is a fact that each function $f: X \to U(W)$ can be extended to a unique linear transformation $f': V_X \to W$ with $Uf' \circ j = f$. This familiar fact states exactly that j is a universal arrow from X to U.

Free Categories from Graphs. Theorem II.7.1 for the free category C on a graph G states exactly that the functor $P: G \to UC$ is universal. The same observation applies to the free monoid on a given set of generators, the free group on a given set of generators, the free R-module (over a given ring R) on a given set of generators, the polynomial algebra over a given commutative ring in a given set of generators, and so on in many cases of free algebraic systems.

Fields of Quotients. To any integral domain D a familiar construction gives a field $Q(D)$ of quotients of D together with a monomorphism $j: D \to Q(D)$ (which is often formulated by making D a subdomain of $Q(D)$). This field of quotients is usually described as the smallest field containing D, in the sense that for each $D \subset K$ with K a field there is a monomorphism $f: Q(D) \to K$ of fields which is the identity on the common subdomain D. However, this inclusion $D \subset K$ may readily be replaced by any monomorphism $D \to K$ of domains. Hence our statement means that the pair $\langle Q(D), j\rangle$ is universal for the forgetful functor $\mathbf{Fld} \to \mathbf{Dom}_m$ from the category of fields to that of domains – provided we take arrows of $\mathbf{Dom}_m$ to be the monomorphisms of integral domains (note that a homomorphism of *fields* is necessarily a monomorphism). However, for the larger category **Dom** with arrows all homomorphisms of integral domains there does not exist a universal arrow from each domain to a field. For instance, for the domain $\mathbf{Z}$ of integers there is for each prime p a homomorphism $\mathbf{Z} \to \mathbf{Z}_p$; the reader should observe that this makes impossible the construction of a universal arrow from $\mathbf{Z}$ to the functor $\mathbf{Fld} \to \mathbf{Dom}$.

Complete Metric Spaces. Let **Met** be the category of all metric spaces $X, Y, \ldots$, with arrows $X \to Y$ those functions which preserve the metric (and which therefore are necessarily injections). The complete metric spaces form (the objects of) a full subcategory. The familiar completion $\bar{X}$ of a metric space X provides an arrow $X \to \bar{X}$ which is universal for the evident forgetful functor (from complete metric spaces to metric spaces).

In many other cases, the function embedding a mathematical object in a suitably completed object can be interpreted as a universal arrow. The general fact of the uniqueness of the universal arrow implies the uniqueness of the completed object, up to a unique isomorphism (who wants more?).

The idea of universality is sometimes expressed in terms of "universal elements". If D is a category and $H : D \to \mathbf{Set}$ a functor, a *universal element* of the functor H is a pair $\langle r, e \rangle$ consisting of an object $r \in D$ and an element $e \in Hr$ such that for every pair $\langle d, x \rangle$ with $x \in Hd$ there is a unique arrow $f: r \to d$ of D with $(Hf)e = x$.

Many familiar constructions are naturally examples of universal elements. For instance, consider an equivalence relation E on a set S, the corresponding quotient set S/E consisting of the equivalence classes of elements of S under E, and the projection $p: S \to S/E$ which sends each $s \in S$ to its E-equivalence class. Now S/E has the familiar property that any function f on S which respects the equivalence relation can be regarded as a function on S/E. More formally, this means that if $f: S \to X$ has $fs = fs'$ whenever sEs', then f can be written as a composite $f = f'p$ for a unique function $f': S/E \to X$:

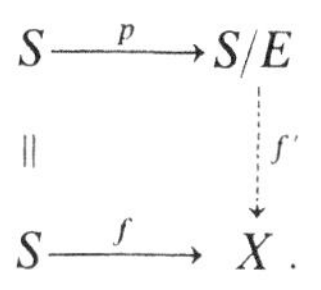

This states exactly that $\langle S/E, p \rangle$ is a universal element for that functor $H : \mathbf{Set} \to \mathbf{Set}$ which assigns to each set X the set HX of all those functions $f: S \to X$ for which sEs' implies $fs = fs'$.

Again, let N be a normal subgroup of a group G. The usual projection $p: G \to G/N$ which sends each $g \in G$ to its coset $pg = gN$ in the quotient group G/N is a universal element for that functor $H : \mathbf{Grp} \to \mathbf{Set}$ which assigns to each group G' the set HG' of all those homomorphisms $f: G \to G'$ which kill N (have $fN = 1$). Indeed, every such homomorphism factors as $f = f'p$, for a unique $f': G/N \to G'$. Now the quotient group is usually described as a group whose elements are cosets. However, once the cosets are used to prove this *one* "universal" property of $p: G \to G/N$, *all* other properties of quotient groups – for example, the isomorphism theorems – can be proved with no further mention of cosets (see Mac Lane-Birkhoff [1967]). All that is needed is the existence of a universal element

p of the functor H. For that matter, even this existence could be proved without using cosets (see the adjoint functor theorem stated in § V.6).

Tensor products provide another example of universal elements. Given two vector spaces V and V' over the field K, the function H which assigns to each vector space W the set $HW = \mathrm{Bilin}(V, V'; W)$ of all bilinear functions $V \times V' \to W$ is the object function of a functor $H: \mathbf{Vect}_K \to \mathbf{Set}$, and the usual construction of the tensor product provides both a vector space $V \otimes V'$ and a bilinear function $\otimes: V \times V' \to V \otimes V'$, usually written $\langle v, v' \rangle \mapsto v \otimes v'$, so that the pair $\langle V \otimes V', \otimes \rangle$ is a universal element for the functor $H = \mathrm{Bilin}(V, V'; -)$. This applies equally well when the field K is replaced by a commutative ring (and vector spaces by K-modules).

The notion "universal element" is a special case of the notion "universal arrow". Indeed, if $*$ is the set with one point, then any element $e \in Hr$ can be regarded as an arrow $e: * \to Hr$ in **Ens**. Thus a universal element $\langle r, e \rangle$ for H is exactly a universal arrow from $*$ to H. Conversely, if C has small hom-sets, the notion "universal arrow" is a special case of the notion "universal element". Indeed, if $S: D \to C$ is a functor and $c \in C$ is an object, then $\langle r, u: c \to Sr \rangle$ is a universal arrow from c to S if and only if the pair $\langle r, u \in C(c, Sr) \rangle$ is a universal element of the functor $H = C(c, S-)$. This is the functor which acts on objects d and arrows h of D by

$$d \mapsto C(c, Sd), \qquad h \mapsto C(c, Sh).$$

Hitherto we have treated universal arrows from an object $c \in C$ to a functor $S: D \to C$. The dual concept is also useful. A *universal arrow* from S to c is a pair $\langle r, v \rangle$ consisting of an object $r \in D$ and an arrow $v: Sr \to c$ with codomain c such that to every pair $\langle d, f \rangle$ with $f: Sd \to c$ there is a unique $f': d \to r$ with $f = v \circ Sf'$, as in the commutative diagram

$$\begin{array}{ccc} d & \qquad & Sd \xrightarrow{\ f\ } c \\ \vdots\, f' & & Sf' \vdots \qquad \| \\ \downarrow & & \downarrow \qquad\quad \\ r, & & Sr \xrightarrow{\ v\ } c. \end{array}$$

The projections $p: a \times b \to a$, $q: a \times b \to b$ of a product in C (for $C = \mathbf{Grp}$, **Set**, **Cat**, ...) are examples of such a universal. Indeed, given any other pair of arrows $f: c \to a$, $g: c \to b$ to a and b, there is a unique $h: c \to a \times b$ with $ph = f$, $qh = g$. Therefore $\langle p, q \rangle$ is a "universal pair". To make it a universal arrow, introduce the diagonal functor $\Delta: C \to C \times C$, with $\Delta c = \langle c, c \rangle$. Then the pair f, g above becomes an arrow $\langle f, g \rangle: \Delta c \to \langle a, b \rangle$ in $C \times C$, and $\langle p, q \rangle$ is a universal arrow from Δ to the object $\langle a, b \rangle$.

Similarly, the kernel of a homomorphism (in **Ab**, **Grp**, **Rng**, R-**Mod**, ...) is a universal, more exactly, a universal for a suitable contravariant functor.

Note that we say "universal arrow to S" and "universal arrow from S" rather than "universal" and "couniversal".

Exercises

1. Show how each of the following familiar constructions can be interpreted as a universal arrow:
   (a) The integral group ring of a group (better, of a monoid).
   (b) The tensor algebra of a vector space.
   (c) The exterior algebra of a vector space.
2. Find a universal element for the contravariant power set functor $\mathscr{P}: \mathbf{Set}^{\mathrm{op}} \to \mathbf{Set}$.
3. Find (from any given object) universal arrows to the following forgetful functors: $\mathbf{Ab} \to \mathbf{Grp}$, $\mathbf{Rng} \to \mathbf{Ab}$ (forget the multiplication), $\mathbf{Top} \to \mathbf{Set}$, $\mathbf{Set}_* \to \mathbf{Set}$.
4. Use only universality (of projections) to prove the following isomorphisms of group theory:
   (a) For normal subgroups M, N of G with $M \subset N$, $(G/M)/(N/M) \cong G/M$.
   (b) For subgroups S and N of G, N normal, with join SN, $SN/N \cong S/S \cap N$.
5. Show that the quotient K-module A/S (S a submodule of A) has a description by universality. Derive isomorphism theorems.
6. Describe quotients of a ring by a two-sided ideal by universality.
7. Show that the construction of the polynomial ring $K[x]$ in an indeterminate x over a commutative ring K is a universal construction.

2. The Yoneda Lemma

Next we consider some conceptual properties of universality. First, universality can be formulated with hom-sets, as follows:

Proposition 1. *For a functor $S: D \to C$ a pair $\langle r, u: c \to Sr \rangle$ is universal from c to S if and only if the function sending each $f': r \to d$ into $Sf' \circ u: c \to Sd$ is a bijection of hom-sets*

$$D(r, d) \cong C(c, Sd). \tag{1}$$

This bijection is natural in d. Conversely, given r and c, any natural isomorphism (1) is determined in this way by a unique arrow $u: c \to Sr$ such that $\langle r, u \rangle$ is universal from c to S.

Proof. The statement that $\langle r, u \rangle$ is universal is exactly the statement that $f' \mapsto Sf' \circ u = f$ is a bijection. This bijection is natural in d, for if $g': d \to d'$, then $S(g'f') \circ u = Sg' \circ (Sf' \circ u)$.

Conversely, a natural isomorphism (1) gives for each object d of D a bijection $\varphi_d: D(r, d) \to C(c, Sd)$. In particular, choose the object d to be r;

the identity $1_r \in D(r, r)$ then goes by φ_r to an arrow $u: c \to Sr$ in C. For any $f': r \to d$ the diagram

$$\begin{array}{ccc} D(r, r) & \xrightarrow{\varphi_r} & C(c, Sr) \\ {\scriptstyle D(r, f')}\downarrow & & \downarrow{\scriptstyle C(c, Sf')} \\ D(r, d) & \xrightarrow{\varphi_d} & C(c, Sd) \end{array} \qquad (2)$$

commutes because φ is natural. But in this diagram, $1_r \in D(r, r)$ is mapped (top and right) to $Sf' \circ u$ and (left and bottom) to $\varphi_d(f')$. Since φ_d is a bijection, this states precisely that each $f: c \to Sd$ has the form $f = Sf' \circ u$ for a unique f'. This is precisely the statement that $\langle r, u \rangle$ is universal.

If C and D have small hom-sets, this result (1) states that the functor $C(c, S-)$ to **Set** is naturally isomorphic to a covariant hom-functor $D(r, -)$. Such isomorphisms are called representations:

Definition. *Let D have small hom-sets. A representation of a functor $K: D \to \mathbf{Set}$ is a pair $\langle r, \psi \rangle$, with r an object of D and*

$$\psi : D(r, -) \cong K \qquad (3)$$

a natural isomorphism. The object r is called the representing object. The functor K is said to be representable when such a representation exists.

Up to isomorphism, a representable functor is thus just a covariant hom-functor $D(r, -)$. This notion can be related to universal arrows as follows.

Proposition 2. *Let $*$ denote any one-point set and let D have small hom-sets. If $\langle r, u: * \to Kr \rangle$ is a universal arrow from $*$ to $K: D \to \mathbf{Set}$, then the function ψ which for each object d of D sends the arrow $f': r \to d$ to $K(f')(u*) \in Kd$ is a representation of K. Every representation of K is obtained in this way from exactly one such universal arrow.*

Proof. For any set X, a function $f: * \to X$ from the one-point set $*$ to X is determined by the element $f(*) \in X$. This correspondence $f \mapsto f(*)$ is a bijection $\mathbf{Set}(*, X) \xrightarrow{\sim} X$, natural in $X \in \mathbf{Set}$. Composing with K yields a natural isomorphism $\mathbf{Set}(*, K-) \xrightarrow{\sim} K$. This plus the representation ψ of (3) gives

$$\mathbf{Set}(*, K-) \cong K \cong D(r, -).$$

Therefore a representation of K amounts to a natural isomorphism $\mathbf{Set}(*, K-) \cong D(r, -)$. The proposition thus follows from the previous one.

A direct proof is equally easy: Given the universal arrow u, the correspondence $f' \mapsto K(f')(u(*))$ is a representation; given a representation ψ as in (3), ψ_r maps $1: r \to r$ to an element of Kr, which is a universal element, hence also a universal arrow $* \to Kr$.

Observe that each of the notions "universal arrow", "universal element", and "representable functor" subsumes the other two. Thus, a universal arrow from c to $S: D \to C$ amounts (Proposition 1) to a natural isomorphism $D(r, d) \cong C(c, Sd)$ and hence to a representation of the functor $C(c, S-): D \to \mathbf{Set}$ or equally well to a universal element for the same functor.

The argument for Proposition 1 rested on the observation that each natural transformation $\varphi: D(r, -) \dot{\to} K$ is completely determined by the image under φ_r of the identity $1: r \to r$. This fact may be stated as follows:

Lemma *(Yoneda). If $K: D \to \mathbf{Set}$ is a functor from D and r an object in D (for D a category with small hom-sets), there is a bijection*

$$y: \mathrm{Nat}(D(r, -), K) \cong Kr \tag{4}$$

which sends each natural transformation $\alpha: D(r, -) \dot{\to} K$ to $\alpha_r 1_r$, the image of the identity $r \to r$.

The proof is indicated by the following commutative diagram:

$$\begin{array}{ccccc} D(r, r) & \xrightarrow{\alpha_r} & K(r) & & r \\ {\scriptstyle f_* = D(r, f)} \downarrow & & \downarrow {\scriptstyle K(f)} & & \downarrow {\scriptstyle f} \\ D(r, d) & \xrightarrow{\alpha_d} & K(d), & & d. \end{array} \tag{5}$$

Corollary. *For objects $r, s \in D$, each natural transformation $D(r, -) \dot{\to} D(s, -)$ has the form $D(h, -)$ for a unique arrow $h: s \to r$.*

The Yoneda map y of (4) is natural in K and r. To state this fact formally, we must consider K as an object in the functor category $\mathbf{Set}^D$, regard both domain and codomain of the map y as functors of the pair $\langle K, r \rangle$, and consider this pair as an object in the category $\mathbf{Set}^D \times D$. The codomain for y is then the evaluation functor E, which maps each pair $\langle K, r \rangle$ to the value Kr of the functor K at the object r; the domain is the functor N which maps the object $\langle K, r \rangle$ to the set $\mathrm{Nat}(D(r, -), K)$ of all natural transformations and which maps a pair of arrows $F: K \to K'$, $f: r \to r'$ to $\mathrm{Nat}(D(f, -), F)$. With these observations we may at once prove an addendum to the *Yoneda Lemma:*

Lemma. *The bijection of* (4) *is a natural isomorphism $y: N \dot{\to} E$ between the functors $E, N: \mathbf{Set}^D \times D \to \mathbf{Set}$.*

The object function $r \mapsto D(r, -)$ and the arrow function

$$(f: s \to r) \mapsto D(f, -): D(r, -) \dot{\to} D(s, -)$$

for f an arrow of D together define a full and faithful functor

$$Y: D^{\mathrm{op}} \to \mathbf{Set}^D \tag{6}$$

called the *Yoneda functor*. Its dual is another such functor

$$Y' : D \to \mathbf{Set}^{D^{\mathrm{op}}} \tag{7}$$

(also faithful) which sends $f: s \to r$ to the natural transformation

$$D(-, f): D(-, s) \dot{\to} D(-, r): D^{\mathrm{op}} \to \mathbf{Set}.$$

D must have small hom-sets if these functors are to be defined (because **Set** is the category of all *small* sets). For larger D, the Yoneda lemmas remain valid if **Set** is replaced by any category **Ens** whose objects are sets $X, Y, \ldots$, and for which $\mathbf{Ens}(X, Y)$ is the set of all functions from X to Y, provided of course that D has hom-sets which are objects in **Ens**. (The meaning of naturality is not altered by further enlargement of **Ens**; see Exercise 4.)

Exercises

1. Let functors $K, K' : D \to \mathbf{Set}$ have representations $\langle r, \psi\rangle$ and $\langle r', \psi'\rangle$, respectively. Prove that to each natural transformation $\tau : K \dot{\to} K'$, there is a unique morphism $h: r' \to r$ of D such that

$$\tau \circ \psi = \psi' \circ D(h, -): D(r, -) \dot{\to} K'.$$

2. State the dual of the Yoneda Lemma (D replaced by D^{op}).
3. (Kan; the coyoneda lemma.) For $K : D \to \mathbf{Set}$, $(* \downarrow K)$ is the category of elements $x \in Kd$, $Q : (* \downarrow K) \to D$ is the projection $x \in Kd \mapsto d$ and for each $a \in D$, $a : (* \downarrow K) \to D$ is the diagonal functor sending everything to the constant value a. Establish a natural isomorphism

$$\mathrm{Nat}(K, D(a, -)) \cong \mathrm{Nat}(a, Q).$$

4. (Naturality is not changed by enlarging the codomain category.) Let E be a full subcategory of E'. For functors $K, L: D \to E$, with $J : E \to E'$ the inclusion, prove that $\mathrm{Nat}(K, L) \cong \mathrm{Nat}(JK, JL)$.

3. Coproducts and Colimits

We introduce colimits by a variety of special cases, each of which is a universal.

Coproducts. For any category C, the diagonal functor $\Delta : C \to C \times C$ is defined on objects by $\Delta(c) = \langle c, c\rangle$, on arrows by $\Delta(f) = \langle f, f\rangle$. A universal arrow from an object $\langle a, b\rangle$ of $C \times C$ to the functor Δ is called a *coproduct diagram*. It consists of an object c of C and an arrow $\langle a, b\rangle \to \langle c, c\rangle$ of $C \times C$; that is, a pair of arrows $i: a \to c$, $j: b \to c$ from a and b to a common codomain c. This pair has the familiar universal property: For any pair of arrows $f: a \to d$, $g: b \to d$ there is a unique $h: c \to d$ with $f = h \circ i$, $g = h \circ j$. When such a coproduct diagram exists, the object c is necessarily unique (up to isomorphism in C); it is written $c = a \amalg b$ or $c = a + b$ and is called a *coproduct object*. The coproduct diagram then is

$$a \xrightarrow{i} a \amalg b \xleftarrow{j} b \, ;$$

the arrows i and j are called the *injections* of the coproduct $a \amalg b$ (though they are not required to be injective as functions). The universality of this diagram states that any diagram of the following form can be filled in uniquely (at h) so as to be commutative:

$$\begin{array}{ccccc} a & \xrightarrow{i} & a \amalg b & \xleftarrow{j} & b \\ & {\scriptstyle f} \searrow & \downarrow {\scriptstyle h} & \swarrow {\scriptstyle g} & \\ & & d & & \end{array} \qquad (1)$$

Hence the assignment $\langle f, g \rangle \mapsto h$ is a bijection

$$C(a, d) \times C(b, d) \cong C(a \amalg b, d) \qquad (2)$$

natural in d, with inverse $h \mapsto \langle hi, hj \rangle$. If every pair of objects a, b in C has a coproduct then, choosing a coproduct diagram for each pair, the coproduct $\amalg : C \times C \to C$ is a bifunctor, with $h \amalg k$ defined for arrows $h : a \to a'$, $k : b \to b'$ as the unique arrow $h \amalg k : a \amalg b \to a' \amalg b'$ with $(h \amalg k) i = i' h$, $(h \amalg k) j = j' k$ (draw the diagram!).

The diagram (1) is more familiar in other guises. For example, in **Set** take $a \amalg b$ to be a disjoint union of the sets a and b (i.e., a union of disjoint copies of a and b), while i and j are the inclusion maps $a \subset a \amalg b$, $b \subset a \amalg b$. Now a function h on a disjoint union is uniquely determined by independently giving its values on a and on b; i.e., by giving the composites hi and hj. This says exactly that diagram (1) can be filled in uniquely at h. To be sure, a disjoint union is not unique, but it is unique up to a bijection, as befits a universal.

The coproduct of any two objects exists in many of the familiar categories, where it has a variety of names as indicated in the following list:

Set	disjoint union of sets,
Top	disjoint union of spaces,
Top$_*$	wedge product (join two spaces at the base points),
Ab, R-**Mod**	direct sum $A \oplus B$,
Grp	free product,
CRng	tensor product $R \otimes S$.

In a preorder P, a least upper bound $a \cup b$ of two elements a and b, if it exists, is an element $a \cup b$ with the properties (i) $a \leqq a \cup b$, $b \leqq a \cup b$; and (ii) if $a \leqq c$ and $b \leqq c$, then $a \cup b \leqq c$. These properties state exactly that $a \cup b$ is a coproduct of a and b in P, regarded as a category.

Infinite Coproducts. In the description of the coproduct, replace $C \times C = C^2$ by C^X for any set X. Here the set X is regarded as a discrete category, so the functor category C^X has as its objects the X-indexed families $a = \{a_x \mid x \in X\}$ of objects of C. The corresponding diagonal functor $\Delta : C \to C^X$ sends each c to the constant family (all $c_x = c$). A universal arrow from a to Δ is an X-fold coproduct diagram; it consists of a coproduct object $\coprod_x a_x \in C$ and arrows (coproduct injections) $i_x : a_x \to \coprod_x a_x$ of C with the requisite universal property. This universal property states that the assignment $f \mapsto \{f i_x \mid x \in X\}$ is a bijection

$$C(\coprod_x a_x, c) \cong \prod_{x \in X} C(a_x, c), \tag{3}$$

natural in c. In **Set**, a coproduct is an X-fold disjoint union.

Copowers. If the factors in a coproduct are all equal ($a_x = b$ for all x), the coproduct $\coprod_x b$ is called a *copower* and is written $X \cdot b$, so that

$$C(X \cdot b, c) \cong C(b, c)^X, \tag{4}$$

natural in c. For example, in **Set**, with $b = Y$ a set, the copower $X \cdot Y = X \times Y$ is the cartesian product of the sets X and Y.

Cokernels. Suppose that C has a null object z, so that for any two objects $b, c \in C$ there is a zero arrow $0 : b \to z \to c$. The cokernel of $f : a \to b$ is then an arrow $u : b \to e$ such that (i) $uf = 0 : a \to e$; (ii) if $h : b \to c$ has $hf = 0$, then $h = h'u$ for a unique arrow $h' : e \to c$. The picture is

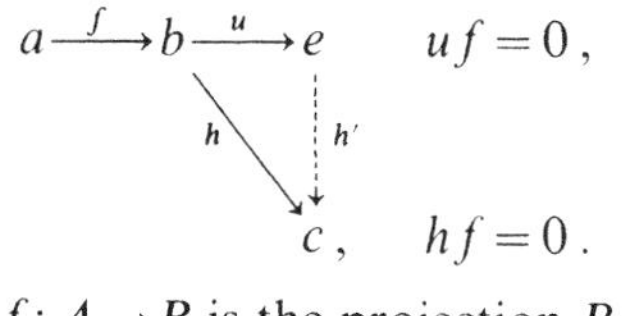

$$\begin{array}{ccccc} a \xrightarrow{f} & b & \xrightarrow{u} & e & \quad uf = 0, \\ & & {\scriptstyle h}\searrow & \downarrow {\scriptstyle h'} & \\ & & & c, & \quad hf = 0. \end{array} \tag{5}$$

In **Ab**, the cokernel of $f : A \to B$ is the projection $B \to B/fA$ to a quotient group of B, and in many other such categories a cokernel is essentially a suitable quotient object. However, in categories without a null object cokernels are not available. Hence we consider more generally certain "coequalizers".

Coequalizers. Given in C a pair $f, g : a \to b$ of arrows with the same domain a and the same codomain b, a *coequalizer* of $\langle f, g \rangle$ is an arrow $u : b \to e$ (or, a pair $\langle e, u \rangle$) such that (i) $uf = ug$; (ii) if $h : b \to c$ has $hf = hg$, then $h = h'u$ for a unique arrow $h' : e \to c$. The picture is

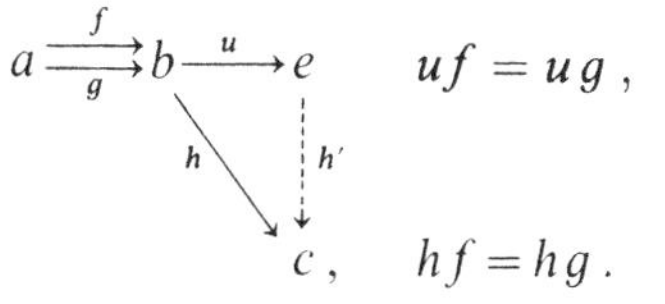

$$\begin{array}{ccccc} a \underset{g}{\overset{f}{\rightrightarrows}} & b & \xrightarrow{u} & e & \quad uf = ug, \\ & & {\scriptstyle h}\searrow & \downarrow {\scriptstyle h'} & \\ & & & c, & \quad hf = hg. \end{array} \tag{6}$$

A coequalizer u can be interpreted as a universal arrow as follows. Let $\downarrow\downarrow$ denote the category which has precisely two objects and two non-identity arrows from the first object to the second; thus the category is $\cdot \rightrightarrows \cdot$. Form the functor category $C^{\downarrow\downarrow}$. An object in $C^{\downarrow\downarrow}$ is then a functor from $\cdot \rightrightarrows \cdot$ to C; that is, a pair $\langle f, g\rangle : a \to b$ of parallel arrows $a \rightrightarrows b$ in C. An arrow in $C^{\downarrow\downarrow}$ from one such pair $\langle f, g\rangle$ to another $\langle f', g'\rangle$ is a natural transformation between the corresponding functors; this means that it is a pair $\langle h, k\rangle$ of arrows $h: a \to a'$ and $k: b \to b'$ in C

$$\begin{array}{ccc} a & \overset{f}{\underset{g}{\rightrightarrows}} & b \\ {\scriptstyle h}\downarrow & & \downarrow{\scriptstyle k} \\ a' & \overset{f'}{\underset{g'}{\rightrightarrows}} & b', \end{array} \qquad \begin{array}{c} kg = g'h, \\ \\ kf = f'h, \end{array}$$

which make the f-square *and* the g-square commute. There is also a diagonal functor $\Delta: C \to C^{\downarrow\downarrow}$, defined on objects c and arrows r of C as

$$\begin{array}{c} c \\ \downarrow{\scriptstyle r} \\ c' \end{array} \quad \mapsto \quad \begin{array}{ccc} c & \overset{1}{\underset{1}{\rightrightarrows}} & c \\ {\scriptstyle r}\downarrow & & \downarrow{\scriptstyle r} \\ c' & \overset{1}{\underset{1}{\rightrightarrows}} & c'; \end{array}$$

in symbols, $\Delta c = \langle 1_c, 1_c\rangle$ and $\Delta r = \langle r, r\rangle$. Now given the pair $\langle f, g\rangle : a \to b$, an arrow $h: b \to c$ with $hf = hg$ is the same thing as an arrow $\langle hf = hg, h\rangle : \langle f, g\rangle \to \langle 1_c, 1_c\rangle$ in the functor category $C^{\downarrow\downarrow}$:

$$\begin{array}{ccc} a & \overset{f}{\underset{g}{\rightrightarrows}} & b \\ {\scriptstyle hf}\downarrow & & \downarrow{\scriptstyle h} \\ c & \overset{1}{\underset{1}{\rightrightarrows}} & c, \end{array} \qquad hf = hg.$$

In other words, the arrows h which "coequalize" f and g are the arrows from $\langle f, g\rangle$ to Δ. Therefore a coequalizer $\langle e, u\rangle$ of the pair $\langle f, g\rangle$ is just a universal arrow from $\langle f, g\rangle$ to the functor Δ.

Coequalizers of any set of maps from a to b are defined in the same way.

In **Ab**, the coequalizer of two homomorphisms $f, g: A \to B$ is the projection $B \to B/(f-g)A$ on a quotient group of B (by the image of the difference homomorphism). In **Set**, the coequalizer of two functions $f, g: X \to Y$ is the projection $p: Y \to Y/E$ on the quotient set of Y by the least equivalence relation $E \subset Y \times Y$ which contains all pairs $\langle fx, gx\rangle$ for $x \in X$. The same construction, using the quotient topology, gives coequalizers in **Top**.

Pushouts. Given in C a pair $f: a \to b$, $g: a \to c$ of arrows with a common domain a, a *pushout* of $\langle f, g\rangle$ is a commutative square, such as that on the left below

$$\begin{array}{ccc} a & \xrightarrow{f} & b \\ {\scriptstyle g}\downarrow & & \downarrow{\scriptstyle u} \\ c & \xrightarrow{v} & r, \end{array} \qquad \begin{array}{ccc} a & \xrightarrow{f} & b \\ {\scriptstyle g}\downarrow & & \downarrow{\scriptstyle h} \\ c & \xrightarrow{k} & s, \end{array} \tag{7}$$

such that to every other commutative square (right above) built on f, g there is a unique $t: r \to s$ with $tu = h$ and $tv = k$. In other words, the pushout is the universal way of filling out a commutative square on the sides f, g. It may be interpreted as a universal arrow. Let $\cdot \leftarrow \cdot \rightarrow \cdot$ denote the category which looks just like that. An object in the functor category $C^{\cdot \leftarrow \cdot \rightarrow \cdot}$ is then a pair of arrows $\langle f, g \rangle$ in C with a common domain, while $\Delta(c) = \langle 1_c, 1_c \rangle$ is the object function of an evident "diagonal" functor $\Delta: C \to C^{\leftarrow \cdot \rightarrow}$. A commutative square $hf = kg$ as on the right above can then be read as an arrow

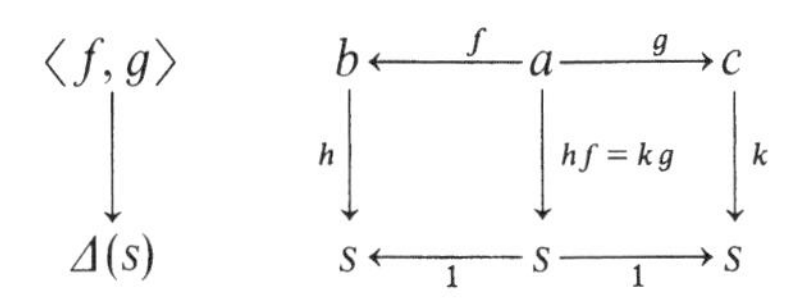

in $C^{\leftarrow \cdot \rightarrow}$ from $\langle f, g \rangle$ to Δs. The pushout is a universal such arrow. Its vertex r, which is uniquely determined up to (a unique) isomorphism, is often written as a coproduct "over a"

$$r = b \amalg_a c = b \amalg_{\langle f, g \rangle} c,$$

and called a "fibered sum" or (the vertex of) a "cocartesian square". In **Set**, the pushout of $\langle f, g \rangle$ always exists; it is the disjoint union $b \amalg c$ with the elements fx and gx identified for each $x \in a$. A similar construction gives pushouts in **Top** – they include such useful constructions as adjunction spaces. Pushouts exist in **Grp**; in particular, if f and g above are monic in **Grp**, the arrows u and v of the pushout square are also monic, and the vertex r is called the "amalgamated product" of b with c.

Cokernel Pair. Given an arrow $f: a \to b$ in C, the pushout of f with f is called the cokernel pair of f. Thus the cokernel pair of f consists of an object r and a parallel pair of arrows $u, v: b \to r$, with domain b, such that $uf = vf$ and such that to any parallel pair $h, k: b \to s$ with $hf = kf$ there is a unique $t: r \to s$ with $tu = h$ and $tv = k$:

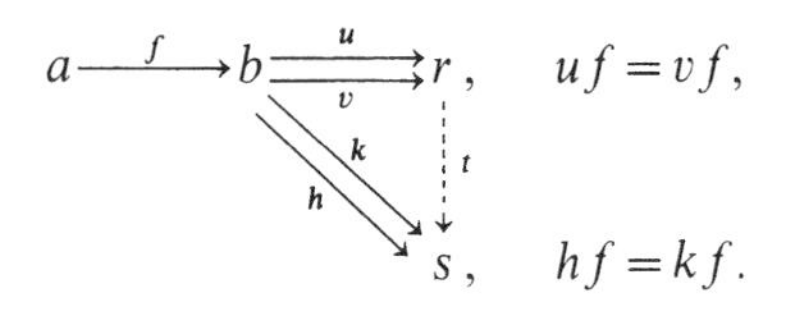

Colimits. The preceding cases all deal with particular functor categories and have the following pattern. Let C and J be categories (J for index category, usually small and often finite). The *diagonal functor*

$$\Delta : C \to C^J$$

sends each object c to the constant functor Δc – the functor which has the value c at each object $i \in J$ and the value 1_c at each arrow of J. If $f : c \to c'$ is an arrow of C, Δf is the natural transformation $\Delta f : \Delta c \dot{\to} \Delta c'$ which has the same value f at each object i of J. Each functor $F : J \to C$ is an object of C^J. A universal arrow $\langle r, u \rangle$ from F to Δ is called a *colimit* (a "direct limit" or "inductive limit") diagram for the functor F. It consists of an object r of C, usually written $r = \varinjlim F$ or $r = \operatorname{Colim} F$, together with a natural transformation $u : F \dot{\to} \Delta r$ which is universal among natural transformations $\tau : F \dot{\to} \Delta c$. Since Δc is the constant functor, the natural transformation τ consists of arrows $\tau_i : F_i \to c$ of C, one for each object i of J, with $\tau_j \quad F u = \tau_i$ for each arrow $u : i \to j$ of J. Pictorially, all the squares in the following schematic diagram (for a special choice of J)

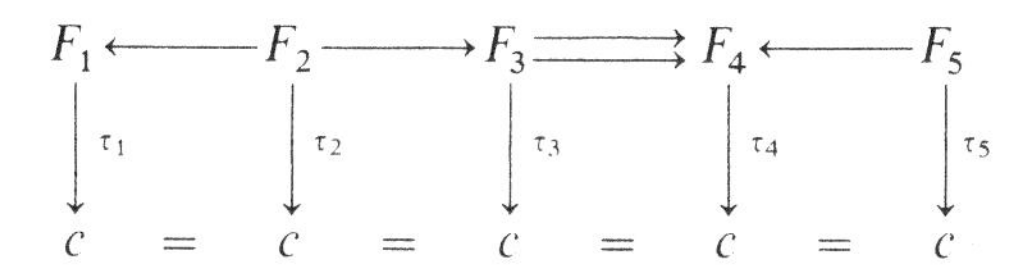

must commute. It is convenient to visualize these diagrams with all the "bottom" objects identified. For this reason, a natural transformation $\tau : F \dot{\to} \Delta c$, often written as $\tau : F \dot{\to} c$, omitting Δ, is called a *cone* from the base F to the vertex c, as in the figure

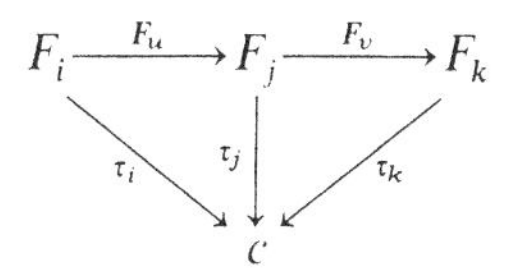

(all triangles commutative). In this language, a colimit of $F : J \to C$ consists of an object $\varinjlim F \in C$ and a cone $\mu : F \dot{\to} \Delta(\varinjlim F)$ from the base F to the vertex $\varinjlim F$ which is universal: For any cone $\tau : F \dot{\to} \Delta c$ from the base F there is a unique arrow $t' : \varinjlim F \to c$ with $\tau_i = t' \mu_i$ for every index $i \in J$. We call μ the *limiting cone* or the *universal cone* (from F).

For example, let $J = \omega = \{0 \to 1 \to 2 \to 3 \to \cdots\}$ and consider a functor $F : \omega \to \mathbf{Set}$ which maps every arrow of ω to an inclusion (subset in set). Such a functor F is simply a nested sequence of sets $F_0 \subset F_1 \subset F_2 \subset \cdots$. The union U of all sets F_n, with the cone given by the inclusion maps

$F_n \to U$, is $\varinjlim F$. The same interpretation of unions as special colimits applies in **Grp**, **Ab**, and other familiar categories. The reader may wish to convince himself now of what we shall soon prove (Exercise V.1.8): For J small, any $F : J \to \mathbf{Set}$ has a colimit.

Exercises

1. In the category of commutative rings, show that $R \to R \otimes S \leftarrow S$, with maps $r \mapsto r \otimes 1$, $1 \otimes s \leftarrow\!\shortmid s$, is a coproduct diagram.
2. If a category has (binary) coproducts and coequalizers, prove that it also has pushouts. Apply to **Set**, **Grp**, and **Top**.
3. In the category $\mathbf{Matr}_K$ of § I.2, describe the coequalizer of two $m \times n$ matrices A, B (i.e., of two arrows $n \to m$ in $\mathbf{Matr}_K$).
4. Describe coproducts (and show that they exist) in **Cat**, in **Mon**, and in **Grph**.
5. If E is an equivalence relation on a set X, show that the usual set X/E of equivalence classes can be described by a coequalizer in **Set**.
6. Show that a and b have a coproduct in C if and only if the following functor is representable: $C(a, -) \times C(b, -) : C \to \mathbf{Set}$, by $c \mapsto C(a, c) \times C(b, c)$.
7. (Every abelian group is a colimit of its finitely generated subgroups.) If A is an abelian group, and J_A the preorder with objects all finitely generated subgroups $S \subset A$ ordered by inclusion, show that A is the colimit of the evident functor $J_A \to \mathbf{Ab}$. Generalize.

4. Products and Limits

The limit notion is dual to that of a colimit. Given categories C, J, and the diagonal functor $\Delta : C \to C^J$, a *limit* for a functor $F : J \to C$ is a universal arrow $\langle r, v \rangle$ from Δ to F. It consists of an object r of C, usually written $r = \varprojlim F$ or $\operatorname{Lim} F$ and called the *limit* object (the "inverse limit" or "projective limit") of the functor F, together with a natural transformation $v : \Delta r \dot{\to} F$ which is universal among natural transformations $\tau : \Delta c \dot{\to} F$, for objects c of C. Since $\Delta c : J \to C$ is the functor constantly c, this natural transformation τ consists of one arrow $\tau_i : c \to F_i$ of C for each object i of J such that for every arrow $u : i \to j$ of J one has $\tau_j = Fu \circ \tau_i$. We may call $\tau : c \dot{\to} F$ a *cone* to the base F from the vertex c. (We say "cone to the base F" rather than "cocone"). The universal property of v is this: It is a cone to the base F from the vertex $\varprojlim F$; for any cone τ to F from an object c, there is a unique arrow $t : c \to \varprojlim F$ such that $\tau_i = v_i t$ for all i. The situation may be pictured as

$$\begin{array}{ccc} c & \xrightarrow{\quad t \quad} & \varprojlim F = \varprojlim_j F_j \\ {\scriptstyle \tau_i}\downarrow & \tau_j \searrow \;\; \swarrow v_i & \downarrow{\scriptstyle v_j} \\ F_i & \xrightarrow[\quad Fu \quad]{} & F_j \end{array} \qquad v = \text{limiting cone},$$

each cone is represented by a commuting triangle (just one of many), with vertex up; there is a unique arrow t which makes all the added (vertex down) triangles commute. As with any universal, the object $\varprojlim F$ and its *limiting cone* $v : \varprojlim F \to F$ are determined uniquely by the functor F, up to isomorphism in C.

The properties of $\varprojlim$ and $\varinjlim$ are summarized in the diagram

$$\begin{array}{ccccc} \operatorname{Lim} F = \varprojlim F & \xrightarrow{v} & F & \xrightarrow{\mu} & \varinjlim F = \operatorname{Colim} F \\ \uparrow & & \| & & \downarrow \\ c & \xrightarrow[\tau]{} & F & \xrightarrow[\sigma]{} & c\,, \end{array} \tag{1}$$

where the horizontal arrows are cones, the vertical arrows are arrows in C. When the limits exist, there are natural isomorphisms

$$C(c, \varprojlim F) \cong \operatorname{Nat}(\Delta c, F) = \operatorname{Cone}(c, F), \tag{2}$$

$$\operatorname{Cone}(F, c) = \operatorname{Nat}(F, \Delta c) \cong C(\varinjlim F, c). \tag{3}$$

There are familiar names for various special limits, dual to those for colimits:

Products. If J is the discrete category $\{1,2\}$, a functor $F : \{1, 2\} \to C$ is a pair of objects $\langle a, b\rangle$ of C. The limit object is called a *product* of a and b, and is written $a \times b$ or $a \amalg b$; the limit diagram consists of $a \times b$ and two arrows p, q (or sometimes pr_1, pr_2),

$$a \xleftarrow{p} a \times b \xrightarrow{q} b\,,$$

called the *projections* of the product. They constitute a cone from the vertex $a \times b$, so by the definition above of a limit, there is a bijection of sets

$$C(c, a \times b) \cong C(c, a) \times C(c, b) \tag{4}$$

natural in c, which sends each $h: c \to a \times b$ to the pair of composites $\langle ph, qh\rangle$. Conversely, given arrows $f: c \to a$ and $g: c \to b$, there is a unique $h: c \to a \times b$ with $ph = f$ and $qh = g$. We write

$$h = (f, g) : c \to a \times b$$

and call h the arrow with *components* f and g. We have already observed (in § II.3) that the product of any two objects exists in **Cat**, in **Grp**, in **Top**, and in **Mon**; in these cases (and in many others) it is called the *direct product*. In a preorder, a product is a greatest lower bound.

Infinite products. If J is a set ($=$ discrete category $=$ category with all arrows identities), then a functor $F: J \to C$ is simply a J-indexed family of objects $a_j \in C$, while a cone with vertex c and base a_j is just a J-indexed family of arrows $f_j: c \to a_j$. A universal cone $p_j: \Pi_j a_j \to a_j$ thus consists of an object $\Pi_j a_j$, called the *product* of the factors a_j, and of arrows p_j, called the *projections* of the product, with the following universal property: To each J-indexed family (=cone) $f_j: c \to a_j$ there is a unique f

$$f: c \to \Pi_j a_j, \quad \text{with} \quad p_j f = f_j, \quad j \in J.$$

The arrow f uniquely determined by this property is called the map (to the product) with *components* f_j, $j \in J$. Also $\{f_j \mid j \in J\} \mapsto f$ is a bijection

$$\Pi_j C(c, a_j) \cong C(c, \Pi_j a_j), \tag{5}$$

natural in c. Here the right hand product is that in C, while the left-hand product is taken in **Set** (where we assume that C has small hom-sets). Observe that the hom-functor $C(c, -)$ carries products in C to products in **Set** (see § V.4). Products over any small set J exist in **Set**, in **Top**, and in **Grp**; in each case they are just the familiar cartesian products.

Powers. If the factors in a product are all equal ($a_j = b \in C$ for all j) the product $\Pi_j a_j = \Pi_j b$ is called a *power* and is written $\Pi_j b = b^J$, so the

$$C(c, b)^J \cong C(c, b^J), \tag{6}$$

natural in c. The power on the left is that in **Set**, where every small power X^J exists (and is the set of all functions $J \to X$).

Equalizers. If $J = \downarrow\downarrow$, a functor $F: \downarrow\downarrow \to C$ is a pair $f, g: b \to a$ of parallel arrows of C. A limit object d of F, when it exists, is called an *equalizer* (or, a "difference kernel") of f and g. The limit diagram is

$$d \xrightarrow{e} b \underset{g}{\overset{f}{\rightrightarrows}} a, \quad fe = ge \tag{7}$$

(the limit arrow e amounts to a cone $a \leftarrow d \to b$ from the vertex d). The limit arrow is often called the equalizer of f and g; its universal property reads: To any $h: c \to b$ with $fh = gh$ there is a unique $h': c \to d$ with $eh' = h$.

In **Set**, the equalizer always exists; d is the set $\{x \in b \mid fx = gx\}$ and $e: d \to b$ is the injection of this subset of b into b. In **Top**, the equalizer has the same description (d has the subspace topology). In **Ab** the equalizer d of f and g is the usual kernel of the difference homomorphism $f - g: b \to a$.

Equalizers for any set of arrows from b to a are described similarly. Any equalizer e is necessarily a monic.

Pullbacks. If $J = (\rightarrow \cdot \leftarrow)$, a functor $F : (\rightarrow \cdot \leftarrow) \rightarrow C$ is a pair of arrows $b \xrightarrow{f} a \xleftarrow{g} d$ of C with a common codomain a. A cone over such a functor is a pair of arrows from a vertex c such that the square (on the left)

$$\begin{array}{ccc} c & \xrightarrow{k} & d \\ {\scriptstyle h}\downarrow & & \downarrow{\scriptstyle g} \\ b & \xrightarrow[f]{} & a, \end{array} \qquad \begin{array}{ccc} b \times_a d & \xrightarrow{q} & d \\ {\scriptstyle p}\downarrow & & \downarrow{\scriptstyle g} \\ b & \xrightarrow{f} & a \end{array} \tag{8}$$

commutes. A universal cone is then a commutative square of this form, with new vertex written $b \times_a d$ and arrows p, q as shown on the right, such that for any square with vertex c there is a unique $r : c \rightarrow b \times_a d$ with $k = qr$, $h = pr$. The square formed by this universal cone is called a *pullback square* or a "cartesian square" and the vertex $b \times_a d$ of the universal cone is called a pullback, a "fibered product", or a product over (the object) a. This construction, possible in many categories, first became prominent in the category **Top**. If $g : d \rightarrow a$ is a "fiber map" (of some type) with "base" a and f is a continuous map into the base, then the projection p of the pullback is the "induced fiber map" (of the type considered).

The pullback of a pair of equal arrows $f : b \rightarrow a \leftarrow b : f$, when it exists, is called the *kernel pair* of f. It is an object d and a pair of arrows $p, q : d \rightarrow b$ such that $fp = fq : d \rightarrow a$ and such that any pair $h, k : c \rightarrow a$ with $fh = fk$ can be written as $h = pr$, $k = qr$ for a unique $r : c \rightarrow d$.

If $J = \mathbf{0}$ is the empty category, there is exactly one functor $\mathbf{0} \rightarrow C$; namely, the empty functor; a cone over this functor is just an object $c \in C$ (i.e., just a vertex). Hence a universal cone on $\mathbf{0}$ is an object t of C such that each object $c \in C$ has a unique arrow $c \dashrightarrow t$. In other words, a limit of the empty functor to C is a terminal object of C.

Limits are sometimes defined for diagrams rather than for functors. In detail, let C be a category, UC the underlying graph of C, and G any graph. Then a *diagram* in C of shape G is a morphism $D : G \rightarrow UC$ of graphs. Now define a *cone* $\mu : c \xrightarrow{\cdot} D$ to be a function assigning to each object $i \in G$ an arrow $\mu_i : c \rightarrow D_i$ of C such that $Dh \circ \mu_i = \mu_j$ for every arrow $h : i \rightarrow j$ of the graph G. This is just the previous definition of a cone (a natural transformation $\mu : \Delta c \xrightarrow{\cdot} D$), coupled with the observation that this definition uses the composition of arrows in C but *not* in the domain G of D. A limit for the diagram D is now a universal cone $\lambda : c \xrightarrow{\cdot} D$.

This variation on the definition of a limit yields no essentially new information. For, let FG be a free category generated by the graph G, and $P : G \rightarrow U(FG)$ the corresponding universal diagram. Then each diagram $D : G \rightarrow UC$ can be written uniquely as $D = UD' \circ P$ for a (unique) functor $D' : FG \rightarrow C$, and one readily observes that limits (and limiting cones) for D' correspond exactly to those for D.

Exercises

1. In **Set**, show that the pullback of $f: X \to Z$ and $g: Y \to Z$ is given by the set of pairs $\{\langle x, y\rangle \mid x \in X,\ y \in Y,\ fx = gy\}$. Describe pullbacks in **Top**.
2. Show that the usual cartesian product over an index set J, with its projections, is a (categorical) product in **Set** and in **Top**.
3. If the category J has an initial object s, prove that every functor $F: J \to C$ to any category C has a limit, namely $F(s)$. Dualize.
4. In any category, prove that $f: a \to b$ is epi if and only if the following square is a pushout:

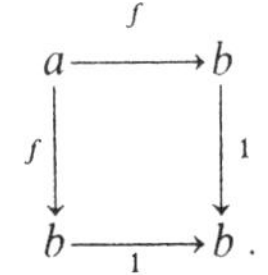

5. In a pullback square (8), show that f monic implies q monic.
6. In **Set**, show that the kernel pair of $f: X \to Y$ is given by the equivalence relation $E = \{\langle x, x'\rangle \mid x, x' \in X \text{ and } fx = fx'\}$, with suitable maps $E \rightrightarrows X$.
7. (Kernel pairs via products and equalizers.) If C has finite products and equalizers, show that the kernel pair of $f: a \to b$ may be expressed in terms of the projections $p_1, p_2: a \times a \to a$ as $p_1 e, p_2 e$, where e is the equalizer of $fp_1, fp_2: a \times a \to b$ (cf. Exercise 6). Dualize.
8. Consider the following commutative diagram

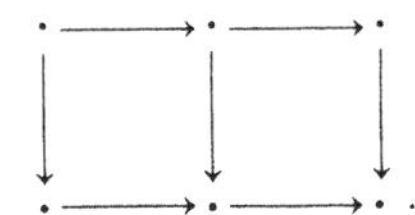

   (a) If both squares are pullbacks, prove that the outside rectangle (with top and bottom edges the evident composites) is a pullback.
   (b) If the outside rectangle and the right-hand square are pullbacks, so is the left-hand square.
9. (Equalizers via products and pullbacks.) Show that the equalizer of $f, g: b \to a$ may be constructed as the pullback of

$$(1_b, f): b \to b \times a \leftarrow b: (1_b, g).$$

10. If C has pullbacks and a terminal object, prove that C has all finite products and equalizers.

5. Categories with Finite Products

A category C is said to *have finite products* if to any finite number of objects $c_1, \ldots, c_n$ of C there exists a product diagram, consisting of a product object $c_1 \times \cdots \times c_n$ and n projections $p_i: c_1 \times \cdots \times c_n \to c_i$, for $i = 1, \ldots, n$, with the usual universal property. In particular, C then has a product of no objects, which is simply a terminal object t in C, as well as a product for any two objects. The *diagonal map* $\delta_c : c \to c \times c$ is defined for each c by $p_1 \delta_c = 1_c = p_2 \delta_c$; it is a natural transformation.

Proposition 1. *If a category C has a terminal object t and a product diagram $a \leftarrow a \times b \to b$ for any two of its objects, then C has all finite products. The product objects provide, by $\langle a, b \rangle \mapsto a \times b$, a bifunctor $C \times C \to C$. For any three objects a, b, and c there is an isomorphism*

$$\alpha = \alpha_{a,b,c} : a \times (b \times c) \cong (a \times b) \times c \tag{1}$$

natural in a, b, and c. For any object a there are isomorphisms

$$\lambda = \lambda_a : t \times a \cong a \qquad \varrho = \varrho_a : a \times t \cong a \tag{2}$$

which are natural in a, where t is the terminal object of C.

Proof. A product of one object c is just the diagram $c \to c$ formed with the identity map of c, so is present in any category. Now suppose that any two objects a_1, a_2 of C have a product. If we choose one such product diagram $a_1 \leftarrow a_1 \times a_2 \to a_2$ for each pair of objects, then $\times$ becomes a functor when $f_1 \times f_2$ is defined on arrows f_i by $p_i(f_1 \times f_2) = f_i p_i$. One may then form a product of three objects a, b, and c by forming the iterated product object $a \times (b \times c)$ with projections as in the diagram

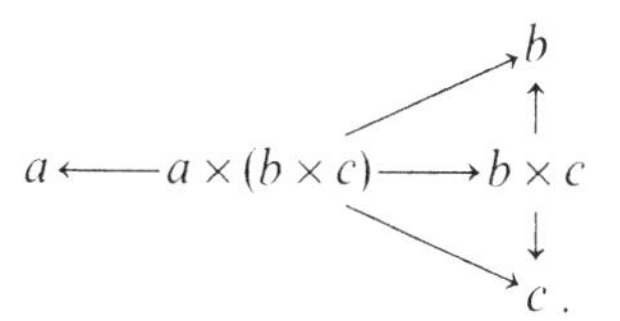

The projections to a and the two indicated composites give three arrows from $a \times (b \times c)$ to a, b, and c respectively. By the universality of the given projections (from two factors) it follows readily that these three arrows form a product diagram for a, b, and c. Product diagrams for more factors can be found by iteration in much the same way. For three factors, one could also form a product diagram by the iteration $(a \times b) \times c$; the uniqueness of the product objects then yields a unique isomorphism $a \times (b \times c) \cong (a \times b) \times c$ commuting with the given projections to a, b, and c. This is the isomorphism α of the proposition, and it is natural. Finally, since every object has a unique arrow to the terminal object t, the diagram $t \leftarrow a \xrightarrow{1} a$ is a product diagram for t and a. The uniqueness of the product object $t \times a$ then yields an isomorphism $\lambda_a : t \times a \to a$, and similarly $\varrho_a : a \times t \to a$. Naturality of λ and ϱ follows. These isomorphisms, α, λ and ρ so constructed are said to be "canonical."

The dual result holds for finite coproducts; in particular a coproduct of no factors is an initial object. For m objects a_j, a coproduct diagram consists of m injections $i_j : a_j \to a_1 \amalg \cdots \amalg a_m$ and any map $f : a_1 \amalg \cdots \amalg a_m \to c$

is uniquely determined by its m *cocomponents* $f\,i_j = f_j : a_j \to c$ for $j = 1, \ldots, m$. In particular, if C has both finite products and finite coproducts, the arrows

$$a_1 \amalg \cdots \amalg a_m \to b_1 \times \cdots \times b_n$$

from a coproduct to a product are determined uniquely by an $m \times n$ matrix of arrows $f_{jk} = p_k f i_j : a_j \to b_k$, where $j = 1, \ldots, m$, $k = 1, \ldots, n$. In categories of finite dimensional vector spaces, where finite coproduct coincides with finite product, this matrix is exactly the usual matrix of a linear transformation relative to given bases in its domain and codomain.

More generally, let C be any category with a null object z (an object z which is both initial and terminal), so that the arrow $a \to z \to b$ through z is the *zero* arrow $0 : a \to b$. If C also has finite products and finite coproducts, there is then a "canonical" arrow

$$a_1 \amalg \cdots \amalg a_n \to a_1 \times \cdots \times a_n$$

of the coproduct to the product – namely, that arrow which has the identity $n \times n$ matrix (identities on the diagonal and zeroes elsewhere). This canonical arrow may be an isomorphism (in **Ab** or R-**Mod**), a proper monic (in **Top**$_*$ or **Set**$_*$) or a proper epi (in **Grp**).

Exercises

1. Prove that the diagonal $\delta_c : c \to c \times c$ is natural in c.
2. In any category with finite products, prove that the following diagrams involving the canonical maps α, ϱ, λ of (1) and (2) always commute:

$$\begin{array}{ccccc} a \times (b \times (c \times d)) & \xrightarrow{\alpha} & (a \times b) \times (c \times d) & \xrightarrow{\alpha} & ((a \times b) \times c) \times d \\ \downarrow{\scriptstyle 1 \times \alpha} & & & & \uparrow{\scriptstyle \alpha \times 1} \\ a \times ((b \times c) \times d) & & \xrightarrow{\qquad\alpha\qquad} & & (a \times (b \times c)) \times d, \end{array}$$

$$\begin{array}{ccc} t \times (b \times c) & \xrightarrow{\alpha} & (t \times b) \times c \\ \downarrow{\scriptstyle \lambda} & & \downarrow{\scriptstyle \lambda \times 1} \\ b \times c & = & b \times c, \end{array} \qquad \begin{array}{ccc} a \times (t \times c) & \xrightarrow{\alpha} & (a \times t) \times c \\ \downarrow{\scriptstyle 1 \times \lambda} & & \downarrow{\scriptstyle \varrho \times 1} \\ a \times c & = & a \times c. \end{array}$$

3. (a) Prove that **Cat** has pullbacks (cf. Exercise II.6.5).
   (b) Show that the comma categories $(b \downarrow C)$ and $(C \downarrow a)$ are pullbacks in **Cat**.
4. Prove that **Cat** has all small coproducts.
5. If B has (finite) products show that any functor category B^C also has (finite) products (calculated "pointwise").

6. Groups in Categories

We return to the ideas of the introduction about expressing algebraic identities by diagrams. Let C be a category with finite products and a terminal object t. Then a *monoid* in C is a triple $\langle c, \mu : c \times c \to c, \eta : t \to c\rangle$, such that the following diagrams commute:

$$\begin{array}{ccccc} c \times (c \times c) & \xrightarrow{\alpha} & (c \times c) \times c & \xrightarrow{\mu \times 1} & c \times c \\ \downarrow{\scriptstyle 1 \times \mu} & & & & \downarrow{\scriptstyle \mu} \\ c \times c & & \xrightarrow{\qquad\mu\qquad} & & c, \end{array} \tag{1}$$

$$\begin{array}{ccccc} t \times c & \xrightarrow{\eta \times 1} & c \times c & \xleftarrow{1 \times \eta} & c \times t \\ \downarrow{\scriptstyle \lambda} & & \downarrow{\scriptstyle \mu} & & \downarrow{\scriptstyle \varrho} \\ c & = & c & = & c\,. \end{array} \tag{2}$$

(This is exactly the definition of the introduction, except for the explicit use in the first diagram of the associativity isomorphism α of (5.1).) We now define a *group* in C to be a monoid $\langle c, \mu, \eta\rangle$ together with an arrow $\zeta : c \to c$ which makes the diagram (with δ_c the diagonal)

$$\begin{array}{ccccc} c & \xrightarrow{\delta_c} & c \times c & \xrightarrow{1 \times \zeta} & c \times c \\ \downarrow & & & & \downarrow{\scriptstyle \mu} \\ t & & \xrightarrow{\qquad\eta\qquad} & & c \end{array} \tag{3}$$

commute (this suggests that ζ sends each $x \in c$ to its right inverse).

By similar diagrams, one may define rings in C, lattices in C, etc.; the process applies to any type of algebraic system defined by operations and identities between them.

It is a familiar fact that if G is an (ordinary) group, so is the function set G^X for any X; indeed the product of two functions f, f' in G^X is defined pointwise, as $(f \cdot f')(x) = fx \cdot f'x$. In the present context this construction takes the following form.

Proposition 1. *If C is a category with finite products, then an object c is a group (or, a monoid) in C if and only if the hom functor $C(-, c)$ is a group (respectively, a monoid) in the functor category* $\mathbf{Set}^{C^{\mathrm{op}}}$.

Proof. Each multiplication μ for c determines a corresponding multiplication $\bar{\mu}$ for the hom-set $C(-, c) : C^{\mathrm{op}} \to \mathbf{Set}$, as the composite

$$C(-, c) \times C(-, c) \xrightarrow{\theta} C(-, c \times c) \xrightarrow{\nu} C(-, c)$$

where $\nu = \mu_* = C(-, \mu)$, while the first natural isomorphism is that given (cf. (4.4)) by the definition of the product object $c \times c$. Conversely, given any natural ν as above, the Yoneda lemma proves that there is a unique $\mu : c \times c \to c$ with $\nu = \mu_*$. A "diagram chase" shows that μ is associative if and only if $\bar{\mu}$ is; the chase uses the definition of the associativity isomorphism α by its commutation with the projections of the three-fold product. The rest of the proof is left as an exercise.

Since the functor category $\mathbf{Set}^{C^{\text{op}}}$ always has finite products (Exercise 5.5) we can consider objects c in C such that $C(-, c)$ is a group in this functor category even if the category C does not have finite products; however, I know no real use of this added generality.

Exercises

(Throughout, C is a category with finite products and a terminal object t.)

1. Describe the category of monoids in C, and show that it has finite products.
2. Show that the category of groups in C has finite products.
3. Show that a functor $T : B \to \mathbf{Set}$ is a group in $\mathbf{Set}^B$ if and only if each Tb, for b an object of B, is an (ordinary) group and each Tf, f in B, is a morphism of groups.
4. (a) If A is an abelian group (in **Set**) show that its multiplication $A \times A \to A$, its unit $1 \to A$, and its inverse $A \to A$ are all morphisms of groups (where $A \times A$ is regarded as the direct product group). Deduce that A with these structure maps is a group in **Grp**.
   (b) Prove that every group in **Grp** has this form.

7. Colimits of Representable Functors

The utility of representable functors $\hom(d, -)$ is emphasized by the following basic result about set-valued functors.

Theorem 1. *Any functor $K : D \to$ **Sets** from a small category D to the category of sets can be represented (in a canonical way) as a colimit of a diagram of representable functors* $\hom(d, -)$ *for objects d in D.*

Proof. First, given K, we construct the needed diagram category (for the colimit) J as the so-called "category of elements" of K; that is, as the comma category $1 \downarrow K$ (see §II.6.(3)) with objects pairs (d, x) of elements $x \in K(d)$ for $d \in D$ and with arrows $f : (d, x) \to (d', x')$ those arrows $f : d \to d'$ of D for which $K(f)x = x'$ (more briefly, $f * x = x'$). We then claim that the given functor K is the colimit of the diagram on $1 \downarrow K$ given by the functor

$$M : J^D \to \mathbf{Sets}^D$$

which sends each object (d, x) to the hom-functor $D(d, -)$ and each arrow f to the induced natural transformation $f^* : D(d', -) \to D(d, -)$. Then the Yoneda isomorphism,

$$y^{-1} : K(d) \to \mathrm{Nat}(D(d, -), K),$$

yields a cone in $\mathbf{Sets}^D$ over the base M to K, as displayed by the arrows to

K at the lower left of the following figure:

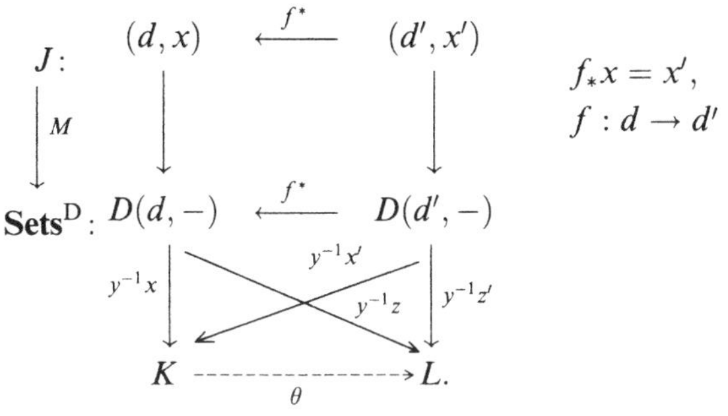

$$
\begin{array}{ccccc}
J: & (d,x) & \xleftarrow{f^*} & (d',x') & \qquad f_* x = x', \\
\Big\downarrow M & \Big\downarrow & & \Big\downarrow & \qquad f : d \to d' \\
\mathbf{Sets}^{\mathrm{D}}: & D(d,-) & \xleftarrow{f^*} & D(d',-) & \\
 & \Big\downarrow y^{-1}x & \searrow y^{-1}z \quad \swarrow y^{-1}x' & \Big\downarrow y^{-1}z' & \\
 & K & \xrightarrow[\theta]{\quad\quad} & L. &
\end{array}
\tag{1}
$$

We claim that this cone to K is a colimiting cone over $D(d,-)$. First, consider any other cone over $D(d,-)$ to the vertex L, some functor $L : D \to \mathbf{Sets}$. The arrows of this cone (arrows in $\mathbf{Sets}^D$) are natural transformations $D(d,-) \to L$, hence are given by the Yoneda lemma in the form $y^{-1}z : D(d,-) \to L$ for some $z \in L(d)$ as well as $y^{-1}z'$: $D(d',-) \to L$, where (since it a cone) $z' = f\, z$.

To show that this cone to the vertex K is universal, we must construct a unique natural transformation $\theta : K \to L$ which carries the first cone into the second one. So for each $x \in K(d)$, we start from the object (d,x) of J^{op}, as at the top in the diagram (1), and set

$$\theta_d x = z$$

for the $z \in L(d)$ present in the natural transformation $y^{-1}z$ in the cone for L. To show θ natural, consider any $f : d \to d'$ with $f\, x = x'$. Then also $f\, z = z'$, and since y^{-1} is natural, $f(y^{-1}z) = y^{-1}(f\, z) = y^{-1}z'$. Therefore, θ is natural. It is evidently unique, q.e.d.

A dual argument will show that any contravariant functor $D^{\mathrm{op}} \to$ **Sets** can be represented as a colimit of a diagram of representable contravariant functors $\hom(-,d)$.

For C a small category, a contravariant functor $F : C^{\mathrm{op}} \to \mathbf{Sets}$ is often called a **presheaf**. The intuition comes from the case where C is the category of open sets U of some topological space and $F(U)$ is the set of smooth (in some sense) functions defined on U, while an inclusion $V \subset U$ gives the map $F(U) \to F(V)$ which restricts a function on U to one on V. The functor category $\mathbf{Sets}^{C^{\mathrm{op}}}$ of all these functors (presheaves) is often written as $\hat{C}$. Certain of these functors (with a "matching" property) are called sheaves; see Mac Lane-Moerdijk [1992].

Notes.

The Yoneda Lemma made an early appearance in the work of the Japanese pioneer N. Yoneda (private communication to Mac Lane) [1954]; with time, its importance has grown.

Representable functors probably first appeared in topology in the form of "universal examples", such as the universal examples of cohomology operations (for instance, in J. P. Serre's 1953 calculations of the cohomology, modulo 2, of Eilenberg-Mac Lane spaces).

Universal arrows are unique only up to isomorphism; perhaps this lack of absolute uniqueness is why the notion was slow to develop. Examples had long been present; the bold step of really formulating the general notion of a universal arrow was taken by Samuel in 1948; the general notion was then lavishly popularized by Bourbaki. The idea that the ordinary cartesian products could be described by universal properties of their projections was formulated about the same time (Mac Lane [1948, 1950]). On the other hand the notions of limit and colimit have a long history in various concrete examples. Thus colimits were used in the proofs of theorems in which infinite abelian groups are represented as unions of their finitely generated subgroups. Limits (over ordered sets) appear in the p-adic numbers of Hensel and in the construction of Čech homology and cohomology by limit processes as formalized by Pontrjagin. An adequate treatment of the natural isomorphisms occurring for such limits was a major motivation of the first Eilenberg-Mac Lane paper on category theory [1945]. E. H. Moore's general analysis (about 1913) used limits over certain directed sets. In all these classical cases, limits appeared only for functors $F: J \to C$ with J a linearly or partly ordered set. Then Kan [1960] took the step of considering limits for all functors, while Freyd [1964] for the general case used the word "root" in place of "limit". His followers have chosen to extend the original word "limit" to this general meaning. Properties special to limits over directed sets will be studied in Chapter IX.

IV. Adjoints

1. Adjunctions

We now present a basic concept due to Kan, which provides a different formulation for the properties of free objects and other universal constructions. As motivation, we first reexamine the construction (§ III.1) of a vector space V_X with basis X. For a fixed field K consider the functors

$$\mathbf{Set} \underset{U}{\overset{V}{\rightleftarrows}} \mathbf{Vct}_K ,$$

where, for each vector space W, $U(W)$ is the set of all vectors in W, so that U is the forgetful functor, while, for any set X, $V(X)$ is the vector space with basis X. The vectors of $V(X)$ are thus the formal finite linear combinations $\Sigma\, r_i x_i$ with scalar coefficients $r_i \in K$ and with each $x_i \in X$, with the evident vector operations. Each function $g: X \to U(W)$ extends to a unique linear transformation $f: V(X) \to W$, given explicitly by $f(\Sigma\, r_i x_i) = \Sigma\, r_i(g x_i)$ (i.e., formal linear combinations in $V(X)$ to actual linear combinations in W). This correspondence $\psi: g \mapsto f$ has an inverse $\varphi: f \mapsto f \mid X$, the restriction of f to X, hence is a bijection

$$\varphi: \mathbf{Vct}_K(V(X), W) \cong \mathbf{Set}(X, U(W)) .$$

This bijection $\varphi = \varphi_{X,W}$ is defined "in the same way" for all sets X and all vector spaces W. This means that the $\varphi_{X,W}$ are the components of a natural transformation φ when both sides above are regarded as functors of X and W. It suffices to verify naturality in X and in W separately. Naturality in X means that for each arrow $h: X' \to X$ the diagram

$$\begin{array}{ccc} \mathbf{Vct}_K(V(X), W) & \xrightarrow{\varphi} & \mathbf{Set}(X, U(W)) \\ {\scriptstyle (Vh)^*}\downarrow & & \downarrow{\scriptstyle h^*} \\ \mathbf{Vct}_K(V(X'), W) & \xrightarrow{\varphi} & \mathbf{Set}(X', U(W)) , \end{array}$$

where $h^* g = g \circ h$, will commute. This commutativity follows from the definition of φ by a routine calculation, as does also the naturality in W.

Note next several similar examples.

The free category $C = FG$ on a given (small) graph G is a functor $\mathbf{Grph} \to \mathbf{Cat}$: it is related to the forgetful functor $U : \mathbf{Cat} \to \mathbf{Grph}$ by the fact (§ II.7) that each morphism $D : G \to UB$ of graphs extends to a unique map $D' : FG \to B$ of categories; moreover, $D \mapsto D'$ is a natural isomorphism

$$\mathbf{Cat}(FG, B) \cong \mathbf{Grph}(G, UB).$$

In the category of small sets, each function $g : S \times T \to R$ of two variables can be treated as a function $\varphi g : S \to \hom(T, R)$ of one variable (in S) whose values are functions of a second variable (in T); explicitly, $[(\varphi g)s]t = g(s, t)$ for $s \in S$, $t \in T$. This describes φ as a bijection

$$\varphi : \hom(S \times T, R) \cong \hom(S, \hom(T, R)) .$$

It is natural in S, T, and R. If we hold the set T fixed and define functors $F, G : \mathbf{Set} \to \mathbf{Set}$ by $F(S) = S \times T$, $G(R) = \hom(T, R)$, the bijection takes the form

$$\hom(F(S), R) \cong \hom(S, G(R))$$

natural in S and R, and much like the previous examples.

For modules A, B, and C over a commutative ring K there is a similar isomorphism

$$\hom(A \otimes_K B, C) \cong \hom(A, \hom_K(B, C))$$

natural in all three arguments.

Definition. *Let A and X be categories. An adjunction from X to A is a triple $\langle F, G, \varphi \rangle : X \rightharpoonup A$, where F and G are functors*

$$X \underset{G}{\overset{F}{\rightleftarrows}} A ,$$

while φ is a function which assigns to each pair of objects $x \in X$, $a \in A$ a bijection of sets

$$\varphi = \varphi_{x,a} : A(Fx, a) \cong X(x, Ga) \tag{1}$$

which is natural in x and a.

Here the left hand side $A(Fx, a)$ is the bifunctor

$$X^{\mathrm{op}} \times A \xrightarrow{F^{\mathrm{op}} \times \mathrm{Id}} A^{\mathrm{op}} \times A \xrightarrow{\hom} \mathbf{Set}$$

which sends each pair of objects $\langle x, a \rangle$ to the hom-set $A(Fx, a)$, and the right hand side is a similar bifunctor $X^{\mathrm{op}} \times A \to \mathbf{Set}$. Therefore the naturality of the bijection φ means that for all $k : a \to a'$ and $h : x' \to x$ both the diagrams:

$$\begin{array}{ccc} A(Fx, a) & \xrightarrow{\varphi} & X(x, Ga) \\ \downarrow{\scriptstyle k_*} & & \downarrow{\scriptstyle (Gk)_*} \\ A(Fx, a') & \xrightarrow{\varphi} & X(x, Ga') \end{array} \qquad \begin{array}{ccc} A(Fx, a) & \xrightarrow{\varphi} & X(x, Ga) \\ \downarrow{\scriptstyle (Fh)^*} & & \downarrow{\scriptstyle h^*} \\ A(Fx', a) & \xrightarrow{\varphi} & X(x', Ga) \end{array} \tag{2}$$

will commute. Here k_* is short for $A(Fx, k)$, the operation of composition with k, and $h^* = X(h, Ga)$.

This discussion assumes that all the hom-sets of X and A are small. If not, we just replace **Set** above by a suitable larger category **Ens** of sets.

An adjunction may also be described without hom-sets directly in terms of arrows. It is a bijection which assigns to each arrow $f: Fx \to a$ an arrow $\varphi f = \operatorname{rad} f: x \to Ga$, the *right adjunct of f*, in such a way that the naturality conditions of (2),

$$\varphi(k \circ f) = Gk \circ \varphi f, \quad \varphi(f \circ Fh) = \varphi f \circ h, \tag{3}$$

hold for all f and all arrows $h: x' \to x$ and $k: a \to a'$. It is equivalent to require that φ^{-1} be natural; i.e., that for every h, k and $g: x \to Ga$ one has

$$\varphi^{-1}(gh) = \varphi^{-1}g \circ Fh, \quad \varphi^{-1}(Gk \circ g) = k \circ \varphi^{-1}g. \tag{4}$$

Given such an adjunction, the functor F is said to be a *left-adjoint* for G, while G is called a *right adjoint* for F. (Some authors write $F \dashv G$; others say that F is the "adjoint" of G and G the "coadjoint" of F, but other authors say the opposite; therefore we shall stick to the language of "left" and "right" adjoints.)

Every adjunction yields a universal arrow. Specifically, set $a = Fx$ in (1). The left hand hom-set of (1) then contains the identity $1: Fx \to Fx$; call its φ-image η_x. By Yoneda's Proposition III.2.1, this η_x is a universal arrow

$$\eta_x: x \to GFx, \quad \eta_x = \varphi(1_{Fx}),$$

from $x \in X$ to G. The adjunction gives such a universal arrow η_x for *every* object x. Moreover, the function $x \mapsto \eta_x$ is a natural transformation $I_X \to GF$ because every diagram

$$\begin{array}{ccc} x' & \xrightarrow{\eta_{x'}} & GFx' \\ {\scriptstyle h}\downarrow & & \downarrow{\scriptstyle GFh} \\ x & \xrightarrow{\eta_x} & GFx \end{array}$$

is commutative. This one proves by the calculation

$$GFh \circ \varphi(1_{Fx'}) = \varphi(Fh \circ 1_{Fx'}) = \varphi(1_{Fx} \circ Fh) = \varphi(1_{Fx}) \circ h,$$

based on the Eq. (3) describing the naturality of φ. This calculation may also be visualized by the commutative diagram

$$\begin{array}{ccccc} A(Fx', Fx') & \xrightarrow{(Fh)_*} & A(Fx', Fx) & \xleftarrow{(Fh)^*} & A(Fx, Fx) \\ \downarrow{\scriptstyle\varphi} & & \downarrow{\scriptstyle\varphi} & & \downarrow{\scriptstyle\varphi} \\ X(x', GFx') & \xrightarrow[(GFh)_*]{} & X(x', GFx) & \xleftarrow[h^*]{} & X(x, GFx), \end{array}$$

where $h^* = X(h, 1)$ and $h_* = X(1, h)$.

The bijection φ can be expressed in terms of the arrows η_x as

$$\varphi(f) = G(f)\eta_x \quad \text{for} \quad f : Fx \to a ; \tag{5}$$

indeed, by the naturality (3) of φ we may compute that

$$\varphi(f) = \varphi(f \circ 1_{Fx}) = Gf \circ \varphi 1_{Fx} = Gf \circ \eta_x .$$

This computation may be visualized by chasing 1 around the commutative square

$$\begin{array}{ccc} A(Fx, Fx) & \xrightarrow{\varphi} & X(x, GFx) \\ \downarrow f_* & & \downarrow (Gf)_* \\ A(Fx, a) & \xrightarrow{\varphi} & X(x, Ga) \end{array} \qquad \begin{array}{ccc} 1 & \longmapsto & \eta_x \\ \downarrow & & \downarrow \\ f \circ 1 & \longmapsto & \varphi f = Gf \circ \eta_x . \end{array}$$

Dually, the adjunction gives a universal arrow from F. Indeed, set $x = Ga$ in the adjunction (1). The identity arrow $1 : Ga \to Ga$ is now present in the right-hand hom-set; its image under φ^{-1} is called ε_a,

$$\varepsilon_a : FGa \to a , \quad \varepsilon_a = \varphi^{-1}(1_{Ga}) , \quad a \in A ,$$

and is a universal arrow from F to a. As before, ε is a natural transformation $\varepsilon : FG \xrightarrow{\cdot} I_A$, and

$$\varphi^{-1}(g) = \varepsilon_a \circ Fg \quad \text{for} \quad g : x \to Ga .$$

Finally, take $x = Ga$. Then $\varepsilon_a = \varphi^{-1}(1_{Ga})$ gives, by the formula (5) for φ,

$$1_{Ga} = \varphi(\varepsilon_a) = G(\varepsilon_a) \circ \eta_{Ga} .$$

This asserts that the composite natural transformation

$$G \xrightarrow{\eta G} GFG \xrightarrow{G\varepsilon} G$$

is the identity transformation.

To summarize, we have proved

Theorem 1. *An adjunction $\langle F, G, \varphi \rangle : X \rightharpoonup A$ determines*

(i) *A natural transformation $\eta : I_X \xrightarrow{\cdot} GF$ such that for each object x the arrow η_x is universal to G from x, while the right adjunct of each $f : Fx \to a$ is*

$$\varphi f = Gf \circ \eta_x : x \to Ga ; \tag{6}$$

(ii) *A natural transformation $\varepsilon : FG \xrightarrow{\cdot} I_A$ such that each arrow ε_a is universal to a from F, while each $g : x \to Ga$ has left adjunct*

$$\varphi^{-1} g = \varepsilon_a \circ Fg : Fx \to a . \tag{7}$$

Moreover, both the following composites are the identities (of G, resp. F).

$$G \xrightarrow{\eta G} GFG \xrightarrow{G\varepsilon} G , \qquad F \xrightarrow{F\eta} FGF \xrightarrow{\varepsilon F} F . \tag{8}$$

We call η the *unit* and ε the *counit* of the adjunction. (Formerly, we called η a "front adjunction" and ε a "back adjunction".)

The given adjunction is actually already determined by various portions of all these data, in the following sense.

Theorem 2. *Each adjunction $\langle F, G, \varphi \rangle : X \rightharpoonup A$ is completely determined by the items in any one of the following lists:*

(i) *Functors F, G, and a natural transformation $\eta : 1_X \dot{\to} GF$ such that each $\eta_x : x \to GFx$ is universal to G from x. Then φ is defined by* (6).

(ii) *The functor $G : A \to X$ and for each $x \in X$ an object $F_0 x \in A$ and a universal arrow $\eta_x : x \to GF_0 x$ from x to G. Then the functor F has object function F_0 and is defined on arrows $h : x \to x'$ by $GFh \circ \eta_x = \eta_{x'} \circ h$.*

(iii) *Functors F, G, and a natural transformation $\varepsilon : FG \dot{\to} I_A$ such that each $\varepsilon_a : FGa \to a$ is universal from F to a. Here φ^{-1} is defined by* (7).

(iv) *The functor $F : X \to A$ and for each $a \in A$ an object $G_0 a \in X$ and an arrow $\varepsilon_a : FG_0 a \to a$ universal from F to a.*

(v) *Functors F, G and natural transformations $\eta : I_X \dot{\to} GF$ and $\varepsilon : FG \dot{\to} I_A$ such that both composites* (8) *are the identity transformations. Here φ is defined by* (6) *and φ^{-1} by* (7).

Because of (v), we often denote the adjunction $\langle F, G, \varphi \rangle$ by $\langle F, G, \eta, \varepsilon \rangle : X \rightharpoonup A$.

Proof. Ad (i): The statement that η_x is universal means that to each $f : x \to Ga$ there is exactly one g as in the commutative diagram

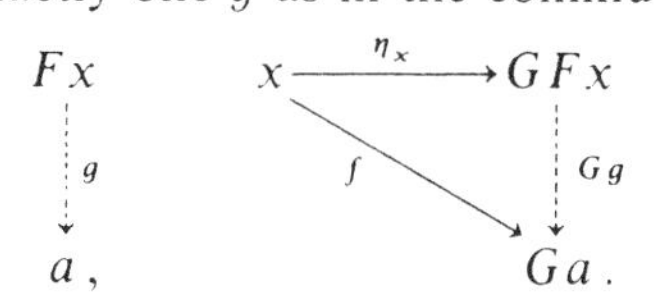

This states precisely that $\theta(g) = Gg \circ \eta_x$ defines a bijection

$$\theta : A(Fx, a) \to X(x, Ga) .$$

This bijection θ is natural in x because η is natural, and natural in a because G is a functor, hence gives an adjunction $\langle F, G, \theta \rangle$. In case η was the unit obtained from an adjunction $\langle F, G, \varphi \rangle$, then $\theta = \varphi$.

The data (ii) can be expanded to (i), and hence determine the adjunction. In (ii) we are given simply a universal arrow $\langle F_0 x, \eta_x \rangle$ for every object $x \in X$; we shall show that there is exactly one way to make F_0 the object function of a functor F for which $\eta : I_X \dot{\to} GF$ will be natural. Specifically, for each $h : x \to x'$ the universality of η_x states that there is exactly one arrow (dotted)

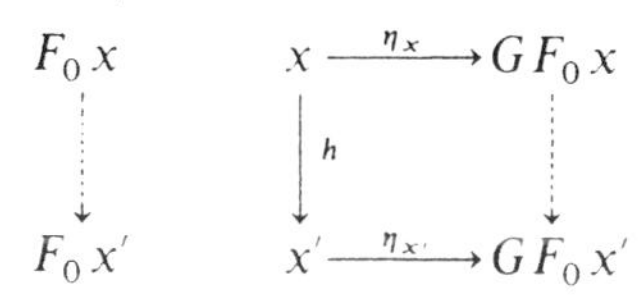

which can make the diagram commute. Choose this arrow as $Fh: F_0 x \to F_0 x'$; the commutativity states that η is now natural, and it is easy to check that this choice of Fh makes F a functor.

The proofs of parts (iii) and (iv) are dual.

To prove part (v) we use η and ε to define functions

$$A(Fx, a) \underset{\theta}{\overset{\varphi}{\rightleftarrows}} X(x, Ga)$$

by $\varphi f = Gf \circ \eta_x$ for each $f: Fx \to a$ and $\theta g = \varepsilon_a \circ Fg$ for each $g: x \to Ga$. Then since G is a functor and η is natural

$$\varphi \theta g = G\varepsilon_a \circ GFg \circ \eta_x = G\varepsilon_a \circ \eta_{Ga} \circ g.$$

But our hypothesis (8) states that $G\varepsilon_a \circ \eta_{Ga} = 1$. Hence $\varphi\theta = \mathrm{id}$. Dually $\theta\varphi = \mathrm{id}$. Therefore φ is a bijection (with inverse θ). It is clearly natural, hence is an adjunction (and, if we started with an adjunction, it is the one from which we started).

This theorem is very useful. For example, parts (ii) and (iv) construct an adjunction whenever we have a universal arrow from (or to) every object of a given category. For example, the category C has finite products when for each pair $\langle a, b\rangle \in C \times C$ there is a universal arrow from $\Delta: C \to C \times C$ to $\langle a, b\rangle$. By the theorem above we conclude that the function $\langle a, b\rangle \to a \times b$ giving the product object is actually a functor $C \times C \to C$, and that this functor is right adjoint to the diagonal functor Δ:

$$\varphi: (C \times C)(\Delta c, \langle a, b\rangle) \cong C(c, a \times b).$$

Using the definition of the arrows in $C \times C$, this is

$$\varphi: C(c, a) \times C(c, b) \cong C(c, a \times b).$$

The counit of this adjunction (set $c = a \times b$ on the right) is an arrow $\langle a \times b, a \times b\rangle \to \langle a, b\rangle$; it is thus just a pair of arrows $a \leftarrow a \times b \to b$; namely, the projections $p: a \times b \to a$ and $q: a \times b \to b$ of the product. The adjunction φ^{-1} sends each $f: c \to a \times b$ to the pair $\langle pf, qf\rangle$; this is the way in which φ is determined by the counit ε.

Similarly, if the category C has coproducts $\langle a, b\rangle \mapsto a \amalg b$, they define the coproduct functor $C \times C \to C$ which is a left adjoint to Δ:

$$C(a \amalg b, c) \cong (C \times C)(\langle a, b\rangle, \Delta c).$$

All the other examples of limits (when they always exist) can be similarly read as examples of adjoints. In many further applications, it turns out that proving universality is an easy way of showing that adjoints are present.

On the other hand, part (v) of the theorem describes an adjunction by two simple identities

$$\begin{array}{ccc} F \xrightarrow{F\eta} FGF & & GFG \xleftarrow{\eta G} G \\ \searrow \quad \downarrow \varepsilon F & & G\varepsilon \downarrow \quad \swarrow \\ F & & G \end{array} \tag{9}$$

on the unit and counit of the adjunction. These *triangular identities* make no explicit use of the objects of the categories A and X, and so are easy to manipulate. As we shall soon see, this is convenient for discussing properties of adjunctions. (For some authors, these identities are said to make η a "quasi-inverse" to ε.)

Corollary 1. *Any two left-adjoints F and F' of a functor $G: A \to X$ are naturally isomorphic.*

The proof is just an application of the fact that a universal arrow, like an initial object, is unique up to isomorphism. Explicitly, adjunctions $\langle F, G, \varphi \rangle$ and $\langle F', G, \varphi' \rangle$ give to each x two universal arrows $x \to GFx$ and $x \to GF'x$; hence there is a unique isomorphism $\theta_x : Fx \to F'x$ with $G\theta_x \cdot \eta_x = \eta'_x$; it is easy to verify that $\theta: F \dot{\to} F'$ is natural.

Corollary 2. *A functor $G: A \to X$ has a left adjoint if and only if, for each $x \in X$, the functor $X(x, Ga)$ is representable as a functor of $a \in A$. If $\varphi: A(F_0 x, a) \cong X(x, Ga)$ is a representation of this functor, then F_0 is the object function of a left-adjoint of G for which the bijection φ is natural in a and gives the adjunction.*

This is just a restatement of part (ii) of the theorem. Equivalently, G has a left-adjoint if and only if there is a universal arrow to G from every $x \in X$.

We leave the reader to state the duals.

Adjoints of additive functors are additive.

Theorem 3. *If the additive functor $G: A \to M$ between Ab-categories A and M has a left adjoint $F: M \to A$, then F is additive and the adjunction bijections*

$$\varphi : A(Fm, a) \cong M(m, Ga)$$

are isomorphisms of abelian groups (for all $m \in M$, $a \in A$).

Proof. If $\eta: I \dot{\to} GF$ is the unit of the adjunction, then φ may be written as $\varphi f = Gf \cdot \eta_m$ for any $f: Fm \to a$. If also $f': Fm \to a$, the additivity of G gives

$$\varphi(f + f') = G(f + f')\eta_m = (Gf + Gf')\eta_m = Gf \cdot \eta_m + Gf' \cdot \eta_m = \varphi f + \varphi f'.$$

Therefore φ is a morphism of abelian groups. Next take $g, g' : m \to n$ in M. Since η is natural,

$$GF(g+g') \circ \eta_m = \eta_n(g+g') = \eta_n g + \eta_n g' .$$

On the other hand, since G is additive,

$$G(Fg+Fg') \circ \eta_m = (GFg+GFg')\eta_m = GFg \circ \eta_m + GFg' \circ \eta_m = \eta_n g + \eta_n g' .$$

The equality of these two results and the universal property of η_m show that $F(g+g') = Fg + Fg'$. Hence F is additive.

Dually, any right adjoint of an additive functor is additive.

Exercises

1. Show that Theorem 2 can have an added clause (and its dual): (vi) A functor $G : A \to X$ and for each $x \in X$ a representation φ_x of the functor $X(x, G-) : A \to \mathbf{Set}$.
2. (Lawvere.) Given functors $G : A \to X$ and $F : X \to A$, show that each adjunction $\langle F, G, \varphi \rangle$ can be described as an isomorphism θ of comma categories such that the following diagram commutes

$$\begin{array}{ccc} \theta : (F \downarrow I_A) & \cong & (I_X \downarrow G) \\ \downarrow & & \downarrow \\ X \times A & = & X \times A . \end{array}$$

Here the vertical maps have components the projection functors P and Q of II.6(5).

3. For the adjunction $\langle \Delta, \times, \varphi \rangle$ – product right adjoint to diagonal – show that the unit $\delta_c : c \to c \times c$ for each object $c \in C$ is the unique arrow such that the diagram

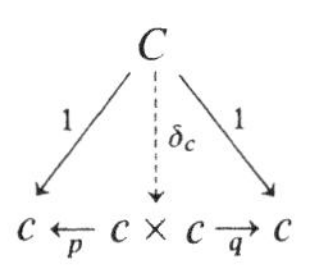

commutes. (This arrow δ_c is often called the *diagonal arrow* of c.) If $C = \mathbf{Set}$, show that $\delta_c x = \langle x, x \rangle$ for $x \in c$.

4. (Paré.) Given functors $G : A \to X$ and $K : X \to A$ and natural transformations $\varepsilon : KG \dot{\to} \mathrm{id}_A$, $\varrho : \mathrm{id}_X \dot{\to} GK$ such that $G\varepsilon \cdot \varrho G = 1_G : G \dot{\to} GKG \dot{\to} G$, prove that $\varepsilon K \cdot K\varrho : K \dot{\to} K$ is an idempotent in A^X and that G has a left adjoint if and only if this idempotent splits; explicitly if $\varepsilon K \cdot K\varrho$ splits as $\alpha \cdot \beta$ with $\beta \cdot \alpha = 1$ and $\beta : K \dot{\to} F$, then F is a left adjoint of G with unit $G\beta \cdot \varrho$ and counit $\varepsilon \cdot \alpha G$.

2. Examples of Adjoints

We now summarize a number of examples of adjoints, beginning with a table of left-adjoints of typical forgetful functors.

Forgetful functor	*Left adjoint F*	*Unit of adjunction*
$U : R\text{-}\mathbf{Mod} \to \mathbf{Set}$	$X \mapsto FX$ Free R-module, basis X	$j : X \to UFX$ (cf. § III.1) "insertion of generators"
$U : \mathbf{Cat} \to \mathbf{Grph}$	$G \mapsto CG$ Free category on graph G	$G \to UCG$ "insertion of generators"
$U : \mathbf{Grp} \to \mathbf{Set}$	$X \mapsto FX$ Free group, generators $x \in X$	$X \to UFX$ "insertion of generators"
$U : \mathbf{Ab} \to \mathbf{Set}$	$X \mapsto F_a X$ Free abelian group on X	"insertion of generators"
$U : \mathbf{Ab} \to \mathbf{Grp}$	$G \mapsto G/[G, G]$ Factor commutator group	$G \to G/[G, G]$ projection on the quotient
$U : R\text{-}\mathbf{Mod} \to \mathbf{Ab}$	$A \mapsto R \otimes A$	$A \to U(R \otimes A)$ $a \mapsto 1 \otimes a$
$U : R\text{-}\mathbf{Mod}\text{-}S \to R\text{-}\mathbf{Mod}$	$A \mapsto A \otimes S$	$A \to U(A \otimes S)$ $a \mapsto a \otimes 1$
$U : \mathbf{Rng} \to \mathbf{Mon}$ (cf. Exercise III.1.1)	$M \mapsto \mathbf{Z}(M)$ (integral) monoid ring	$M \to U\mathbf{Z}M$ $m \mapsto m$
$U : K\text{-}\mathbf{Alg} \to K\text{-}\mathbf{Mod}$	$V \mapsto TV$ Tensor algebra on V	$V \subset TV$ "insertion of generators"
$U : \mathbf{Fld} \to \mathbf{Dom}_m$ (cf. § III.1)	$D \mapsto QD$ Field of quotients	$D \subset UQD$ "insertion of $D : a \mapsto a/1$"
$U : \mathbf{Compmet} \to \mathbf{Met}$	Completion of metric space	(§ III.1)

There is a similar description of counits. For example, in the free R-module FX generated by elements $jx = \langle x \rangle$ for $x \in X$, the elements may be written as finite sums $\Sigma\, r_i \langle x_i \rangle$ with scalars $r_i \in R$. Then for any R-module A the counit $\varepsilon_A : FUA \to A$ is $\Sigma\, r_i \langle a_i \rangle \mapsto \Sigma\, r_i a_i$ (linear combinations in A). In other words ε_A is the epimorphism appearing in the standard representation of an arbitrary R-module as a quotient of a free module (the free module on its own elements as generators).

Next, we list some left and right adjoints (which need not exist in every category C) for diagonal functors; with the unit when C is **Set**.

Diagonal functor	*Adjoint*	*Unit*	*Counit*
$\Delta : C \to C \times C$	Left: Coproduct $\amalg : C \times C \to C$ $\langle a, b \rangle \mapsto a \amalg b$	(pair of) injections $i : a \to a \amalg b$ $j : b \to a \amalg b$	"folding" map $c \amalg c \to c$ $ix \mapsto x,\ jx \mapsto x$
	Right: Product $\Pi : C \times C \to C$ $\langle a, b \rangle \mapsto a \times b$	Diagonal arrow $\delta_c : c \to c \times c$ $x \mapsto \langle x, x \rangle$	(pair of) projections $p : a \times b \mapsto a$ $q : a \times b \mapsto b$

Diagonal functor	*Adjoint*	*Unit*	*Counit*
$C \to \mathbf{1}$	Left: Initial object s		$s \to c$
	Right: Terminal object t	$c \to t$	
$\Delta : C \to C^{\downarrow\downarrow}$ (III.3.6)	Left: Coequalizer $\langle f, g\rangle \mapsto$ coeq. object e	Coequalizing arrow $\langle f, g\rangle \xrightarrow{\langle uf, u\rangle} \langle e, e\rangle$	Identity $1 : c \to c$
(III.4.7)	Right: Equalizer d $\langle f, g\rangle \mapsto$ equal. object	Identity	Equalizing arrow $\langle d, d\rangle \to \langle f, g\rangle$
$\Delta : C \to C^{\leftarrow\cdot\to}$	Left: (Vertex of) pushout (III.3.7)		
	Right: (Vertex of) pullback (III.4.8)		
$\Delta : C \to C^J$	Left: Colimit object	Universal cone	
	Right: Limit object		Universal cone

In the case of limits, the form of the unit depends on the number of connected components of J. Here a category J is called *connected* when to any two objects $j, k \in J$ there is a finite sequence of arrows

$$j = j_0 \to j_1 \leftarrow j_2 \to \cdots \to j_{2n-1} \leftarrow j_{2n} = k \qquad \text{(both directions possible)}$$

joining j to k (see Exercises 7, 8).

Duality functors provide further examples. For vector spaces V, W over a field K, the dual $\bar{D}$ is a contravariant functor on **Vct** to **Vct**, given on objects by $\bar{D}V = \mathbf{Vct}(V, K)$ with the usual vector space structure and on arrows $h : V \to W$ as $\bar{D}h : \bar{D}W \to \bar{D}V$, where $(\bar{D}h)f = fh$ for each $f : W \to K$. A function

$$\varphi = \varphi_{V,W} : \mathbf{Vct}(V, \mathbf{Vct}(W, K)) \to \mathbf{Vct}(W, \mathbf{Vct}(V, K)) \tag{1}$$

is defined for $h : V \to \bar{D}W$ by $[(\varphi h)w]v = (hv)w$ for all $v \in V$, $w \in W$. Since $\varphi_{W,V}\varphi_{V,W}$ is the identity, each φ is a bijection. This bijection can be made into an adjuction as follows. The contravariant functor $\bar{D}$ leads to two different (covariant!) functors with the same object function,

$$D : \mathbf{Vct}^{\mathrm{op}} \to \mathbf{Vct}, \qquad D^{\mathrm{op}} : \mathbf{Vct} \to \mathbf{Vct}^{\mathrm{op}},$$

defined (as usual) for arrows $h^{\mathrm{op}} : W \to V$ and $h : V \to W$ by

$$Dh^{\mathrm{op}} = \bar{D}h : \bar{D}W \to \bar{D}V; \qquad D^{\mathrm{op}}h = (\bar{D}h)^{\mathrm{op}} : \bar{D}V \to \bar{D}W.$$

The bijection φ of (1) above may now be written as

$$\mathbf{Vct}^{\mathrm{op}}(D^{\mathrm{op}}W, V) \cong \mathbf{Vct}(W, DV), \tag{2}$$

natural in V and W. Therefore D^{op} is the left adjoint of D. (Warning: It is not a right adjoint of D, see § V.5, Exercise 2.) If $\kappa_W : W \to \bar{D}\bar{D}W$ is the usual canonical map to the double dual, the unit of the adjunction (set $V = D^{op} W$ in (2)) is this map $\eta_W = \kappa_W : W \to D D^{op} W$, and the counit is an arrow $\varepsilon_V : D^{op} D V \to V$ in $\mathbf{Vct}^{op}$ which turns out to be $\varepsilon_V = (\kappa_V)^{op}$ for the *same* κ.

This example illustrates the way in which adjunctions may replace isomorphisms of categories. For finite dimensional vector spaces, D and D^{op} are isomorphisms; for the general case, this is not true, but D is the right adjoint of D^{op}.

This example also bears on adjoints for other contravariant functors. Two contravariant functors $\bar{S}$ from A to X and $\bar{T}$ from X to A are "adjoint on the right" (Freyd) when there is a bijection $A(a, \bar{T}x) \cong X(x, \bar{S}a)$, natural in a and x. We shall not need this terminology, because we can replace $\bar{S}$ and $\bar{T}$ by the covariant functors $S : A^{op} \to X$ and $T : X^{op} \to A$ and form the dual $S^{op} : A \to X^{op}$, also covariant; thus the natural bijection above becomes $X^{op}(S^{op} a, x) \cong A(a, Tx)$, and so states that S^{op} is left adjoint (in our usual sense) to T – or, equivalently, that T^{op} is left adjoint to S. It is not necessarily equivalent to say that $\bar{T}$ and $\bar{S}$ are adjoint "on the left".

The next three sections will be concerned with three other types of adjoints: A left adjoint to an inclusion functor (of a full subcategory) is called a *reflection*; certain other special sorts of adjoints are "equivalences" of categories. Some other amusing examples of adjoints are given in the exercises to follow, some of which require knowledge of the subject matter involved. Goguen [1971] shows for finite state machines that the functor "minimal realization" is left adjoint to the functor "behavior". The reader is urged to find his own examples as well.

Exercises

1. For K a field and V a vector space over K, there is an "exterior algebra" $E(V)$, which is a graded, anticommutative algebra. Show that E is the left adjoint of a suitable forgetful functor (one which is not faithful).
2. Show that the functor $U : R\text{-}\mathbf{Mod} \to \mathbf{Ab}$ has not only a left adjoint $A \mapsto R \otimes A$, but also a right adjoint $A \mapsto \hom_{\mathbf{Z}}(R, A)$.
3. For K a field, let $\mathbf{Lie}_K$ be the category of all (small) Lie algebras L over K, with arrows the morphisms of K-modules which also preserve the Lie bracket operation $\langle a, b \rangle \mapsto [a, b]$. Let $V : \mathbf{Alg}_K \to \mathbf{Lie}_K$ be the functor which assigns to each (associative) algebra A the Lie algebra VA on the same vector space, with bracket $[a, b] = ab - ba$ for $a, b \in A$. Using the Poincaré-Birkhoff-Witt Theorem show that the functor E, where EL is the enveloping associative algebra of L, is a left adjoint for V.
4. Let $\mathbf{Rng}'$ denote the category of rings R which do not necessarily have an identity element for multiplication. Show that the standard process of adding an identity to R provides a left adjoint for the forgetful functor $\mathbf{Rng} \to \mathbf{Rng}'$ (forget the presence of the identity).

5. If a monoid M is regarded as a discrete category, with objects the elements $x \in M$, then the multiplication of M is a bifunctor $\mu : M \times M \to M$. If M is a group, show that the group inverse provides right adjoints for the functors $\mu(x, -)$ and $\mu(-, y) : M \to M$. Conversely, does the presence of such adjoints make a monoid into a group?
6. Describe units and counits for pushout and pullback.
7. If the category J is a disjoint union (coproduct) $\amalg J_k$ of categories J_k, for index k in some set K, with $I_k : J_k \to J$ the injections of the coproduct, then each functor $F : J \to C$ determines functors $F_k = F I_k : J_k \to C$.
   (a) Prove that $\operatorname{Lim} F \cong \Pi_k \operatorname{Lim} F_k$, if the limits on the right exist.
   (b) Show that every category J is a disjoint union of connected categories (called the *connected components* of J).
   (c) Conclude that all limits can be obtained from products and limits over connected categories.
8. (a) If the category J is connected, prove for any $c \in C$ that $\operatorname{Lim} \Delta c = c$ and $\operatorname{Colim} \Delta c = c$, where $\Delta c : J \to C$ is the constant functor.
   (b) Describe the unit for the right adjoint to $\Delta : C \to C^J$.
9. (Smythe.) Show that the functor O : **Cat** $\to$ **Set** assigning to each category C the set of its objects has a left adjoint D which assigns to each set X the discrete category on X, and that D in turn has a left adjoint assigning to each category the set of its connected components. Also show that O has a right adjoint which assigns to each set X a category with objects X and exactly one arrow in every hom-set.
10. If a category C has both cokernel pairs and equalizers, show that the functor $K : C^2 \to C^{\downarrow\downarrow}$ which assigns to each arrow of C its cokernel pair has as right adjoint the functor which assigns to each parallel pair of arrows its equalizing arrow.
11. If C has finite coproducts and $a \in C$, prove that the projection $Q : (a \downarrow C) \to C$ of the comma category $(Q(a \to c) = c)$ has a left adjoint, with $c \mapsto (a \to a \amalg c)$.
12. If X is a set and C a category with powers and copowers, prove that the copower $c \mapsto X \cdot c$ is left adjoint to the power $c \mapsto c^X$.

3. Reflective Subcategories

For many of the forgetful functors $U : A \to X$ listed in § 2, the counit $\varepsilon : FU \dot{\to} I_A$ of the adjunction assigns to each $a \in A$ the epimorphism $\varepsilon_a : F(Ua) \to a$ which gives the standard representation of a as a quotient of a free object. This is a general fact: Whenever a right adjoint G is faithful, every counit ε_a of the adjunction is epi.

Theorem 1. *For an adjunction* $\langle F, G, \eta, \varepsilon \rangle : X \rightharpoonup A$: (i) G *is faithful if and only if every component* ε_a *of the counit* ε *is epi,* (ii) G *is full if and only if every* ε_a *is a split monic. Hence* G *is full and faithful if and only if each* ε_a *is an isomorphism* $FGa \cong a$.

The proof depends on a lemma.

Lemma. *Let $f^*: A(a, -) \dot{\to} A(b, -)$ be the natural transformation induced by an arrow $f: b \to a$ of A. Then f^* is monic if and only if f is epi, while f^* is epi if and only if f is a split monic (i.e., if and only if f has a left inverse).*

Note that $f^* \mapsto f$ is the bijection $\mathrm{Nat}(A(a, -), A(b, -)) \cong A(b, a)$ given by the Yoneda lemma.

Observe, also, that for functors $S, T: C \to B$, a natural transformation $\tau: S \dot{\to} T$ is epi (respectively, monic) in B^C if and only if every component $\tau_c: S_c \to T_c$ is epi (respectively, monic) in B for $B = \mathbf{Set}$; this follows by Exercise III.4.4, computing the pushout pointwise as in Exercise III.5.5.

Proof. For $h \in A(a, c)$, $f^* h = hf$. Hence the first result is just the definition of an epi f. If f^* is epi, there is an $h_0: a \to b$ with $f^* h_0 = h_0 f = 1: b \to b$, so f has a left inverse. The converse is immediate.

Now we prove the theorem. Apply the Yoneda Lemma to the natural transformation (arrow function of G followed by the adjunction)

$$A(a, c) \xrightarrow{G_{a,c}} X(Ga, Gc) \xrightarrow{\varphi^{-1}} A(FGa, c).$$

It is determined (set $c = a$) by the image of $1: a \to a$, which is exactly the definition of the counit $\varepsilon_a: FGa \to a$. But φ^{-1} is an isomorphism, hence this natural transformation is monic or epi, respectively, when every $G_{a,c}$ is injective or surjective, respectively; that is, when G is faithful or full, respectively. The result now follows by the lemma.

A subcategory A of B is called *reflective* in B when the inclusion functor $K: A \to B$ has a left adjoint $F: B \to A$. This functor F may be called a *reflector* and the adjunction $\langle F, K, \varphi \rangle = \langle F, \varphi \rangle : B \rightharpoonup A$ a *reflection* of B in its subcategory A. Since the inclusion functor K is always faithful, the counit ε of a reflection is always epi. A reflection can be described in terms of the composite functor $R = KF: B \to B$; indeed, $A \subset B$ is reflective in B if and only if there is a functor $R: B \to B$ with values in the subcategory A and a bijection of sets

$$A(Rb, a) \cong B(b, a)$$

natural in $b \in B$ and $a \in A$. A reflection may be described in terms of universal arrows: $A \subset B$ is reflective if and only if to each $b \in B$ there is an object Rb of the subcategory A and an arrow $\eta_b: b \to Rb$ such that every arrow $g: b \to a \in A$ has the form $g = f \circ \eta_b$ for a unique arrow $f: Rb \to a$ of A. As usual, R is then (the object function of) a functor $B \to B$ (with values in A).

If a full subcategory $A \subset B$ is reflective in B, then by Theorem 1 each object $a \in A$ is isomorphic to FKa, and hence $Ra \cong a$ for all a.

Dually, $A \subset B$ is *coreflective* in B when the inclusion functor $A \to B$ has a right adjoint. (Warning: Mitchell [1965] has interchanged the meanings of "reflection" and "coreflection".)

Here are some examples. **Ab** is reflective in **Grp**. For, if $G/[G, G]$ is the usual factor-commutator group of a group G, then $\hom(G/[G, G], A) \cong \hom(G, A)$ for A abelian, and **Ab** is full in **Grp**. Or consider the category of all metric spaces X, with arrows uniformly continuous functions. The (full) subcategory of complete metric spaces is reflective; the reflector sends each metric space to its completion. Again, consider the category of all completely regular Hausdorff spaces (with arrows all continuous functions). The (full) subcategory of all compact Hausdorff spaces is reflective; the reflector sends each completely regular space to its Stone-Čech compactification.

A coreflective subcategory of **Ab** is the full subcategory of all torsion abelian groups (a group is torsion if all elements have finite order); the coreflector sends each abelian group A to the subgroup TA of all elements of finite order in A.

Exercises

1. Show that the table of dual statements (§ II.1) extends as follows:

Statement	*Dual statement*
$S, T: C \to B$ are functors	$S, T: C \to B$ are functors
T is full	T is full
T is faithful	T is faithful
$\eta: S \dot{\to} T$ is a natural transformation.	$\eta: T \dot{\to} S$ is a natural transformation.
$\langle F, G, \varphi \rangle: X \rightharpoonup A$ is an adjunction	$\langle G, F, \varphi^{-1} \rangle: A \rightharpoonup X$ is an adjunction
η is the unit of $\langle F, G, \varphi \rangle$.	η is the counit of $\langle G, F, \varphi^{-1} \rangle$.

2. Show that the torsion-free abelian groups form a full reflective subcategory of **Ab**.
3. If $\langle G, F, \varphi \rangle: X \rightharpoonup A$ is an adjunction with G full and every unit η_x a monic, then every η_x is also epi.
4. Show the following subcategories to be reflective:
   (a) The full subcategory of all partial orders in the category **Preord** of all preorders, with arrows all monotone functions.
   (b) The full subcategory of T_0-spaces in **Top**.
5. Given an adjunction $\langle F, G, \varphi \rangle: X \rightharpoonup A$, prove that G is faithful if and only if φ^{-1} carries epis to epis.
6. Given an adjunction $\langle F, G, \eta, \varepsilon \rangle$ with either F or G full, prove that $G\varepsilon: GFG \to G$ is invertible with inverse $\eta G: G \to GFG$.
7. If A is a full and reflective subcategory of B, prove that every functor $S: J \to A$ with a limit in B has a limit in A.

4. Equivalence of Categories

A functor $S: A \to C$ is an *isomorphism* of categories when there is a functor $T: C \to A$ (backwards) such that $ST = I: C \to C$ and $TS = I: A \to A$. In this case, the identity natural transformations

$\eta: I \dot{\to} ST$ and $\varepsilon: TS \dot{\to} I$ make $\langle T, S; \eta, \varepsilon\rangle: C \rightharpoonup A$ an adjunction. In other words, a two-sided inverse T of a functor S is a left-adjoint of S – and for that matter, T is also a right-adjoint of S.

There is a more general (and more useful) notion:

A functor $S: A \to C$ is an *equivalence of categories* (and the categories A and C are *equivalent*) when there is a functor $T: C \to A$ (backwards) and natural isomorphisms $ST \cong I: C \to C$ and $TS \cong I: A \to A$. In this case $T: C \to A$ is also an equivalence of categories. We shall soon see that T is then both a left adjoint and a right adjoint of S.

Here is an example. In any category C a *skeleton* of C is any full subcategory A such that each object of C is isomorphic (in C) to exactly one object of A. Then A is equivalent to C and the inclusion $K: A \to C$ is an equivalence of categories. For, select to each $c \in C$ an isomorphism $\theta_c: c \cong Tc$ with Tc an object of A. Then we can make T a functor $T: C \to A$ in exactly one way so that θ will become a natural isomorphism $\theta: I \cong KT$. Moreover $TK \cong I$, so K is indeed an equivalence: *A category is equivalent to (any one of) its skeletons.* For example, the category of all finite sets has as a skeleton the full subcategory with objects all finite ordinal numbers 0, 1, 2, ..., n, ... (Here 0 is the empty set and each $n = \{0, 1, \ldots, n-1\}$.)

A category is called *skeletal* when any two isomorphic objects are identical; i.e., when the category is its own skeleton.

An *adjoint equivalence* of categories is an adjunction $\langle T, S; \eta, \varepsilon\rangle: C \rightharpoonup A$ in which both the unit $\eta: I \dot{\to} ST$ and the counit $\varepsilon: TS \dot{\to} I$ are natural isomorphisms: $I \cong ST$, $TS \cong I$. Then η^{-1} and ε^{-1} are also natural isomorphisms, and the triangular identities $\varepsilon T \cdot T\eta = 1$, $S\varepsilon \cdot \eta S = 1$ can be written as $T\eta^{-1} \cdot \varepsilon^{-1} T = 1$, $\eta^{-1} S \cdot S\varepsilon^{-1} = 1$, respectively. These identities then state that $\langle S, T, \varepsilon^{-1}, \eta^{-1}\rangle: A \rightharpoonup C$ is an adjunction with $\varepsilon^{-1}: I \dot{\to} TS$ as unit and $\eta^{-1}: ST \dot{\to} I$ as counit. Thus in an adjoint equivalence $\langle T, S, -, -\rangle$ the functor $T: C \to A$ is the left adjoint of $S: A \to C$ with unit η and at the same time T is the right adjoint of S, with unit ε^{-1}.

We can now state the main facts about equivalence.

Theorem 1. *The following properties of a functor $S: A \to C$ are logically equivalent:*

(i) *S is an equivalence of categories,*

(ii) *S is part of an adjoint equivalence $\langle T, S; \eta, \varepsilon\rangle: C \rightharpoonup A$,*

(iii) *S is full and faithful, and each object $c \in C$ is isomorphic to Sa for some object $a \in A$.*

Proof. Trivially, (ii) implies (i). To prove that (i) implies (iii), note that $ST \cong I$ shows that each $c \in C$ has the form $c \cong S(Tc)$ for an $a = Tc \in A$. The natural isomorphism $\theta: TS \cong I$ gives for each $f: a \to a'$ the commutative square

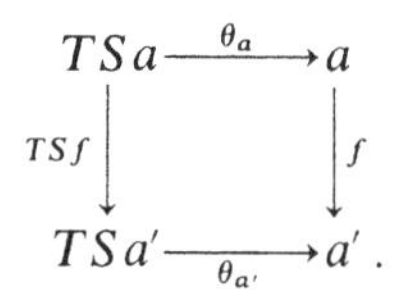

Hence $f = \theta_{a'} \circ TSf \circ \theta_a^{-1}$; it follows that S is faithful. Symmetrically, $ST \cong I$ proves T faithful. To show S full, consider any $h: Sa \to Sa'$ and set $f = \theta_{a'} \circ Th \circ \theta_a^{-1}$. Then the square above commutes also with Sf replaced by h, so $TSf = Th$. Since T is faithful, $Sf = h$, which means that S is full.

To prove that (iii) implies (ii) we must construct from S a (left) adjoint T. For each $c \in C$ we can choose some object $a_0 = T_0 c \in A$ and an isomorphism η_c:

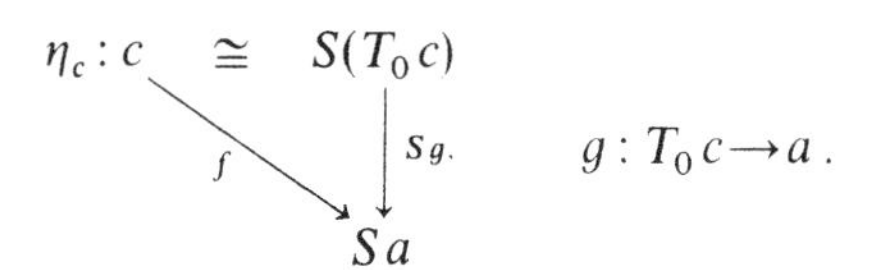

For every arrow $f: c \to Sa$, the composite $f \circ \eta_c^{-1}$ has the form Sg for some g because S is full; this g is unique because S is faithful. In other words, $f = Sg \circ \eta_c$ for a unique g, so η_c is universal from c to S. Therefore T_0 can be made a functor $T: C \to A$ in exactly one way so that $\eta: I \dot{\to} ST$ is natural, and then T is the left adjoint of S with unit the isomorphism η. As with any adjunction, $S\varepsilon_a \cdot \eta_{Sa} = 1$ (put $c = Sa$, $f = 1$ in the diagram above). Thus $S\varepsilon_a = (\eta_{Sa})^{-1}$ is invertible. Since S is full and faithful, the counit ε_a is also invertible. Therefore $\langle T, S; \eta, \varepsilon \rangle : C \rightharpoonup A$ is an adjoint equivalence, and the proof is complete.

In this proof, suppose that A is a full subcategory of C and that $S = K : A \to C$ is the insertion. For objects $a \in A \subset C$ we can then choose $a_0 = a = Ka$ and η_{Ka} the identity. Then $K\varepsilon_a = 1$, hence $\varepsilon_a = 1$ for all a. This proves

Proposition 2. *If A is a full subcategory of C and every $c \in C$ is isomorphic (in C) to some object of A, then the insertion $K: A \to C$ is an equivalence and is part of an adjoint equivalence $\langle T, K; \eta, 1 \rangle : C \rightharpoonup A$ with counit the identity. Therefore A is reflective in C.*

This includes in particular the case already noted, when A is a skeleton of C.

A functor $F: X \to A$ is said to be a *left-adjoint-left-inverse* of $G: A \to X$ when there is an adjunction $\langle F, G; \eta, 1 \rangle : X \rightharpoonup A$ with counit the identity. This means (Exercise 4) that G is an isomorphism of A to a reflective subcategory of X. In the case of the Proposition 2 just above, we have shown that the insertion $A \to C$ has a left-adjoint-left-inverse.

Duality theorems in functional analysis are often instances of equivalences. For example, let **CAb** be the category of compact topological abelian groups, and let P assign to each such group G its character group PG, consisting of all continuous homomorphisms $G \to \mathbf{R}/\mathbf{Z}$. The Pontrjagin duality theorem asserts that $P: \mathbf{CAb} \to \mathbf{Ab}^{\mathrm{op}}$ is an equivalence of categories. Similarly, the Gelfand-Naimark theorem states that the functor C which assigns to each compact Hausdorff space X its abelian C^*-algebra of continuous complex-valued functions is an equivalence of categories (see Negrepontis [1971]).

Exercises

1. Prove: (a) Any two skeletons of a category C are isomorphic.
   (b) If A_0 is a skeleton of A and C_0 a skeleton of C, then A and C are equivalent if and only if A_0 and C_0 are isomorphic.
2. (a) Prove: the composite of two equivalences $D \to C$, $C \to A$ is an equivalence.
   (b) State and prove the corresponding fact for adjoint equivalences.
3. If $S: A \to C$ is full, faithful, and surjective on objects (each $c \in C$ is $c = Sa$ for some $a \in A$), prove that there is an adjoint equivalence $\langle T, S; 1, \varepsilon \rangle : C \rightharpoonup A$ with unit the identity (and thence that T is a left-adjoint-right-inverse of S).
4. Given a functor $G: A \to X$, prove the three following conditions logically equivalent:
   (a) G has a left-adjoint-left-inverse.
   (b) G has a left adjoint, and is full, faithful, and injective on objects.
   (c) There is a full reflective subcategory Y of X and an isomorphism $H: A \cong Y$ such that $G = KH$, where $K: Y \to X$ is the insertion.
5. If J is a connected category and $\Delta: C \to C^J$ has a left adjoint (colimit), show that this left adjoint can be chosen to be a left-adjoint-left-inverse.

5. Adjoints for Preorders

Recall that a preorder P is a set $P = \{p, p', \ldots\}$ equipped with a reflexive and transitive binary relation $p \leqq p'$, and that preorders may be regarded as categories so that order-preserving functions become functors. An order-reversing function $\bar{L}$ on P to Q is then a functor $L: P \to Q^{\mathrm{op}}$.

Theorem 1 *(Galois connections are adjoint pairs). Let P, Q be two preorders and $L: P \to Q^{\mathrm{op}}$, $R: Q^{\mathrm{op}} \to P$ two order-preserving functions. Then L (regarded as a functor) is a left adjoint to R if and only if, for all $p \in P$ and $q \in Q$,*

$$Lp \geqq q \text{ in } Q \text{ if and only if } p \leqq Rq \text{ in } P. \tag{1}$$

When this is the case, there is exactly one adjunction φ making L the left adjoint of R. For all p and q, $p \leqq RLp$ and $LRq \geqq q$; hence also

$$Lp \geqq LRLp \geqq Lp, \quad Rq \leqq RLRq \leqq Rq. \tag{2}$$

Proof. Recall that P becomes a category in which there is (exactly) one arrow $p \to p'$ whenever $p \leqq p'$. Thus the condition (1) states precisely that there is a bijection $\hom_{Q^{\mathrm{op}}}(Lp, q) \cong \hom_P(p, Rq)$; since each hom-set has at most one element, this bijection is automatically natural. The unit of the adjunction is the inequality $p \leqq RLp$ for all p, while the counit is $LRq \geqq q$ for all q. The two Eqs. (2) are the triangular identities connecting unit and counit. In the convenient case when both P and Q are posets (i.e., when both the relations $\leqq$ are antisymmetric) these conditions become $L = LRL$, and $R = RLR$ (each three passages reduce to one!).

A pair of order-preserving functions L and R which satisfy (1) is called a *Galois connection* from P to Q. Here is the fundamental example, for a group G acting on a set U, by $\langle \sigma, x \rangle \mapsto \sigma \cdot x$ for $\sigma \in G$, $x \in U$. Take $P = \mathscr{P}(U)$, the set of all subsets $X \subset U$, ordered by inclusion, while $Q = \mathscr{P}(G)$ is the set of subsets $S \subset G$ also ordered by inclusion ($S \leqq S'$ if and only if $S \subset S'$). Let $LX = \{\sigma \mid x \in X \text{ implies } \sigma \cdot x = x\}$, $RS = \{x \mid \sigma \in S \text{ implies } \sigma \cdot x = x\}$; in other words, LX is the subgroup of G which fixes all points $x \in X$ and RS is the set of fixed points of the automorphisms of S. Then $LX \geqq S$ in Q if and only if $\sigma \cdot x = x$ for all $\sigma \in S$ and all $x \in X$, which in turn holds if and only if $X \leqq RS$ in P. Therefore, L and R form an adjoint pair (a Galois connection). The original instance is that with G a group of automorphisms of a field U, as in the classical Galois theory.

If U and V are sets, the set $\mathscr{P}(U)$ of all subsets of U is a preorder under inclusion. For each function $f: U \to V$ the direct image f_*, defined by $f_*(X) = \{f(x) \mid x \in X\}$ is an order-preserving function and hence a functor $f_*: \mathscr{P}(U) \to \mathscr{P}(V)$. The inverse image $f^*(Y) = \{x \mid fx = y$ for some $y \in Y\}$ defines a functor $f^*: \mathscr{P}(V) \to \mathscr{P}(U)$ in the opposite direction. Since $f_* X \subset Y$ if and only if $X \subset f^* Y$, the direct image functor f_* is left adjoint to the inverse image functor f^*.

Certain adjoints for Boolean algebras are closely related to the basic connectives in logic. We again regard $\mathscr{P}(U)$ as a preorder, and hence as a category. The diagonal functor $\Delta: \mathscr{P}(U) \to \mathscr{P}(U) \times \mathscr{P}(U)$ has (as we have already noted) a right adjoint $\cap$, sending subsets X, Y to their intersection $X \cap Y$, and a left adjoint $\cup$, with $\langle X, Y \rangle \mapsto X \cup Y$, the union. If X is a fixed subset of U, then intersection with X is a functor $X \cap - : \mathscr{P}(U) \to \mathscr{P}(U)$. Since $X \cap Y \leqq Z$ if and only if $Y \leqq X' \cup Z$, where X' is the complement of X in U, the right adjoint of $X \cap -$ is $X' \cup -$. Thus the construction of suitable adjoints yields the Boolean operations $\cap$, $\cup$, and $'$ corresponding to "and", "or", and "not". Now consider the first projection $P: U \times V \to U$ from the product of two sets U and V. Each subset $S \subset U \times V$ defines two corresponding subsets of U by

$$P_* S = \{x \mid \exists y, y \in V \quad \text{and} \quad \langle x, y \rangle \in S\},$$

$$P_\# S = \{x \mid \forall y, y \in V \quad \text{implies} \quad \langle x, y \rangle \in S\};$$

they arise from $\langle x, y \rangle \in S$ by applying the existential quantifier $\exists y$, "there exists a y" and the universal quantifier $\forall y$, "for all y", respectively to $\langle x, y\rangle \in S$. Also $P_* S$ is the direct image of S under the projection P. Now for all subsets $X \subset U$ one has

$$S \leqq P^* X \Leftrightarrow P_* S \leqq X\,; \qquad P^* X \leqq S \Leftrightarrow X \leqq P_\# S\,,$$

where "$\Leftrightarrow$" means "if and only if". These state that P^*, which is the inverse image operation, has both a left adjoint P_* and a right adjoint $P_\#$. In this sense, both quantifiers $\exists$ and $\forall$ can be interpreted as adjoints.

There is also a geometric interpretation: $P^* X$ is the cylinder $X \times V \subset U \times V$ over the base $X \subset U$, $P_* S$ is the projection of $S \subset U \times V$ on the base U, and $P_\# S$ is the largest subset X of U such that the cylinder on X is wholly contained in S. This analysis has revealed several basic concepts of logic (and, or, not, $\forall y$, $\exists y$) to be adjoints. This illustrates the slogan "adjoints are everywhere".

Exercises

1. Let H be a space with an inner product (e.g., Hilbert space). If $P = Q$ is the set of all subsets S of H, ordered by inclusion, show that $LS = RS =$ the orthogonal complement of S gives a Galois connection.
2. In a Galois connection between posets, show that the subset $\{p \mid p = RLp\}$ of P equals $\{p \mid p = Rq$ for some $q\}$ and give a bijection from this set to the subset $\{q \mid q = LRq\}$ of Q. What are these sets in the case of a group of automorphisms of a field? Does this generalize to an arbitrary adjunction?
3. For C a category with pullbacks, each arrow $f: a \to a'$ defines a functor $(C \downarrow f) = f_*: (C \downarrow a) \to (C \downarrow a')$ which carries each object $x \to a$ of $(C \downarrow a)$ to the composite $x \to a \to a'$. Show that f_* has a right adjoint f^* with $f^*(x' \to a') = y \to a$, where y is the vertex of the pullback of $a \to a' \leftarrow x'$.

6. Cartesian Closed Categories

Much of the force of category theory will be seen to reside in using categories with specified additional structures. One basic example will be the closed categories (§ VII. 7); at present we can define readily one useful special case, "cartesian closed".

To assert that a category C has all finite products and coproducts is to assert that products, terminal, initial and coproducts exist, thus the functors $C \to \mathbf{1}$ and $\Delta: C \to C \times C$ have both left and right adjoints. Indeed, the left adjoints give initial object and coproduct, respectively, while the right adjoints give terminal object and product, respectively.

Using just adjoints we will now define "cartesian closed category". A category C with all finite products specifically given is called *cartesian closed* when each of the following functors

$$C \to \mathbf{1}, \qquad C \to C \times C, \qquad C \xrightarrow{-\times b} C,$$
$$c \mapsto 0, \qquad c \mapsto \langle c, c\rangle, \qquad a \mapsto a \times b,$$

has a *specified* right adjoint (with a specified adjunction). These adjoints are written as follows

$$t \dashv 0\,, \qquad a \times b \dashv \langle a, b \rangle\,, \qquad c^b \dashv c\,.$$

Thus to specify the first is to specify a terminal object t in C, and specifying the second is specifying for each pair of objects $a, b \in C$ a product object $a \times b$ together with its projections $a \leftarrow a \times b \rightarrow b$. These projections determine the adjunction (they constitute the counit of the adjunction); as already noted, $\times$ is then a bifunctor. The third required adjoint specifies for each functor $- \times b : C \rightarrow C$ a right adjoint, with the corresponding bijection

$$\hom(a \times b, c) \cong \hom(a, c^b)$$

natural in a and in c. By the parameter theorem (to be proved in the next section), $\langle b, c \rangle \mapsto c^b$ is then (the object function of) a bifunctor $C^{\mathrm{op}} \times C \rightarrow C$. Specifying the adjunction amounts to specifying for each c and b an arrow e

$$e : c^b \times b \rightarrow c$$

which is natural in c and universal from $- \times b$ to c. We call this $e = e_{b,c}$ the *evaluation* map. It amounts to the ordinary evaluation $\langle f, x \rangle \mapsto f x$ of a function f at an argument x in both of the following cases:

Set is a cartesian closed category, with $c^b = \hom(b, c)$.

Cat is cartesian closed, with exponent C^B the functor category.

A closely related example of adjoints is the functor

$$- \otimes_K B : K\text{-}\mathbf{Mod} \rightarrow K\text{-}\mathbf{Mod}$$

which has a right adjoint $\hom_K(B, -)$; the adjunction is determined by a counit $\hom_K(B, A) \otimes_K B \rightarrow A$ given by evaluation.

Exercises

1. (a) If U is any set, show that the preorder $\mathscr{P}(U)$ of all subsets of U is a cartesian closed category.
   (b) Show that any Boolean algebra, regarded as a preorder, is cartesian closed.
2. In some elementary theory T, consider the set $S = \{p, q, \ldots\}$ of sentences of T as a preorder, with $p \leqq q$ meaning "p entails q" (i.e., q is a consequence of p on the basis of the axioms of T). Prove that S is a cartesian closed category, with product given by conjunction and exponential q^p given by "p implies q".
3. In any cartesian closed category, prove $c^t \cong c$ and $c^{b \times b'} \cong (c^b)^{b'}$.
4. In any cartesian closed category obtain a natural transformation $c^b \times b^a \rightarrow c^a$ which agrees in **Set** with composition of functions. Prove it (like composition) associative.
5. Show that A cartesian closed need not imply A^J cartesian closed.

7. Transformations of Adjoints

We next study maps comparing different adjunctions. Given two adjunctions

$$\langle F, G, \varphi, \eta, \varepsilon\rangle : X \rightharpoonup A\,, \qquad \langle F', G', \varphi', \eta', \varepsilon'\rangle : X' \rightharpoonup A' \tag{1}$$

we define a *map of adjunctions* (from the first to the second adjunction) to be a pair of functors $K : A \to A'$ and $L : X \to X'$ such that both squares

$$\begin{array}{ccccc} A & \xrightarrow{G} & X & \xrightarrow{F} & A \\ {\scriptstyle K}\downarrow & & \downarrow{\scriptstyle L} & & \downarrow{\scriptstyle K} \\ A' & \xrightarrow[G']{} & X' & \xrightarrow[F']{} & A' \end{array} \tag{2}$$

of functors commute, and such that the diagram of hom-sets and adjunctions

$$\begin{array}{ccc} A(Fx, a) & \xrightarrow{\varphi} & X(x, Ga) \\ {\scriptstyle K = K_{Fx,a}}\downarrow & & \downarrow{\scriptstyle L = L_{x,Ga}} \\ A'(KFx, Ka) & & X'(Lx, LGa) \\ \| & & \| \\ A'(F'Lx, Ka) & \xrightarrow{\varphi'} & X'(Lx, G'Ka) \end{array} \tag{3}$$

commutes for all objects $x \in X$ and $a \in A$. Here $K_{Fx,a}$ is the map $f \mapsto Kf$ given by the functor K applied to each $f : Fx \to a$.

Proposition 1. *Given adjunctions* (1) *and functors* K *and* L *satisfying* (2), *the condition* (3) *on hom-sets is equivalent to* $L\eta = \eta' L$ *and also to* $\varepsilon' K = K\varepsilon$.

Proof. Given (3) commutative, set $a = Fx$ and chase the identity arrow $1 : Fx \to Fx$ around (3) to get the units η, η' and the equality

$$\langle L\eta : L \to LGF\rangle = \langle \eta' L : L \to G'F'L\rangle\,,$$

where $LGF = G'F'L$ by (2). Conversely, given the equality $L\eta = \eta' L$ of natural transformations, the definition of the adjunctions φ and φ' by their units gives (3). The case of the counits is dual to this one.

Next, given two adjunctions

$$\langle F, G, \varphi, \eta, \varepsilon\rangle\,, \qquad \langle F', G', \varphi', \eta', \varepsilon'\rangle : X \rightharpoonup A \tag{4}$$

between the same two categories, two natural transformations

$$\sigma : F \dot{\to} F'\,, \qquad \tau : G' \dot{\to} G$$

are said to be *conjugate* (for the given adjunctions) when the diagram

$$\begin{array}{ccc} A(F'x,a) & \overset{\varphi'}{\cong} & X(x,G'a) \\ {\scriptstyle (\sigma_x)^* = A(\sigma_x, a)}\downarrow & & \downarrow{\scriptstyle X(x,\tau_a) = (\tau_a)_*} \\ A(Fx,a) & \underset{\varphi}{\cong} & X(x,Ga) \end{array} \tag{5}$$

commutes for every pair of objects $x \in X$, $a \in A$.

Theorem 2. *Given the two adjunctions* (4), *the natural transformations σ and τ are conjugate if and only if any one of the four following diagrams (of natural transformations) commutes*

$$\begin{array}{ccc} G' & \xrightarrow{\tau} & G \\ {\scriptstyle \eta G'}\downarrow & & \uparrow{\scriptstyle G\varepsilon'} \\ GFG' & \xrightarrow[G\sigma G']{} & GF'G', \end{array} \qquad \begin{array}{ccc} F & \xrightarrow{\sigma} & F' \\ {\scriptstyle F\eta'}\downarrow & & \uparrow{\scriptstyle \varepsilon F'} \\ FG'F' & \xrightarrow[F\tau F']{} & FGF', \end{array} \tag{6}$$

$$\begin{array}{ccc} FG' & \xrightarrow{F\tau} & FG \\ {\scriptstyle \sigma G'}\downarrow & & \downarrow{\scriptstyle \varepsilon} \\ F'G' & \xrightarrow{\varepsilon'} & I_A, \end{array} \qquad \begin{array}{ccc} I_X & \xrightarrow{\eta} & GF \\ {\scriptstyle \eta'}\downarrow & & \downarrow{\scriptstyle G\sigma} \\ G'F' & \xrightarrow[\tau F']{} & GF'. \end{array} \tag{7}$$

Also, given the adjunctions (4) *and the natural transformation $\sigma : F \dot{\rightarrow} F'$, there is a unique $\tau : G' \dot{\rightarrow} G$ such that the pair $\langle \sigma, \tau \rangle$ is conjugate. Dually, given* (4) *and τ, there is a unique σ with $\langle \sigma, \tau \rangle$ conjugate.*

Proof. First, (5) implies (6) and (5) implies (7). For, put $x = G'a$ in (5), start with the identity arrow $1 : G'a \rightarrow G'a$ in the upper right and use the description of φ and φ' by unit and counit to chase this element 1 around the diagram as follows

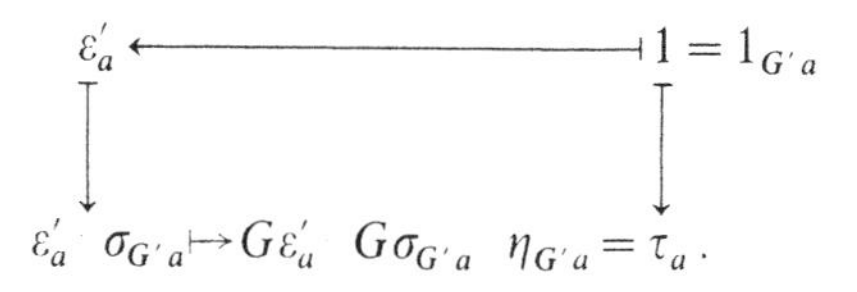

The result (lower right) is the first equality of (6). A slightly different chase yields

$$\begin{array}{ccc} \varepsilon'_a & \longleftarrow\!\!\!\dashv & 1 \\ \top\!\!\downarrow & & \top\!\!\downarrow \\ \varepsilon'_a \circ \sigma_{G'a} = \varepsilon_a \circ F\tau_a & \longleftarrow\!\!\!\dashv & \tau_a. \end{array}$$

The resulting equality is the first diagram of (7). The second halves of (6) and (7) are duals.

Next, suppose σ but not τ given. Then the Yoneda Lemma applied to the composite transformation $\varphi \circ (\sigma_x)^* \circ \varphi'^{-1}$ (three legs of (5)) shows that there is a unique family of arrows τ'_a for which (5) commutes, and this family is a natural transformation. Since each $\varepsilon_a : FGa \to a$ is universal from F to a, there is also a unique family of arrows $\tau''_a : G'a \to Ga$ for which the first of (7) commutes. Since (5) implies (7), $\tau'_a = \tau''_a$. In other words, if $\tau = \tau''$ makes the first square of (7) commute, it also makes (5) commute. Therefore the first square of (7) implies (5). Given σ, there is immediately a unique natural transformation $\tau : G' \stackrel{\bullet}{\to} G$ for which the first of (6) commutes; since (5) implies (6), $\tau'_a = \tau_a$, and hence the solutions τ'_a of (5) are necessarily natural; moreover (6) implies (5).

The reader may also show that (6) implies (5) or (7) by constructing suitable diagrams of natural transformations.

We now regard a conjugate pair $\langle \sigma, \tau \rangle$ of natural transformations as a *transformation* (or morphism) from the first to the second adjunction. The "vertical" composite of two such

$$\langle F, G, \eta, \varepsilon \rangle \xrightarrow{\langle \sigma, \tau \rangle} \langle F', G', \eta', \varepsilon' \rangle \xrightarrow{\langle \sigma', \tau' \rangle} \langle F'', G'', \eta'', \varepsilon'' \rangle \tag{8}$$

is evidently (say by condition (5)) a transformation $\langle \sigma', \tau' \rangle \circ \langle \sigma, \tau \rangle = \langle \sigma' \cdot \sigma, \tau \cdot \tau' \rangle$ from the first to the third adjunction. For the two given categories X and A we thus have a new category $A^{(\mathrm{adj})X}$, the *category of adjunctions* from X to A; its objects are the adjunctions $\langle F, G; \eta, \varepsilon \rangle$; its arrows are the transformations (conjugate pairs) $\langle \sigma, \tau \rangle$, with the composition just noted. Also there are two evident "forgetful" functors to the ordinary functor categories, as follows:

$$\begin{array}{ccccc} A^X & \longleftarrow & A^{(\mathrm{adj})X}, & [A^{(\mathrm{adj})X}]^{\mathrm{op}} \longrightarrow & X^A, \\ F & \longleftarrow\!\!\shortmid & \langle F, G, \eta, \varepsilon \rangle & \shortmid\!\!\longrightarrow & G \\ {\scriptstyle \sigma}\downarrow & & \downarrow{\scriptstyle \langle \sigma, \tau \rangle} & & \uparrow{\scriptstyle \tau} \\ F' & \longleftarrow\!\!\shortmid & \langle F', G', \eta', \varepsilon' \rangle & \shortmid\!\!\longrightarrow & G'. \end{array}$$

A typical example for **Set** is the bijection

$$\hom(S \times T, R) \cong \hom(S, \hom(T, R)) \tag{9}$$

discussed in § 1 as an example of an adjunction (for each fixed set T). If $t : T \to T'$ is a function between two such sets, then $- \times t$ is a natural transformation of functors $- \times T \stackrel{\bullet}{\to} - \times T'$. Its conjugate is the natural transformation $\hom(t, -) : \hom(T', -) \stackrel{\bullet}{\to} \hom(T, -)$; this is, as it should be, in the reverse direction, corresponding to the fact that $S \times T$ is covariant and $\hom(T, R)$ contravariant in the argument T. We may call (9) an adjunction with a "parameter" $T \in \mathbf{Set}$. For a commutative ring

K the adjunction $\mathbf{Mod}_K(A \otimes_K B, C) \cong \mathbf{Mod}_K(A, \mathrm{Hom}_K(B, C))$ has a parameter $B \in \mathbf{Mod}_K$. Here is general statement:

Theorem 3 *(Adjunctions with a parameter). Given a bifunctor $F: X \times P \to A$, assume for each object $p \in P$ that $F(-, p): X \to A$ has a right adjoint $G(p, -): A \to X$, via an adjunction*

$$\hom(F(x, p), a) \cong \hom(x, G(p, a)), \tag{10}$$

natural in x and a. There is then a unique way to assign to each arrow $h: p \to p'$ of P and each object $a \in A$ an arrow $G(h, a): G(p', a) \to G(p, a)$ of X so that G becomes a bifunctor $P^{\mathrm{op}} \times A \to X$ for which the bijection of the adjunction (10) *is natural in all three variables x, p, and a. This assignment of arrows $G(h, a)$ to $\langle h, a \rangle$ may also be described as the unique way to make $G(h, -)$ a natural transformation conjugate to $F(-, h)$.*

Proof. The condition that the adjunction (10) be natural in $p \in P$ is the commutativity of the square

$$\begin{array}{ccc}
\hom(F(x, p), a) & \cong & \hom(x, G(p, a)) \\
\uparrow{\scriptstyle F(x,h)^*} & & \uparrow{\scriptstyle G(h,a)_*} \\
\hom(F(x, p'), a) & \cong & \hom(x, G(p', a)).
\end{array}$$

This commutativity (for all a) states precisely that $G(h, -): G(p', -) \dot{\to} G(p, -)$ must be chosen as the conjugate to $F(-, h): F(-, p) \dot{\to} F(-, p')$. By the previous theorem, there exists a unique choice of $G(h, -)$ to realize this – and the condition of conjugacy may be expressed in *any of the five equivalent ways* stated there. For a second arrow $h': p' \to p''$, the uniqueness of the choice of conjugates shows for $h'h$ that $G(h'h, -) = G(h, -) \circ G(h', -)$, so that $G(-, a)$ is a functor and G a bifunctor, as required.

Dually, given a bifunctor $G: P^{\mathrm{op}} \times A \to X$ where each $G(p, -)$ has a right adjoint $F(-, p)$, there is a unique way to make F a bifunctor $X \times P \to A$.

Exercises

1. Interpret the definition $C(X \cdot a, c) \cong \mathbf{Set}(X, C(c, a))$ of copowers $X \cdot a$ in C as an adjunction with parameter a.
2. Let $\eta_x: x \to G(p, F(x, p))$ be the unit of an adjunction with parameter. It is natural in x, but what property of η corresponds to the naturality of the adjunction (10) in p?
3. In the functor category A^X let S be that full subcategory with objects those functors $F: X \to A$ which have a right adjoint $RF: A \to X$. Make R a functor $S^{\mathrm{op}} \to X^A$ by choosing one RF for each F, with $R\sigma$ the conjugate of σ.

4. (Kelly.) An *adjoint square* is an array of categories, functors, adjunctions, and natural transformations

$$\begin{array}{ccc} X & \xrightarrow{\langle F, G, \varphi\rangle} & A \\ \downarrow H & & \downarrow K \\ X' & \xrightarrow{\langle F', G', \varphi'\rangle} & A', \end{array} \qquad \begin{array}{l} \sigma : F'H \dot{\to} KF, \\ \\ \tau : HG \dot{\to} G'K, \end{array}$$

such that the following diagram of hom-sets always commutes

$$\begin{array}{ccccc} A(Fx, a) & \xrightarrow[K]{} & A'(KFx, Ka) & \xrightarrow{(\sigma x)^*} & A'(F'Hx, Ka) \\ \downarrow \varphi & & & & \downarrow \varphi' \\ X(x, Ga) & \xrightarrow{H} & X'(Hx, HGa) & \xrightarrow{(\tau a)_*} & X'(Hx, G'Ka). \end{array}$$

Express this last condition variously in terms of unit and counit of the adjunctions and prove that each of σ, τ determines the other. (The case $H = K =$ identity functor is that treated in the text above.)

5. (Palmquist.) Given H, K, and the two adjunctions as in Exercise 4, establish a bijection between natural transformations $\alpha : F'HG \dot{\to} K$ and natural transformations $\beta : H \dot{\to} G'KF$.

8. Composition of Adjoints

Two successive adjunctions compose to give a single adjunction, in the following sense:

Theorem 1. *Given two adjunctions*

$$\langle F, G, \eta, \varepsilon\rangle : X \rightharpoonup A, \qquad \langle \bar{F}, \bar{G}, \bar{\eta}, \bar{\varepsilon}\rangle : A \rightharpoonup D$$

the composite functors yield an adjunction

$$\langle \bar{F}F, G\bar{G}, G\bar{\eta}F \cdot \eta, \bar{\varepsilon} \cdot \bar{F}\varepsilon\bar{G}\rangle : X \rightharpoonup D.$$

Proof. With hom-sets, the two given adjunctions yield a composite isomorphism, natural in $x \in X$ and $d \in D$:

$$D(\bar{F}Fx, d) \cong A(Fx, \bar{G}d) \cong X(x, G\bar{G}d).$$

This makes the composite $\bar{F}F$ left adjoint to $G\bar{G}$. Setting $d = \bar{F}Fx$, and applying these two isomorphisms to the identity $1 : \bar{F}Fx \to \bar{F}Fx$, we find that the unit of the composite adjunction is $x \xrightarrow{\eta x} GFx \xrightarrow{G\bar{\eta}Fx} G\bar{G}\bar{F}Fx$, so is $G\bar{\eta}F \cdot \eta$, as asserted. By the dual argument, the counit is $\bar{\varepsilon} \cdot \bar{F}\varepsilon\bar{G}$, q.e.d. One can also calculate directly that these last formulas give natural transformations $I \dot{\to} G\bar{G}\bar{F}F$ and $\bar{F}FG\bar{G} \dot{\to} I$ which satisfy the triangular identities.

Using this composition, we may form a category **Adj**, whose objects are all (small) categories $X, A, D, \ldots$ and whose arrows are the adjunctions $\langle F, G, \eta, \varepsilon\rangle : X \rightharpoonup A$, composed as above; the identity arrow for each category A is the identity adjunction $A \rightharpoonup A$.

This category has additional structure. Each hom-set $\mathbf{Adj}(X, A)$ may be regarded as a category; to wit, the category $A^{(\mathrm{adj})X}$ of adjunctions from X to A as described in the last section. Its objects are these adjunctions and its arrows are the conjugate pairs $\langle\sigma, \tau\rangle$, under "vertical" composition defined in (7.8).

Theorem 2. *Given two conjugate pairs*

$$\langle\sigma, \tau\rangle : \langle F, G, \eta, \varepsilon\rangle \dot{\rightarrow} \langle F', G', \eta', \varepsilon'\rangle : X \rightharpoonup A\,,$$
$$\langle\bar{\sigma}, \bar{\tau}\rangle : \langle\bar{F}, \bar{G}, \bar{\eta}, \bar{\varepsilon}\rangle \dot{\rightarrow} \langle\bar{F}', \bar{G}', \bar{\eta}', \bar{\varepsilon}'\rangle : A \rightharpoonup D$$

the (horizontal) composite natural transformations $\bar{\sigma}\sigma$ and $\tau\bar{\tau}$ yield a conjugate pair $\bar{\sigma}\sigma : \bar{F}F \dot{\rightarrow} \bar{F}'F', \tau\bar{\tau} : G'\bar{G}' \dot{\rightarrow} G\bar{G}$ of natural transformations for the composite adjunctions.

The proof may be visualized by the diagram of hom-sets

$$\begin{array}{ccccc} D(\bar{F}'F'x, d) & \cong & A(F'x, \bar{G}'d) & \cong & X(x, G'\bar{G}'d) \\ \downarrow{\scriptstyle (\bar{\sigma}\sigma x)^*} & & \downarrow{\scriptstyle (\sigma x)^*(\bar{\tau}d)_*} & & \downarrow{\scriptstyle (\tau\bar{\tau}_d)_*} \\ D(\bar{F}Fx, d) & \cong & A(Fx, \bar{G}d) & \cong & X(x, G\bar{G}d)\,. \end{array}$$

Moreover, this operation of (horizontal) composition is a bifunctor

$$\mathbf{Adj}(A, D) \times \mathbf{Adj}(X, A) \rightarrow \mathbf{Adj}(X, D)\,. \qquad (1)$$

This means that **Adj** is a "two-dimensional" category, as is **Cat** (see § II.5).

There is additional discussion in Chapter XII.

Exercises

1. Prove that horizontal composition is a bifunctor, as in (1), and that this implies an interchange law between horizontal and vertical composition of conjugate pairs.
2. Show that the adjunction with right adjoint the forgetful functor **Rng**→**Set** can be obtained as a composite adjunction in two ways, **Rng**→**Ab**→**Set** and **Rng**→**Mon**→**Set**.
3. Let R, S, and T be rings.
   (a) For a bimodule ${}_RE_S$, show that $-\otimes_R E : \mathbf{Mod}_R \rightarrow \mathbf{Mod}_S$ has a right adjoint $\hom_S(E, -)$.
   (b) Show that this is an adjunction with parameter $E \in R$-**Mod**-S.
   (c) Describe the composite of this adjunction with a similar adjunction $\mathbf{Mod}_S \rightharpoonup \mathbf{Mod}_T$.

9. Subsets and Characteristic Functions

The characteristic function of a subset $S \subset X$ is the two-valued function $\psi_s : X \to \{0, 1\}$ on X with the values

$$\psi_s x = 0 \quad \text{if } x \in S\,; \qquad \psi_s x = 1 \quad \text{if } x \in X \text{ but } x \notin S\,. \tag{1}$$

Put differently, $\{0\} \subset \{0, 1\}$ represents the simplest non-trivial subset. An arbitrary subset $S \subset X$ can be mapped into this simple subset by ψ_s, as defined. This map produces a pullback square

$$\begin{array}{ccc} S & \longrightarrow & \{0\} \\ \downarrow & & \downarrow \\ X & \xrightarrow{\psi_s} & \{0,1\}\,. \end{array} \tag{2}$$

Such characteristic functions are often used in probability theory; in logic, $\{0, 1\}$ is the set of two "truth values" with 0 the value "truth". One says that the monomorphism (the typical subset) $t : \{0\} \to \{0, 1\}$ is a "subobject classifier" for the category of sets.

It turns out that there are similar classifiers for subobjects in other categories. In general, a **subobject classifier** for a category C with a terminal object 1 is defined to be a monomorphism $t : 1 \rightarrowtail \Omega$ such that every monomorphism m in C is a pullback of t in an unique way. In other words, for each m there exists a unique pullback square

$$\begin{array}{ccc} S & \longrightarrow & 1 \\ {\scriptstyle m}\downarrow & & \downarrow{\scriptstyle t} \\ X & \xrightarrow{\psi} & \Omega\,. \end{array} \tag{3}$$

In the resulting pullback square (3), the top horizontal arrow is the unique map to the terminal object 1, the lower horizontal arrow acts as the "characteristic function" of the given subobject S, while the "universal" monomorphism $t : 1 \to \Omega$ may be called "truth".

For example, take C to be the category of functions $f : X \to Y$. Here, a monomorphism $g \rightarrowtail f$ is a function $g : S \to T$ between a pair of subsets $S \subset X$ and $T \subset Y$ such that $g(s) = f(s)$ for all $s \in S$. This means that the diagram

$$\begin{array}{ccc} S & \xrightarrow{g} & T \\ \downarrow & & \downarrow \\ X & \xrightarrow{f} & Y \end{array}$$

commutes. In this case, there are three types of elements of X: those x in

S, those x not in S but with gx in T, and, finally, those x not in S with gx not in T. We may then define a characteristic function with three values by setting

$$\psi_S x = 0 \quad \text{if } x \in S\,,$$

$$\psi_S x = 1 \quad \text{if } x \notin S \text{ but } f\,x \in T\,,$$

$$\psi_S x = 2 \quad \text{if } f\,x \notin T (\text{and hence, } x \notin S)\,.$$

Again this prescription provides a pullback

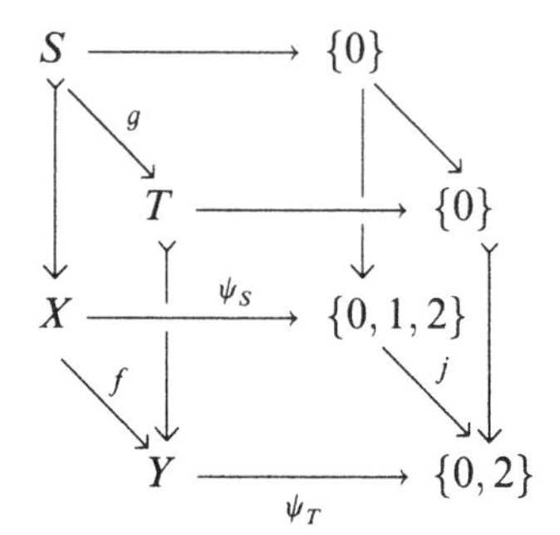

of objects $X \to Y$ and $j : \{0, 1, 2\} \to \{0,\ 1\}$ in the category of functions, where the function j on the right is given by $j0 = 0, j1 = 0, j2 = 2$. Thus, in this case, the inclusion j on the right is a subobject classifier for the category of functions.

There are many other examples of subobject classifiers. First, recall that the arrow category **2** is the category with only two objects 0 and 1 and only one non-identity arrow $a : 0 \to 1$. Thus, an ordinary function f is the same thing as a functor $\mathbf{2} \to \mathbf{Sets}$. Hence, we have constructed above the subobject classifier for the functor category $\mathbf{Sets}^{\mathbf{2}}$. For any category C, there is a subobject classifier (find it!) for the functor category $\mathbf{Sets}^{\mathbf{C}}$.

10. Categories Like Sets

An (elementary) **topos** is defined to be a category E with the following properties:

(i) E has all finite limits;
(ii) E has a subobject classifier;
(iii) E is cartesian closed.

We recall that requiring E to be cartesian closed means requiring that each functor "product with b" (i.e., $a \mapsto a \times b$) has for all b in E a right adjoint $c \mapsto c^b$, so that

$$\hom(a \times b,\ c) \cong \hom(a, c^b)\ .$$

In other words, E has exponentials.

The category **Sets** of all (small) sets is a topos and so is the category $\mathbf{Sets}^{C^{\mathrm{op}}}$ of all set-valued contravariant functors on a small category C. Such a functor $F : C^{\mathrm{op}} \mapsto \mathbf{Sets}$ is also called a presheaf. This is in reference to topology, where C is the set of all open sets U of a topological space X. In this case, a presheaf F assigns to each open set U a set $F(U)$, with functorial properties for continuous maps $U \to V$. For example, $F(U)$ might be the set of all continuous real-valued functions on U. In this case, F is said to be a sheaf (think of the sheaf of coefficients for a cohomology theory!). This and other categories of sheaves play a central role in algebraic geometry and in algebraic topology; the word "topos" is evidently a derivative of the word "topology", suggesting that a topological structure is essentially described by its topos of sheaves of **Sets**.

This study of categories of sheaves on topological spaces and their generalization has led to the study of toposes (see Mac Lane-Moerdijk [1992]). In particular, various logical properties are reflected in the subobject classifiers Ω of a topos. Under many circumstances, a topos provides an alternative view of the foundations of mathematics; for example, the use of "forcing" to prove the independence of the continuum hypothesis can be well organized in terms of constructions on toposes. (see Mac Lane-Moerdijk [1992], Chapter VI). Also, suitable toposes can replace the category of sets as a foundation for mathematics.

The axioms for a topos have many useful consequences. For examples, every topos has all finite colimits.

Notes.

The multiple examples, here and elsewhere, of adjoint functors tend to show that adjoints occur almost everywhere in many branches of Mathematics. It is the thesis of this book that a systematic use of all these adjunctions illuminates and clarifies these subjects. Nevertheless, the notion of an adjoint pair of functors was developed only very recently. The word "adjoint" seems to have arisen first (and long ago) to describe certain linear differential operators. About 1930 the concept was carried over to a Hilbert space H, where the adjoint T^* of a given linear transformation T on H is defined by equality of the inner products

$$(T^*x, y) = (x, Ty)$$

for all vectors $x, y \in H$. Clearly, there is a formal analogy to the definition of adjoint functor.

Daniel Kan in [1958] was the first to recognize and study adjoint functors. He needed them for the study of simplicial objects, and he developed the basic properties such as units and counits, limits as adjoints, adjunctions with a parameter, and conjugate transformations, as well as an important existence theorem (the Kan extension – see Chapter X). Note that his discovery came ten years after the exact formulation of universal constructions. Initially, the idea of adjunctions took on slowly, and the relation to universal arrows was not clear. Freyd in his 1960 Princeton thesis (unpublished but widely circulated) and in his book [1964] and Lawvere [1963, 1964] emphasized the dominant position of adjunctions. One must pause to ask if there are other basic general notions still to be discovered.

One may also speculate as to why the discovery of adjoint functors was so delayed. Ideas about Hilbert space or universal constructions in general topology might have suggested adjoints, but they did not; perhaps the 1939–1945 war interrupted this development. During the next decade 1945–55 there were very few studies of categories, category theory was just a language, and possible workers may have been discouraged by the widespread pragmatic distrust of "general abstract nonsense" (category theory). Bourbaki just missed ([1948], Appendix III). His definition of universal construction was clumsy, because it avoided categorical language, but it amounted to studying a bifunctor $W : X^{\mathrm{op}} \times A \rightarrow \mathbf{Set}$ and asking for a universal element of $W(x, -)$ for each x. This amounts to asking for objects $Fx \in A$ and a natural isomorphism $W(x, a) \cong A(Fx, a)$; it *includes* the problem of finding a left adjoint F to a functor $G: A \rightarrow X$, with $W(x, a) = \hom_X(x, Ga)$. It also includes the problem of finding a tensor product for two modules A and B, with $W(\langle A, B \rangle, C)$ taken to be the set of bilinear functions $A \times B \rightarrow C$. Moreover, the tensor product $A \otimes B$ is not in this way an example of a left adjoint (though it is an example of our universal arrows). In other words Bourbaki's idea of universal construction was devised to be so general as to include more – and in particular, to include the ideas of multilinear algebra which were important to French Mathematical traditions. In retrospect, this added generality seems mistaken; Bourbaki's construction problem emphasized representable functors, and asked "Find Fx so that $W(x, a) \cong A(Fx, a)$". This formulation lacks the symmetry of the adjunction problem, "Find Fx so that $X(x, Ga) \cong A(Fx, a)$" – and so missed a basic discovery; this discovery was left to a younger man, perhaps one less beholden to tradition or to fashion. Put differently, good general theory does not search for the maximum generality, but for the right generality.

V. Limits

This chapter examines the construction and properties of limits, as well as the relation of limits to adjoints. This relation is then used in the basic existence theorems for adjoint functors, which give universals and adjoints in a wide variety of cases. The chapter closes with some indications of the uses of adjoint functors in topology.

1. Creation of Limits

A category C is called *small-complete* (usually just *complete*) if all small diagrams in C have limits in C; that is, if every functor $F: J \to C$ to C from a small category J has a limit. We shall show that **Set**, **Grp**, **Ab**, and many other categories of algebras are small-complete.

The construction of limits in **Set** may be illustrated by considering the limit of a functor $F: \omega^{\mathrm{op}} \to \mathbf{Set}$; here ω, the linearly ordered set of all finite ordinals, is the free category generated by the graph

$$\{0 \to 1 \to 2 \to 3 \to \cdots\}.$$

The functor $F: \omega^{\mathrm{op}} \to \mathbf{Set}$ is just a list of sets F_n and of functions f_n, as in the first row of the diagram below:

$$\begin{array}{c} F_0 \xleftarrow{f_0} F_1 \xleftarrow{f_1} F_2 \leftarrow \cdots \leftarrow F_n \xleftarrow{f_n} F_{n+1} \leftarrow \cdots \\ \Pi_i F_i \xleftarrow{\text{incl.}} L = \varprojlim F \end{array} \qquad (1)$$

$$\{x_0, x_1, \ldots \mid x_n \in F_n\} \hookleftarrow \{x_0, x_1, \ldots \mid f_n x_{n+1} = x_n \in F_n\}.$$

Given F, form first the product set $\Pi_i F_i$; it consists of all strings $x = \{x_0, x_1, x_2 \ldots\}$ of elements, with each $x_n \in F_n$, and it has projections $p_n : \Pi_i F_i \to F_n$, but the triangles formed by these projections need not commute ($f_n p_{n+1} \neq p_n$). A limit must be at least a vertex of a set of commuting triangles (a cone). So take the subset L of those strings x which "match" under f, in that $f_n x_{n+1} = x_n$ for all n. Then functions $\mu_n : L \to F_n$ are defined by $\mu_n x = x_n$; since the string x matches, $f_n \mu_{n+1} = \mu_n$

for all n, so $\mu: L \dot{\to} F$ is a cone from the vertex $L \in \mathbf{Set}$ to the base F. If $\tau: M \dot{\to} F$ is any other cone from a set M as vertex, each $m \in M$ determines a string $\{\tau_n m\}$ which matches and hence a function $g: M \to L$, with $gm = \{\tau_n m\}$, so with $\mu g = \tau$. Since g is the unique such function this shows that μ is a universal cone to F, and so that L is the limit set of F.

A string x which "matches" is the same thing as a cone $x: * \dot{\to} F$ to F from the one point set $*$. Hence the limit L above can be described as the set $L = \mathrm{Cone}(*, F)$ of all such cones. The same construction applies for any domain category (in place of ω^{op}).

Theorem 1 *(Completeness of* **Set***). If the category J is small, any functor $F: J \to \mathbf{Set}$ has a limit which is the set* $\mathrm{Cone}(*, F)$ *of all cones $\sigma: * \dot{\to} F$ from the one point set $*$ to F, while the limiting cone v, with*

$$v_j: \mathrm{Cone}(*, F) \to F_j, \qquad \sigma \mapsto \sigma_j, \tag{2}$$

is for each j that function sending each cone σ to the element $\sigma_j \in F_j$.

For example, if J is discrete, the set $\mathrm{Cone}(*, F)$ of J-cones is just the cartesian product $\Pi_j F_j$.

Proof. Since J is small, $\mathrm{Cone}(*, F)$ is a small set, hence an object of **Set**. If $u: j \to k$ is any arrow of J, then $F_u \sigma_j = \sigma_k$ because σ is a cone; hence v as defined in (2) is a cone to the base F. To prove it universal, consider any other cone $\tau: X \dot{\to} F$ to F from some set X. Then for each $x \in X$, τx is a cone to F from one point, so there is a unique function $h: X \to \mathrm{Cone}(*, F)$ sending each x to τx, q.e.d.

The crux of this proof is the (natural) bijection

$$\mathrm{Cone}(X, F) \cong \mathbf{Set}(X, \mathrm{Cone}(*, F)) \tag{3}$$

given by $\tau \mapsto h$, as above. Since a cone is just a natural transformation, this may be rewritten as an adjunction

$$\mathrm{Nat}(\Delta X, F) \cong \mathbf{Set}(X, \mathrm{Cone}(*, F)).$$

By the very definition of limit, this proves that $\mathrm{Lim}\, F \cong \mathrm{Cone}(*, F)$.

Limits in **Grp** and other categories may be constructed from the set of all cones in much the same way. For example, if $F: \omega^{\mathrm{op}} \to \mathbf{Grp}$, as displayed in (1), then each F_n is a group, the set L of all cones (all matching strings x) is also a group under pointwise multiplication ($(xy)_n = x_n y_n$), and, the projection $\mu_n: L \to F_n$ with $x \mapsto x_n$ is a group homomorphism, so that $\mu: L \dot{\to} F$ is a limiting cone in **Grp**.

The p-adic integers $\mathbf{Z}_p$ (with p a prime) illustrate this construction. Take $F: \omega^{\mathrm{op}} \to \mathbf{Rng}$ with $F_n = \mathbf{Z}/p^n\mathbf{Z}$, the ring of integers modulo p^n, and with $F_{n+1} \to F_n$ the canonical projection $\mathbf{Z}/p^{n+1}\mathbf{Z} \to \mathbf{Z}/p^n\mathbf{Z}$. Then $\mathbf{Z}_p = \underleftarrow{\mathrm{Lim}}\, F$ exists. An element λ of $\mathbf{Z}_p$ is a cone from $*$ to F; that is, λ can be written as a sequence $\lambda = \{\lambda_0, \lambda_1, \ldots\}$ of integers with $\lambda_{n+1} \equiv \lambda_n (\mathrm{mod}\, p^n)$ for all n, where $\lambda = \lambda'$ holds when $\lambda_n \equiv \lambda'_n (\mathrm{mod}\, p^n)$

for all n. Two p-adic integers λ and μ can be added and multiplied "termwise", by the formulas

$$(\lambda+\mu)_n=\lambda_n+\mu_n\,,\qquad (\lambda\mu)_n=\lambda_n\mu_n\,.$$

These operations make $\mathbf{Z}_p=\operatorname{Lim}F$ a ring, the ring of p-adic integers, and this description completely determines $\mathbf{Z}_p$. This description is quicker than the classical one, which first defines a p-adic valuation (and thus a topology) in $\mathbf{Z}$, and then observes that each p-adic integer λ is represented by a Cauchy sequence in that topology.

Formal power series rings also can be described as limits (Ex. 7).

Again, in **Top**, take each object F_n to be a circle S^1, and each arrow $f_n: S^1\to S^1$ to be the continuous map wrapping the domain circle S^1 uniformly p times around the codomain circle. The inverse limit set L then becomes a topological space when we introduce just those open sets in L necessary to make all the functions $\mu_n: L\to S^1$ continuous. This L is the limit space in **Top**; it is known as the *p-adic solenoid*.

Here is the general construction for groups.

Theorem 2. *Let U : **Grp**$\to$**Set** be the forgetful functor. If $H: J\to$ **Grp** is such that the composite UH has a limit L and a limiting cone $v: L \dot{\to} UH$ in **Set**, then there is exactly one group structure on the set L for which each arrow $v_j: L\to UH_j$ of the cone v is a morphism of groups; moreover, this group L is a limit of H with v as limiting cone.*

Proof. By Theorem 1, take $L=\operatorname{Cone}(*, UH)$; define the product of two such cones $\sigma,\tau\in\operatorname{Cone}(*, UH)$ by $(\sigma\tau)_j=\sigma_j\tau_j$ (the product in the group H_j) and the inverse by $(\sigma^{-1})_j=\sigma_j^{-1}$ (the inverse in H_j). These definitions make L a group and each component of v a morphism of groups; conversely, if v given by $\tau\mapsto\tau_j$ is to be a morphism of groups for each j, then the product of $\sigma,\tau\in L$ must be given by this formula.

Now if G is any group and $\lambda: G\dot{\to} H$ any cone in **Grp** (consisting of group morphisms $\lambda_j: G\to H_j$ for $j\in J$), then $U\lambda: UG\dot{\to} UH$ is a cone in **Set**, so by universality $U\lambda=(Uv)h$ for a unique function $h: UG\to L$. For any two group elements g_1 and g_2 in G,

$$(h(g_1g_2))_j=\lambda_j(g_1g_2)=(\lambda_jg_1)(\lambda_jg_2)=(hg_1)_j(hg_2)_j=((hg_1)(hg_2))_j;$$

because λ is a morphism of groups, so is h, and therefore L is indeed the limit in **Grp**.

This argument is just a formalization of the familiar termwise construction of the multiplication in cartesian products of groups, in the p-adic numbers, etc. The conclusion of the Theorem constructs limits in **Grp** from the limits in **Set** in a unique way, using U. The same argument will construct all small limits in **Rng**, **Ab**, R-**Mod** and similar algebraic categories, using the forgetful functors U to **Set**. In other words, each forgetful functor "creates" limits in the sense of the following definition:

Definition. *A functor $V: A \to X$ creates limits for a functor $F: J \to A$ if*

(i) *To every limiting cone $\tau: x \dot{\to} VF$ in X there is exactly one pair $\langle a, \sigma \rangle$ consisting of an object $a \in A$ with $Va = x$ and a cone $\sigma: a \dot{\to} F$ with $V\sigma = \tau$, and if, moreover,*

(ii) *This cone $\sigma: a \dot{\to} F$ is a limiting cone in A.*

Similarly, we may define "V creates products" (the above, with J restricted to be discrete); "V creates finite limits" (the above, with J finite), or "V creates colimits" (the above with the arrows in all cones reversed). Note especially that "V creates limits" means only that V produces limits for functors F whose composite VF already *has* a limit.

In this terminology, Theorem 2 now reads

Theorem 3. *The forgetful functor U :* **Grp**$\to$**Set** *creates limits.*

Exercises

1. Prove that the projection $(x \downarrow C) \to C$ of the comma category creates limits.
2. If **Comp Haus** $\subset$ **Top** is the full subcategory of all compact Hausdorff spaces, show that the forgetful functor **Comp Haus**$\to$**Set** creates limits.
3. For any category X, show that the projection $X^2 \to X \times X$ which sends each arrow $f: x \to y$ in X to the pair $\langle x, y \rangle$ creates limits.
4. Prove that the category of all small finite sets is *finitely complete* (i.e., has all finite limits).
5. Prove that **Cat** is small-complete.
6. Show that each p-adic integer λ is determined by a string of integers a_i with all $a_i \in \{0, 1, \ldots, p-1\}$, with each $\lambda_n \equiv a_0 + a_1 p + \cdots + a_{n-1} p^{n-1} \pmod{p^n}$. Show that addition and multiplication of p-adic integers correspond to the usual operations of addition and multiplication of infinite "decimals" $\ldots a_n \ldots a_0$ (with base p, the decimals extending infinitely to the left).
7. Let $K[x]$ be the usual ring of polynomials in x with coefficients in the commutative ring K, while $F: \omega^{\mathrm{op}} \to$ **Rng** is defined by $F_n = K[x]/(x^n)$, with the evident projections, and (x^n) the usual principal ideal. Prove that $\operatorname{Lim} F$ is the ring of formal power series in x, coefficients in K.
8. Show that the category of sets is cocomplete.

2. Limits by Products and Equalizers

The construction of the limit of $F: J \to$ **Set** as the set of all cones

$$\operatorname{Cone}(*, F) \subset \Pi_j F_j$$

can be made in two steps: Each cone σ is an element x of the product $\Pi_j F_j$ with projections p_j; to require that an element x of the product be a cone is to require that $(Fu)x_j = x_k$ for every arrow $u: j \to k$ in J; this amounts to requiring that x lie in the equalizer of $(Fu)\, p_j$ and $p_k: \Pi_j F_j \to F_k$. Here is the general formulation of this process in any category.

Theorem 1. *For categories C and J, if C has equalizers of all pairs of arrows and all products indexed by the sets* $\operatorname{obj}(J)$ *and* $\operatorname{arr}(J)$, *then C has a limit for every functor $F : J \to C$.*

The proof constructs the following diagram in stages, with i denoting an object and $u : j \to k$ an arrow of the index category J. By assumption, the products $\Pi_i F_i$ and $\Pi_u F_k$ and their projections exist, where the second product is taken over all arrows u of J, with argument at each arrow u the value $F_k = F_{\mathrm{cod}\,u}$ of F at the codomain object of u. Since $\Pi_u F_k$ is a product, there is a unique arrow f such that the upper square commutes for every u and a unique arrow g such that the lower square commutes for every u. By hypothesis,

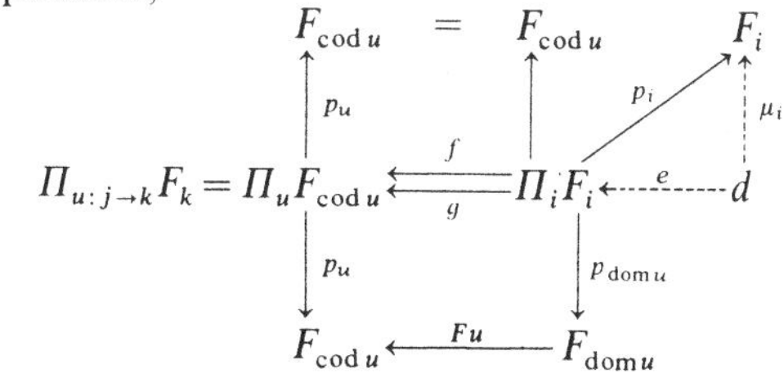

$$
\begin{array}{ccccc}
 & F_{\mathrm{cod}\,u} & = & F_{\mathrm{cod}\,u} & \qquad F_i \\
 & \uparrow{\scriptstyle p_u} & & \uparrow & {\scriptstyle p_i}\nearrow \quad \uparrow{\scriptstyle \mu_i} \\
\Pi_{u:j\to k} F_k = & \Pi_u F_{\mathrm{cod}\,u} & \overset{f}{\underset{g}{\longleftleftarrows}} & \Pi_i F_i & \xleftarrow{\;e\;} d \\
 & \downarrow{\scriptstyle p_u} & & \downarrow{\scriptstyle p_{\mathrm{dom}\,u}} & \\
 & F_{\mathrm{cod}\,u} & \xleftarrow{\;Fu\;} & F_{\mathrm{dom}\,u} &
\end{array}
\qquad (1)
$$

there exists an equalizer e for f and g. Its composite with the projections p_i give arrows $\mu_i = p_i e : d \to F_i$ for each i. Since e equalizes f and g and the two squares above commute, one has $F_u \mu_j = \mu_k$ for every $u : j \to k$; hence $\mu : \Delta d \stackrel{\cdot}{\to} F$ is a cone from the vertex d to the base F. If τ is any other such cone, of vertex c, its maps τ_i combine to yield a unique map $h : c \to \Pi_i F_i$ to the product; τ a cone implies $fh = gh$. Hence h factors uniquely through e and therefore the cone τ factors uniquely through the cone μ. This proves that d and the cone μ provide a limit for F. For the record, much as in the case of **Sets**:

Theorem 2 *(Limits by product and equalizers, continued). The limit of $F : J \to C$ is the equalizer e of $f, g : \Pi_i F_i \to \Pi_u F_{\mathrm{cod}\,u}$* $(u \in \operatorname{arr} J, i \in J)$, *where $p_u f = p_{\mathrm{cod}\,u}$, $p_u g = F_u \cdot p_{\mathrm{dom}\,u}$; the limiting cone μ is $\mu_j = p_j e$, for $j \in J$, all as in* (1).

This theorem has several useful consequences and special cases.

Corollary 1. *If a category C has a terminal object, equalizers of all pairs of arrows, and products of all pairs of objects, then C has all finite limits.*

Here a *finite limit* is a limit of $J \to C$, with the category J finite.

Corollary 2. *If C has equalizers of all pairs of arrows and all small products, then C is small-complete.*

For example, this gives another proof that **Set** is small-complete.

The concept of completeness is useful chiefly for large categories and for preorders. In a preorder P, a product of objects a_j, $j \in J$, is an object d with $d \leqq a_j$ for all j and such that $c \leqq a_j$ for all j implies $c \leqq d$; in other words a product is just a *greatest lower bound* or *meet* of the factors a_j (dually, a coproduct is a *least upper bound* or *join*).

Proposition 3 *(Freyd). A small category C which is small-complete is simply a preorder which has a greatest lower bound for every small set of its elements.*

Proof. Suppose C is not a preorder. Then there are objects $a, b \in C$ with arrows $f \neq g : a \to b$. For any small set J form the product $\Pi_j b$ of factors b_j all equal to b. Then an arrow $h : a \to \Pi_j b$ is determined by its components, which can be f or g. There are thus at least 2^J arrows $a \to \Pi_j b$. If the small set J has cardinal larger than arr$C$, this is a contradiction.

Exercises

1. (Manes.) A parallel pair of arrows $f, g : a \to b$ in C has a *common left inverse* h when there is an arrow $h : b \to a$ with $hf = 1 = hg$.
   (a) Prove that a category C with all small products and with equalizers for all those parallel pairs with a common left inverse is small complete. (Hint: The parallel pair used in the proof of Theorem 1 does in fact have a common left inverse.)
   (b) In **Set**, show that a parallel pair of arrows $f, g : X \to Y$ has a common right inverse if and only if the corresponding function $(f, g) : X \to Y \times Y$ has image containing the diagonal $\{\langle y, y\rangle \mid y \in Y\}$.
2. Prove that C_1, C_2 complete (or cocomplete) imply the same for the product category $C_1 \times C_2$.
3. ($\underleftarrow{\mathrm{Lim}}$ and $\underrightarrow{\mathrm{Lim}}$ **as functors.**) If $F, F' : J \to C$ have limiting cones μ, μ' (or colimiting cones ν, ν'), show that each natural transformation $\beta : F \dot{\to} F'$ determines uniquely arrows $\underleftarrow{\lim}\,\beta$ or $\underrightarrow{\lim}\,\beta$ such that the following diagram commutes, where $\Delta : C \to C^J$ is the diagonal functor:

$$\begin{array}{ccccc} \Delta\,\underleftarrow{\mathrm{Lim}}\,F & \xrightarrow{\mu} & F & \xrightarrow{\nu} & \Delta(\underrightarrow{\mathrm{Lim}}\,F) \\ {\scriptstyle \Delta(\underleftarrow{\lim}\beta)}\downarrow & & \downarrow{\scriptstyle \beta} & & \downarrow{\scriptstyle \Delta(\underrightarrow{\lim}\beta)} \\ \Delta\,\underleftarrow{\mathrm{Lim}}\,F' & \xrightarrow{\mu'} & F' & \xrightarrow{\nu'} & \Delta(\underrightarrow{\mathrm{Lim}}\,F')\,. \end{array}$$

   Conclude: If C is complete, $\underleftarrow{\mathrm{Lim}}$ (or $\underrightarrow{\mathrm{Lim}}$) is a functor $C^J \to C$.
4. (Limits of composites.) Given composable functors

$$J' \xrightarrow{W} J \xrightarrow{F} C \xrightarrow{H} C'$$

   and limiting cones ν for F, ν' for HFW, observe that $\Delta_{J'}(Hc) = H \circ \Delta_J c \circ W : J' \to C'$, and show that there is a unique "canonical" arrow $t : H \circ \underleftarrow{\mathrm{Lim}}\,F \to \underleftarrow{\mathrm{Lim}}\,HFW$ such that the following diagram commutes:

$$\begin{array}{ccccc} \Delta_{J'}(H \circ \underleftarrow{\mathrm{Lim}}\,F) & \xrightarrow{H\nu W} & HFW & \xrightarrow{\mu'} & \Delta_{J'}(\underrightarrow{\mathrm{Lim}}\,HFW) \\ {\scriptstyle \Delta_{J'}(t)}\downarrow & & \| & & \downarrow{\scriptstyle \Delta_{J'}(s)} \\ \Delta_{J'}(\underleftarrow{\mathrm{Lim}}\,HFW) & \xrightarrow{\nu'} & HFW & \xrightarrow{H\mu W} & \Delta_{J'}(H \circ \underrightarrow{\mathrm{Lim}}\,F)\,. \end{array}$$

   Dually, construct $s : \underrightarrow{\mathrm{Lim}}\,HFW \to H \circ \underrightarrow{\mathrm{Lim}}\,F$ as indicated at the right.

5. (Limit as a functor on the comma category of all diagrams in C.)
   (a) Interpret W of Ex. 4 as an arrow in $(\mathbf{Cat}\downarrow C)$ to show (for C complete) that $\underleftarrow{\mathrm{Lim}}$ is a functor $(\mathbf{Cat}\downarrow C)^{\mathrm{op}}\to C$.
   (b) Let $(\mathbf{Cat}\,\ddagger\,C)$ be the ("super-comma") category with objects $F: J\to C$, arrows $\langle\beta, W\rangle: F'\to F$ those pairs consisting of a functor $W: J'\to J$ and a natural transformation $\beta: FW\dot\to F'$. Combine Exercise 3 and Exercise 4 to show (for C complete) that Lim is a functor $(\mathbf{Cat}\,\ddagger\,C)^{\mathrm{op}}\to C$. Dualize.

3. Limits with Parameters

Let $T: J\times P\to X$ be a bifunctor, and suppose for each value $p\in P$ of the "parameter" p that $T(-,p): J\to X$ has a limit. Then these limits for all p form the object function $p\mapsto \mathrm{Lim}_j T(j,p)$ of a functor $P\to X$.

Instead of proving this directly, we replace functors $P\to X$ by objects of the functor category X^P. This replaces $T: J\times P\to X$ by its adjunct $S: J\to X^P$, under the adjunction $\mathbf{Cat}(J\times P, X)\cong \mathbf{Cat}(J, X^P)$. Recall that for each object $p\in P$ there is a functor $E_p: X^P\to X$, "evaluate at p", given for arrows (natural transformations) $\sigma: H\dot\to H'$ of X^P as

$$E_pH=H_p,\qquad E_p\sigma=\sigma_p: H_p\to H'_p. \tag{1}$$

Theorem 1. *If $S: J\to X^P$ is such that for each object $p\in P$ the composite $E_pS: J\to X$ has a limit L_p with a limiting cone $\tau_p: L_p\dot\to E_pS$, then there is a unique functor $L: P\to X$ with object function $p\mapsto L_p$ such that $p\mapsto\tau_p$ is a natural transformation $\tau: \Delta L=\Delta_J L\dot\to S$; moreover, this τ is a limiting cone from the vertex $L\in X^P$ to the base $S: J\to X^P$.*

Proof. Let $h: p\to q$ be any arrow of P. Then, writing E_pS as S_p, the given cones τ_p and τ_q for a typical arrow $u: j\to k$ of J have the form

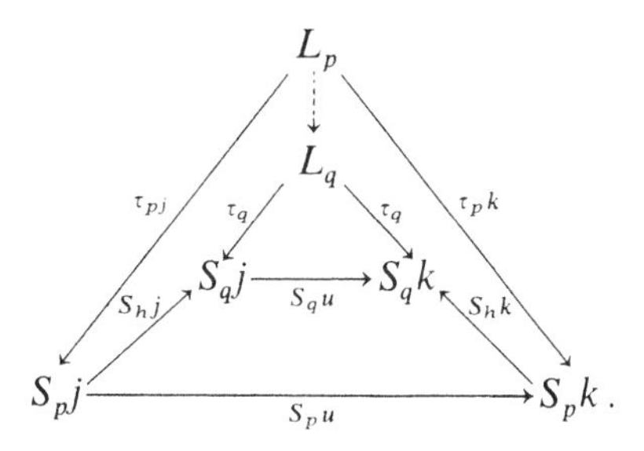

The triangles commute because τ_p and τ_q are cones and the parallelogram because S is a functor. Since the inside cone is universal there is a unique arrow $L_h: L_p\to L_q$ such that $\tau_qj\circ L_h=S_hj\circ\tau_pj$ for all $j\in J$. The assignment $h\mapsto L_h$ makes L a functor (Proof: put another cone outside) and τ a natural transformation $\Delta L\dot\to S$ (a cone from the object $L\in X^P$ to the functor $S: J\to X^P$). It is a limiting cone; for if $\sigma: M\dot\to S$ is any other cone there are unique arrows $M_p \to L_p$ because L_p is a limit; they combine to give a unique natural transformation $M \dashrightarrow L$.

The conclusion may be written

$$E_p(\underleftarrow{\mathrm{Lim}}\, S) = \underleftarrow{\mathrm{Lim}}\,(E_p S):$$

In a functor category, limits may be calculated pointwise (provided the pointwise limits exist).

Corollary. *If X is small-complete, so is every functor category X^P.*

This theorem becomes a case of "creation" of limits, if we write $|P|$ for the discrete subcategory consisting of all objects and identity arrows of P.

Theorem 2. *For any categories X and P, the inclusion functor $i : |P| \to P$ induces a functor $i^* = X^i : X^P \to X^{|P|}$ which creates limits.*

4. Preservation of Limits

A functor $H : C \to D$ is said to *preserve the limits* of functors $F : J \to C$ when every limiting cone $v : b \dashrightarrow F$ in C for a functor F yields by composition with H a limiting cone $Hv : Hb \dashrightarrow HF$ in D; this requires not only that H take each limit object which exists in C to a limit object in D but also that H take limiting cones to limiting cones. A functor is called *continuous* when it preserves all small limits.

Theorem 1. *For any category C with small hom-sets, each hom-functor $C(c, -) : C \to$ **Set** preserves all limits; in particular, all small limits.*

The same proof will give a more general result: If C has hom-sets in **Ens**, any category of sets in which **Ens**(X, Y) consists of *all* functions on X to Y, then each hom-functor $C(c, -) : C \to$ **Ens** preserves all limits which exist in C.

Proof. Let J be any category and $F : J \to C$ a functor with a limiting cone $v : \mathrm{Lim}\, F \dashrightarrow F$ in C. Apply the hom-functor $C(c, -)$; there results a cone $v_* = C(c, v)$, as in the diagram

$$\begin{array}{ccc} C(c, \mathrm{Lim}\, F) & \xrightarrow{v_{*i}} & C(c, F_i), \quad i \in J \\ \uparrow k & & \| \\ X & \xrightarrow{\tau_i} & C(c, F_i) \end{array}$$

in **Set**. For any other cone τ to the same base from a vertex set X, each element $x \in X$ gives a cone $\tau_i x : c \to F_i$ in C and hence, because v is universal, a unique arrow $h_x : c \to \mathrm{Lim}\, F$ with $v_i h_x = \tau_i x$. Then setting $kx = h_x$ for each x defines a function, and hence an arrow k in **Ens** as shown, with $\nu_{*i}k = \tau_i$ for all i. Since k is clearly unique with this property, ν_* is a limiting cone in **Set**, as required.

The same proof, differently stated, might start by noting that the definition of the functor $C(c, F-): J \to \mathbf{Set}$ shows that a cone $\lambda: c \xrightarrow{\cdot} F$ in C is the same thing as a cone $\lambda: * \xrightarrow{\cdot} C(c, F-)$ in **Set**, with vertex a point $*$. Then, because $\mathrm{Cone}(X, -) \cong \mathbf{Set}(X, \mathrm{Cone}(*, -))$ as in (1.3),

$$\begin{aligned}\mathrm{Cone}\big(X, C(c, F-)\big) &\cong \mathbf{Set}\big(X, \mathrm{Cone}(*, C(c, F-))\big) \\ &= \mathbf{Set}\big(X, \mathrm{Cone}(c, F)\big) \cong \mathbf{Set}\big(X, C(c, \mathrm{Lim}\, F)\big)\ ,\end{aligned}$$

where "Cone" means J-cone and where the last step uses the definition of $\mathrm{Lim}\, F$. But $\mathrm{Lim}\, S$, for each $S: J \to \mathbf{Set}$, is defined by the adjunction $\mathrm{Cone}(X, S) \cong \mathbf{Set}(X, \mathrm{Lim}\, S)$. Therefore the above equations determine this $\mathrm{Lim}\, S$ (together with the correct limiting cone) as

$$\mathrm{Lim}\, C(c, F-) \cong C(c, \mathrm{Lim}\, F)\,. \tag{1}$$

Some authors use this equation to define limits in C in terms of limits in **Set**; for example, the product of objects a_i in C is defined by

$$\prod_i C(c, a_i) \cong C(c, \prod_i a_i)\,. \tag{2}$$

The contravariant hom-functor may be written as

$$C(-, c) = C^{\mathrm{op}}(c, -): C^{\mathrm{op}} \to \mathbf{Set}\,;$$

hence the theorem shows that this functor $C(-, c)$ carries small colimits (and their colimiting cones) in C to the corresponding limits and limiting cones in **Set**. For example, the definition of a small coproduct provides an isomorphism (coproduct to product):

$$C(\amalg_j a_j, c) \cong \prod_j C(a_j, c)\,.$$

More generally, the colimit of any $F: J \to C$ is determined by

$$C(\mathrm{Colim}\, F, c) \cong \mathrm{Lim}\, C(F-, c)\,. \tag{3}$$

Creation and preservation are related:

Theorem 2. *If $V: A \to X$ creates limits for $F: J \to A$ and the composite $VF: J \to X$ has a limit, then V preserves the limit of F.*

In particular, if V creates all small limits and X is small-complete, then A is also small-complete, and V is continuous.

Proof. Let $\tau: a \xrightarrow{\cdot} F$ and $\sigma: x \xrightarrow{\cdot} VF$ be limiting cones in A and X, respectively. Since V creates limits, there is a unique cone $\varrho: b \xrightarrow{\cdot} F$ in A with $V\varrho: Vb \xrightarrow{\cdot} VF$ equal to $\sigma: x \xrightarrow{\cdot} VF$; moreover, ϱ is a limiting cone. But limits are unique up to isomorphism, so there is an isomorphism $\theta: b \cong a$ with $\tau\theta = \varrho$. Thus $V\theta: Vb = x \cong Va$, with $V\tau \cdot V\theta = V\varrho = \sigma$, so Va is a limit and V preserves limits, as desired.

In any category an object p is called *projective* if every arrow $h: p \to c$ from p factors through every epi $g: b \to c$, as $h = gh'$ for some h'

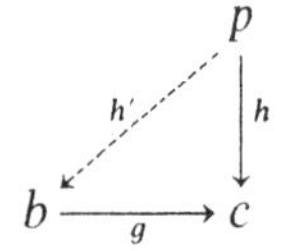

It is equivalent to require that g epi implies $\hom(p,g): \hom(p,b) \to \hom(p,c)$ epi in **Set**. In other words, p is projective exactly when $\hom(p, -)$ preserves epis. Dually, an object q is *injective* when $\hom(-, q)$ carries monics to epis. These notions are especially useful in R-**Mod** and other Ab-categories; in R-**Mod** the projectives are the direct summands of the free modules.

Exercises

1. Prove that the composite of continuous functors is continuous.
2. If C is complete, and $H: C \to D$ preserves all small products and all equalizers (of parallel pairs) prove that H is continuous.
3. Show that the functor $F: \mathbf{Set} \to \mathbf{Ab}$ sending each set X to the free abelian group generated by the set X is not continuous.
4. For any small set X, show that the functor (product with X) $X \times - : \mathbf{Set} \to \mathbf{Set}$ preserves all colimits.
5. (Preservation of Limits.) Given $H: C \to C'$ and a functor $F: J \to C$ such that F and HF have limits, prove that H preserves the limits of F if and only if the canonical arrow $H \cdot \varprojlim F \to \varprojlim HF$ of Exercise 2.4 is an isomorphism (This is a natural way to describe the preservation of limits when both categories C and C' are given with specified limits).

5. Adjoints on Limits

One of the most useful properties of adjoints is this: A functor which is a right adjoint preserves all the limits which exist in its domain:

Theorem 1. *If the functor $G: A \to X$ has a left adjoint, while the functor $T: J \to A$ has a limiting cone $\tau: a \dot{\to} T$ in A, then GT has the limiting cone $G\tau: Ga \dot{\to} GT$ in X.*

Proof. By composition, $G\tau$ is indeed a cone from the vertex Ga in X. If F is a left adjoint to G, and if we apply the adjunction isomorphism to every arrow of a cone $\sigma: x \dot{\to} GT$, we get arrows $(\sigma_i)^\flat: Fx \to Ti$ for $i \in J$ which form a cone $\sigma^\flat: Fx \dot{\to} T$ in A. But $\tau: a \dot{\to} T$ is universal among cones to T in A, so there is a unique arrow $h: Fx \to a$ with $\tau h = \sigma^\flat$. Taking adjuncts again, this gives a unique arrow $h^\sharp: x \to Ga$ with $G\tau \cdot h^\sharp = (\tau h)^\sharp = (\sigma^\flat)^\sharp = \sigma$. The uniqueness of the arrow $h^\sharp$ states precisely that $G\tau: Ga \dot{\to} T$ is universal, q.e.d.

The proof may be illustrated by the following diagrams (where $u : i \to j$ is any arrow of J).

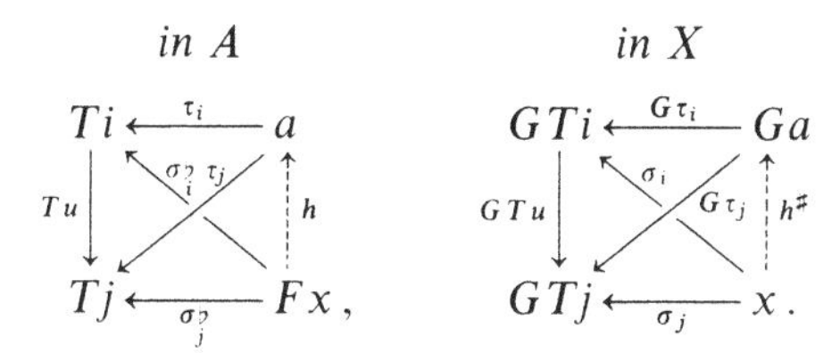

This proof can also be cast in a more sophisticated form by using the fact that Lim is right adjoint to the diagonal functor Δ. In fact, given an adjunction

$$\langle F, G, \eta, \varepsilon \rangle : X \rightharpoonup A$$

and any index category J, one may form the functor categories (from J) and hence the diagram

$$\langle F^J, G^J, \eta^J, \varepsilon^J \rangle : X^J \rightharpoonup A^J,$$

where $F^J(S) = FS$ for each functor $S : J \to X$, and $\eta^J S = \eta S : S \dot{\to} GFS$, etc. The triangular identities for η and ε yield the same identities for η^J and ε^J, so the second diagram is indeed an adjunction (in brief, adjunctions pass to the functor category). Now we have the diagram of adjoint pairs

$$\begin{array}{ccc} X^J & \underset{G^J}{\overset{F^J}{\rightleftarrows}} & A^J \\ \Delta \uparrow\downarrow \underleftarrow{\mathrm{Lim}} & & \Delta \uparrow\downarrow \underleftarrow{\mathrm{Lim}} \\ X & \underset{G}{\overset{F}{\rightleftarrows}} & A . \end{array}$$

The definitions of the diagonal functors Δ show at once that $F^J \Delta = \Delta F$, so the diagram of left adjoints commutes in this square. Since compositions of adjoints give adjoints, it follows that the composites $\underleftarrow{\mathrm{Lim}} \circ G^J$ and $G \circ \underleftarrow{\mathrm{Lim}}$ are both right adjoints to $F^J \circ \Delta = \Delta \circ F$. Since the right adjoint of a given functor is unique up to natural isomorphism, it now follows that $\mathrm{Lim} \circ G^J \cong G \circ \mathrm{Lim}$. This proves again for each functor $T : J \to A$ with limit a (and limiting cone $\tau : a \dot{\to} T$ in A) that $Ga = G \,\mathrm{Lim}\, T = \mathrm{Lim}\, G^J(T) = \mathrm{Lim}\, GT$. The reader should show that the same argument proves that G preserves limiting cones (put units and counits in the square diagram above, and recall that the limiting cone $\tau : a \dot{\to} T$ is just the value of the counit of the adjunction $\langle \Delta, \mathrm{Lim}, \ldots \rangle : A \rightharpoonup A^J$ on the functor T).

The dual of the theorem is equally useful: Any functor P which has a right adjoint (i.e., which *is* a left adjoint) must preserve colimits (coproducts, coequalizers, etc.). This explains why the coproduct (free product) of two free groups is again a free group (on the disjoint union of the sets of generators).

Similarly (by the original theorem) all the typical forgetful functors in algebra preserve products, kernels, equalizers, and other types of limits. Typically, the product of two algebraic systems (groups, rings, etc.) has as underlying set just the (cartesian) product of the two underlying sets. This, and other similar facts, are immediate consequences of this one (easy) theorem. The theorem can also be used to show that certain functors do not have adjoints.

Exercises

1. Show that, for a fixed set X, the functor $X \times - : \mathbf{Set} \to \mathbf{Set}$ cannot have a left adjoint, unless X is a one-point set.
2. For the functor $D : \mathbf{Vct}^{\mathrm{op}} \to \mathbf{Vct}$ of (IV.2.2) show that D has no right adjoint (and hence, in particular, is not the left adjoint of D^{op}).
3. If C is a full and reflective subcategory of a small-cocomplete category D, prove that C is small-cocomplete.
4. Prove that $\mathbf{Set}^{\mathrm{op}}$ is not cartesian closed.

6. Freyd's Adjoint Functor Theorem

To formulate the basic theorem for the existence of a left adjoint to a given functor, we first treat the case of the existence of an initial object in a category and then use the fact that each universal arrow defined by the unit of a left adjoint is an initial object in a suitable comma category.

Theorem 1 *(Existence of an initial object). Let D be a small-complete category with small hom-sets. Then D has an initial object if and only if it satisfies the following*

Solution Set Condition. There exists a small set I and an I-indexed family k_i of objects of D such that for every $d \in D$ there is an $i \in I$ and an arrow $k_i \to d$ of D.

Proof. This solution set condition is necessary: If D has an initial object k, then k indexed by the one-point set realizes the condition, since there is always a (unique) arrow $k \to d$.

Conversely, assume the solution set condition. Since D is small-complete, it contains a product object $w = \Pi k_i$ of the given I-indexed family. For each $d \in D$, there is at least one arrow $w \to d$, for example, a composite $w = \Pi k_i \to k_i \to d$, where the first arrow is a projection of the product. By hypothesis, the set of endomorphisms $D(w, w)$ of w is small and D is complete, so we can construct the equalizer $e : v \to w$ of the set of all the endomorphisms of w. For each $d \in D$, there is by $v \to w \to d$ at least one arrow $v \to d$. Suppose there were two, $f, g : v \to d$, and take their equalizer e_1 as in the figure below

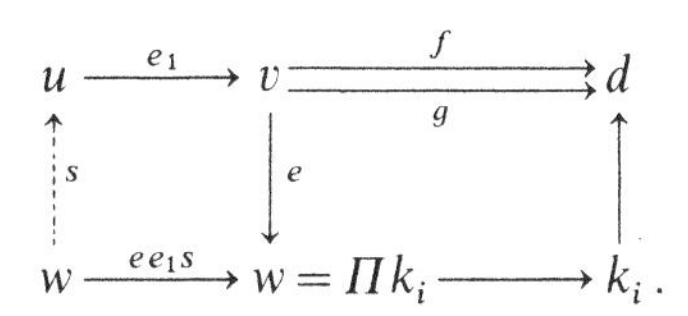

By the construction of w, there is an arrow $s: w \to u$, so the composite $ee_1 s$ is, like 1_w, an endomorphism of w. But e was defined as the equalizer of all endomorphisms of w, so

$$ee_1 se = 1_w e = e 1_v .$$

Now e is an equalizer, hence is monic; cancelling e on the left gives $e_1 se = 1_v$. This states that the equalizer e_1 of f and g has a right inverse. Like any equalizer, e_1 is monic, hence is an isomorphism. Therefore, $f = g$; this conclusion means that v is initial in D.

This proof will be reformulated in § X.2.

Theorem 2 *(The Freyd Adjoint Functor Theorem). Given a small-complete category A with small hom-sets, a functor $G: A \to X$ has a left adjoint if and only if it preserves all small limits and satisfies the following*

Solution Set Condition. For each object $x \in X$ there is a small set I and an I-indexed family of arrows $f_i: x \to Ga_i$ such that every arrow $h: x \to Ga$ can be written as a composite $h = Gt \circ f_i$ for some index i and some $t: a_i \to a$.

Proof. If G has a left adjoint F, then it must preserve all the limits which exist in its domain A; in particular, all the small ones. Moreover, the universal arrow $\eta_x: x \to GFx$ which is the unit of the adjunction satisfies the solution set condition for x, with I the one-point set.

Conversely, given these conditions, it will suffice to construct a universal arrow $x \to Ga$ from each $x \in X$ to G; then G has a left adjoint by the pointwise construction of adjoints. This universal arrow is an initial object in the comma category $(x \downarrow G) = D$, so we need only verify the conditions of the previous theorem for this category. The solution set condition for G clearly gives the condition of the same name for $(x \downarrow G) = D$. Since A has small hom-sets, so does D. To show D small-complete we need only arbitrary small products and equalizers of parallel pairs in D. They may be created as follows:

Lemma. *If $G: A \to X$ preserves all small products (or, all equalizers) then for each $x \in X$ the projection*

$$Q: (x \downarrow G) \to A, \qquad (x \xrightarrow{s} Ga) \mapsto a$$

of the comma category creates all small products (or, all equalizers).

Proof. Let J be a set (a discrete category) and $f_j : x \to Ga_j$ a J-indexed family of objects of $(x \downarrow G)$ such that the product diagram $p_j : \Pi a_j \to a_j$ exists in A. Since G preserves products, $Gp_j : G\Pi a_j \to Ga_j$ is a product diagram in X, so there is a unique arrow $f : x \to G\Pi a_j$ in X with $(Gp_j) f = f_j$ for all j:

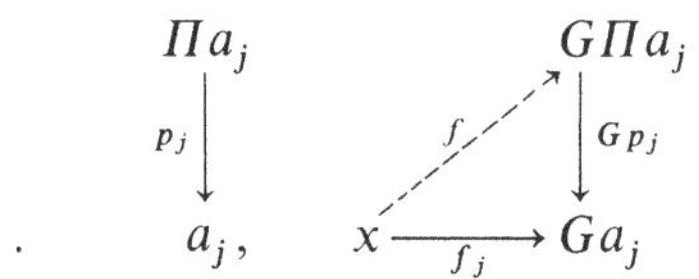

This equation states that $p_j : f \to f_j$ is a cone of arrows in $(x \downarrow G)$; indeed, it is the unique cone there which projects under Q to the given cone $p_j : \Pi a_j \to a_j$. One then verifies that this cone p_j is a product diagram in $(x \downarrow G)$; these two results show that Q creates products.

Similarly, we "create" the equalizer of two arrows $s, t : f \to g$ in $(x \downarrow G)$. As in the figure below, we are given the equalizer e of Qs, Qt; – that is, of s and t as arrows in A. Since G preserves equalizers, Ge is then the equalizer of Gs and Gt. But $Gs \circ f = g = Gt \circ f$, so there is a unique arrow $h : x \to Ga$ making $Ge \circ h = f$, as below. In other words $e : h \to f$ in $(x \downarrow G)$ is the unique arrow of $(x \downarrow G)$ with Q-projection $e : a \to b$.

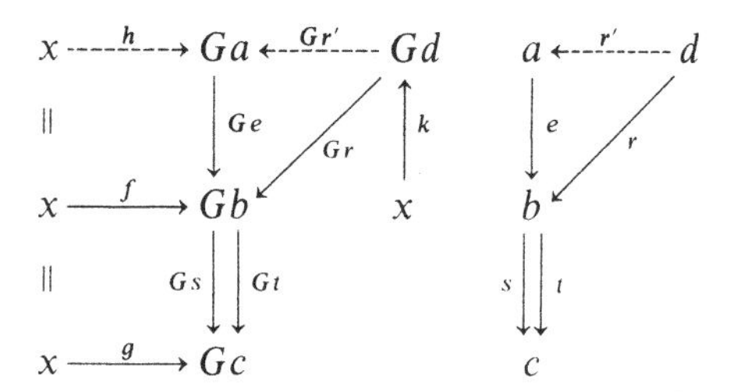

It remains to show that the arrow e is an equalizer in $(x \downarrow G)$. Sc consider another object $k : x \to Gd$ of $(x \downarrow G)$ and an arrow $r : k \to f$ of $(x \downarrow G)$ with $sr = tr$ in $(x \downarrow G)$. Then $sr = tr$ in A, so there is a unique r' in A with $r = er'$. It remains only to show r' an arrow $k \to h$ of $(x \downarrow G)$; but $Ge(Gr' \circ k) = G(er') \circ k = Gr \circ k = f$, so by the unique choice of h, $Gr' \circ k = h$, which states that r' is an arrow of $(x \downarrow G)$.

This line of argument applies not just to products or equalizers, but to the creation of *any* limit (Exercise 1).

Theorem 3 *(The Representability Theorem). Let the category D be small complete with small hom-sets. A functor $K : D \to$* **Set** *is representable if and only if K preserves all small limits and satisfies the following*

Solution Set Condition. There exists a small set S of objects of D such that for any object $d \in D$ and any element $x \in Kd$ there exist an $s \in S$, an element $y \in Ks$ and an arrow $f : s \to d$ with $(Kf)\, y = x$.

Proof. This is another reformulation of the existence Theorem 1 for initial objects. Indeed, a representation of K is a universal arrow from the one-point set $*$ to K (Proposition III.2.2), hence an initial object in the comma category $(* \downarrow K)$, which is small-complete because K is assumed continuous. Conversely, if K is representable, it is necessarily continuous.

The solution set condition (or something like it) is requisite in all three theorems. For an example, let **Ord** be the ordered set of all small ordinal numbers $\alpha, \beta, \ldots$; it is a category with hom-set $\mathbf{Ord}(\alpha, \beta)$ empty or the one-point set according as $\alpha > \beta$ or $\alpha \leqq \beta$. The category $\mathbf{Ord}^{\mathrm{op}}$ is small-complete, because the product of any small set of ordinals is their least upper bound. The functor $K : \mathbf{Ord}^{\mathrm{op}} \to \mathbf{Set}$ with $K\alpha = *$ the one-point set for every α is clearly continuous. However K is not representable: Were $K\alpha \cong \mathbf{Ord}^{\mathrm{op}}(\beta, \alpha)$ for some β, then $\alpha \leqq \beta$ for all α, so β would be a largest small ordinal, which is known to be impossible.

Complete Boolean algebras provide another example to show that some solution set condition is requisite. For a given denumerable set D one can construct an arbitrarily large complete Boolean algebra generated by D (Solovay [1966]); this implies that there is no free complete Boolean algebra generated by D, and hence that the forgetful functor **Comp Bool**$\to$**Set** has no left adjoint – though it is continuous and **Comp Bool** is small-complete.

The adjoint functor theorem has many applications.

For example, it gives a left adjoint to the forgetful functor $U : \mathbf{Grp} \to \mathbf{Set}$. Indeed, we already know that U creates all limits (Theorem 1.3), hence that **Grp** is small-complete and U continuous. It remains to find a solution-set for each $X \in \mathbf{Set}$. Consider any function $f : X \to UG$ for G a group, and take the subgroup S of G generated by all elements fx, for $x \in X$. Every element of S is then a finite product, say $(fx_1)^{\pm 1} (fx_2)^{\pm 1} \cdots (fx_n)^{\pm 1}$, of these generators and their inverses so the cardinal number of S is bounded, given X. Taking one copy of each isomorphism class of such groups S then gives a small set of groups, and the set of all functions $X \to US$ is then a solution set.

This left adjoint $F : \mathbf{Set} \to \mathbf{Grp}$ assigns to each set X the free group FX generated by X, so our theorem has produced this free group without entering into the usual (rather fussy) explicit construction of the elements of FX as equivalence classes of words in letters of X. To be sure, the usual construction also shows that the universal arrow $X \to UFX$ is injective (different elements of X are different as generators of the free group). However, we can also obtain this fact by general arguments and the observation that there does exist a group H with two different elements $h \neq k$. Indeed, for any two elements $x \neq y$ in X we then take a function $f : X \to UH$ with $fx = h$ and $fy = k$. Since f must factor through the universal $X \to UFX$, it follows that this universal must be an injection.

This construction applies not just to **Grp** but to the category of all small algebraic systems of a given type τ. The type τ of an *algebraic system* is given by a set Ω of operators and a set E of identities. The set Ω of operators is a *graded set*; that is, a set Ω with a function which assigns to each element $\omega \in \Omega$ a natural number n, called the *arity* of ω. Thus an operator ω of arity 2 is a binary operator, one of arity 3 a ternary operator, and so on. If S is any set, an *action* of Ω on S is a function A which assigns to each operator ω of arity n an n-ary operation $\omega_A : S^n \to S$ (Here $S^n = S \times \cdots \times S$, with n factors). From the given operators Ω one forms the set Λ of all "derived" operators; given ω of arity n and n derived operators $\lambda_1, \ldots, \lambda_n$ of arities $m_1, \ldots, m_n$, the evident "composite" $\omega(\lambda_1, \ldots, \lambda_n)$ is a derived operator of arity $m_1 + \cdots + m_n$; also, given λ of arity n and $f : n \to m$ any function from $\{1, \ldots, n\}$ to $\{1, \ldots, m\}$, "substitution" of f in λ gives a derived operator θ of arity m, described in terms of variables x_i as $\theta(x_1, \ldots, x_m) = \lambda(x_{f1}, \ldots, x_{fn})$. (This description by variables refers implicitly to the action of Ω on a set; for the abstract formulation of this and of composition, we refer to the standard treatments of universal algebra such as: Cohn [1965], or Grätzer [1968]). At any rate, each action A of Ω on a set S extends uniquely to an action of the set Λ of derived operators on S.

The set E of *identities* for algebraic systems of type τ is a set of ordered pairs $\langle \lambda, \mu \rangle$ of derived operators, where λ and μ have the same arity n. An action A of Ω on S *satisfies* the identity $\langle \lambda, \mu \rangle$ if $\lambda_A = \mu_A : S^n \to S$. An algebra A of type τ – an $\langle \Omega, E \rangle$-algebra – is a set S together with an action A of Ω on S which satisfies all the identities of E; so we call S the *underlying set* of the algebra and often write $|A| = S$. A morphism $g : A \to A'$ of $\langle \Omega, E \rangle$-algebras is a function $g : S \to S'$ on the underlying sets which preserve all the operators of Ω (and hence of Λ) in the sense that

$$g\omega_A(a_1, \ldots, a_n) = \omega_{A'}(ga_1, \ldots, ga_n) \tag{1}$$

for all $a_i \in A$. The collection of all small $\langle \Omega, E \rangle$-algebras, with these morphisms as arrows, is a category $\langle \Omega, E \rangle -$ **Alg**, often called a *variety* or an *equational class* of algebras. This description includes the familiar cases such as **Grp**, **Rng**, **Ab** and many others less familiar (e.g. nilpotent groups of specified class). For example, to describe **Grp**, take three operators in Ω, the product, the inverse, and the assignment of the identity element e, of arities 2, 1, and 0, respectively, and take in E the axioms for the identity ($ex = x = xe$), the axioms for the inverse ($xx^{-1} = e = x^{-1}x$), and the associative law.

For any variety of algebras, the adjoint functor theorem will yield a left adjoint for the forgetful functor $\langle \Omega, E \rangle$-**Alg**$\to$**Set**; the solution set is obtained just as in the case of groups (see also § 7 below). Thus this theorem produces for any set X the free ring, the free abelian group, the free R-module, etc. generated by the elements of the given set X. It does not produce free fields: In defining a field, the inverse to multiplication is not everywhere defined, so fields are not algebraic systems in the sense considered (and, for that matter, free fields do not exist).

Another illustration of the adjoint functor theorem is the construction of the left adjoint to

$$V: \mathbf{Comp\ Haus} \to \mathbf{Set}, \qquad (2)$$

the forgetful functor which sends each compact Hausdorff space to the set of all its points. Given compact Hausdorff spaces X_i, the usual product topology on the cartesian product set $Y = \Pi_i V X_i$ is Hausdorff and compact (the latter by the Tychonoff theorem); hence **Comp Haus** has all small products and V preserves them. For that matter, V creates these products: The product topology is chosen with the fewest open sets to make all the projections $p_i: Y \to X_i$ continuous, so any other compact topology Y' with all p_i continuous would be the same set Y topologized with more open sets; then $\mathrm{id}: Y' \to Y$ is a continuous injection from a compact to a Hausdorff space, hence an isomorphism. By a similar argument, V creates all equalizers, hence all small limits. It remains to find for each set S a solution set of arrows $f: S \to VX$ where each X is compact Hausdorff. Since X may be replaced by the closure $\overline{fS} \subset X$, it is enough to assume fS dense in X. To each point $x \in X$, consider the set $Lx = \{D \mid D \subset S \text{ and } x \in \overline{fD}\}$; thus Lx is a non-void set of subsets of S. If $x \neq x'$ are separated in X by disjoint open sets U and U', then $f^{-1}U \in Lx$ but $f^{-1}U$ is not in Lx', so $Lx \neq Lx'$. Thus L is an injection $X \to \mathscr{P}\mathscr{P}S$ from X to the double power set of S. If we take all subsets X of $\mathscr{P}\mathscr{P}S$, all topologies on each set X and all functions $f: S \to VX$ we obtain a small solution set for S. The adjoint functor theorem then provides a left adjoint to V; it assigns to each set S the Stone-Čech compactification of the discrete topology on S.

Exercises

1. For $G: A \to X$ continuous, show that the projection $(x \downarrow G) \to A$ creates all small limits.
2. Use the adjoint functor theorem to find a left adjoint to each of the forgetful functors **Rng**$\to$**Set**, **Rng**$\to$**Ab**, **Cat**$\to$**Grph**. Compare with the standard explicit construction of these adjoints.
3. Given a pullback diagram in **Cat**,

$$\begin{array}{ccc} A' & \xrightarrow{H'} & A \\ {\scriptstyle G'}\downarrow & & \downarrow{\scriptstyle G} \\ X' & \xrightarrow{H} & X, \end{array}$$

if H creates limits and G preserves them prove that H' creates them.
4. Use Exercise 3 and the fact that $(x \downarrow X) \to X$ creates limits to give a new proof of the result of Exercise 1.

7. Subobjects and Generators

Concepts such as subring, subspace, and subfield will now be treated categorically, using arrows instead of elements. For instance we will regard a subgroup S of a group G not as a set of elements of G, but as the monomorphism $S \to G$ given by insertion.

Let A be any category. If $u: s \to a$ and $v: t \to a$ are two monics with a common codomain a, write $u \leqq v$ when u factors through v; that is, when $u = vu'$ for some arrow u' (which is then necessarily also monic). When both $u \leqq v$ and $v \leqq u$, write $u \equiv v$; this defines an equivalence relation $\equiv$ among the monics with codomain a, and the corresponding equivalence classes of these monics are called the *subobjects* of a. It is often convenient to say that a monic $u: s \to a$ *is* a subobject of a – that is, to identify u with the equivalence class of all $v = u\theta$, for $\theta: s' \to s$ an invertible arrow. These subobjects do correspond to the usual subobjects (defined via elements) in familiar large categories such as **Rng**, **Grp**, **Ab**, and R-**Mod**, but not in **Top**.

Lemma. *In any square pullback diagram*

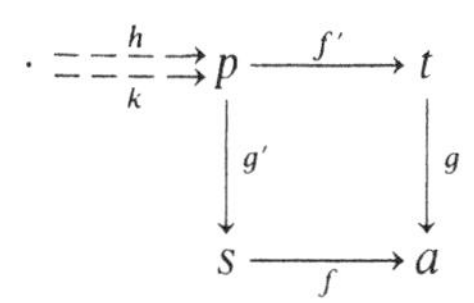

f monic implies f' monic (and g monic implies g' monic).

Briefly, pullbacks of monics are monic.

Proof. Consider a parallel pair h, k, as shown, with $f'h = f'k$. Then $gf'h = gf'k$, so $fg'h = fg'k$. Since f is monic, this gives $g'h = g'k$. But we also have $f'h = f'k$; these two equations, since p is a pullback, imply $h = k$. Therefore f' is monic.

The set of all subobjects of each $a \in A$ is partly ordered by the binary relation $u \leqq v$. If $u: s \to a$ and $v: t \to a$ are two subobjects of a, and A has pullbacks, the pullback of these two arrows gives (Lemma above) another monic $w: p \to a$ with codomain a and with $w \leqq u$, $w \leqq v$; it is the *intersection* (= meet or *greatest lower bound*) of the subobjects u and v in the partly ordered set of all subobjects of $a \in A$. Similarly, if J is any set and $u_i: s_i \to a$ for $i \in J$ any J-indexed set of subobjects of $a \in A$, the pullback of all these arrows, if it exists, gives the intersection of the subobjects u_i of a. The *union* (= join or *least upper bound*) of subobjects can be found under added hypotheses.

Dually, two epis r, s with domain a are equivalent when $r = \theta s$ for some invertible θ. The equivalence classes of such epis are the *quotient objects* of a, partly ordered by the relation $r \leqq s$, which holds when r factors through s as $r = r's$. This definition of quotients by duality is simpler than the usual definition of quotient algebras by equivalence classes, and agrees with the usual definition in those categories where epis are onto. This latter is the case, for example, in **Grp**. Hence every quotient object of a group G in **Grp** is represented by the projection $p: G \to G/N$ of G onto the factor group G/N of G by some normal subgroup N of G, and $G/M \leqq G/N$ holds if and only if $M \supset N$ (in general, the relation $r \leqq s$ for quotients mean that in r "more" is divided out!).

A set S of objects of the category C is said to *generate* C when to any parallel pair $h, h': c \to d$ of arrows of C, $h \neq h'$ implies that there is an $s \in S$ and an arrow $f: s \to c$ with $hf \neq h'f$ (the term "generates" is well established but poorly chosen; "separates" would have been better). This definition includes the case of a single object s generating a category C. For example, any one-point set generates **Set**, **Z** generates **Ab** and **Grp**, and R generates R-**Mod**. The set of finite cyclic groups is a generator for the category of all finite abelian groups (or, of all torsion abelian groups).

Dually, a set Q of objects is a *cogenerating set* for the category C when to every parallel pair $h \neq h': a \to b$ of arrows of C there is an object $q \in Q$ and an arrow $g: b \to q$ with $gh \neq gh'$. A single object q is a *cogenerator* when $\{q\}$ is a cogenerating set. For example, any two-point set is a cogenerator in **Set**.

In terms of subobjects we can examine further the construction of solution sets. Given any functor $G: A \to X$ an arrow $f: x \to Ga$ is said to *span* a when there is no proper monomorphism $s \to a$ in A such that f factors through $Gs \to Ga$.

Lemma. *In the category A, suppose that every set of subobjects of an object $a \in A$ has a pullback. Then if $G: A \to X$ preserves all these pullbacks, every arrow $h: x \to Ga$ factors through an arrow $f: x \to Gb$ which spans b.*

Proof. Consider the set of all those subobjects $u_j: s_j \to a$ such that h factors through Gu_j as $h = Gu_j \circ h_j$. Take the pullback $v: b \to a$ of all the u_j. Then, as in the commutative diagrams

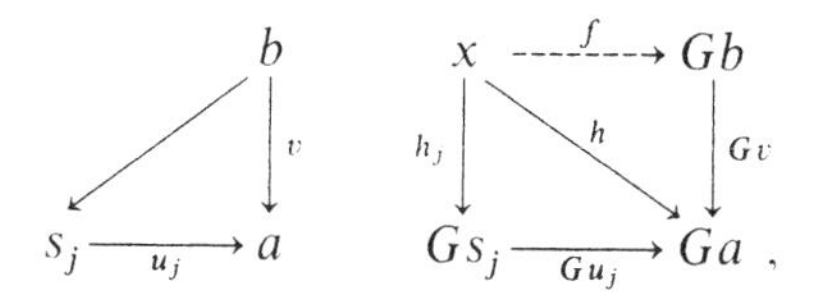

$Gv: Gb \to Ga$ is still a pullback (for the Gu_j), so h factors through Gv via f, as shown. It follows from the construction that f spans b.

This lemma states that a solution set for x can be the set of all arrows from x which span.

As an application consider the category of algebras of given type τ. Given an arrow $f: S \to GA$, the algebra A has a subalgebra consisting of all elements obtained from elements of $f(S)$ by iterated applications of operators $\omega \in \Omega$. The cardinal number of this subalgebra A_f is then bounded by the cardinal of S and that of Ω. Since f factors through $S \to GA_f$, these latter arrows from the set S form a small set which is a solution set for $G: \mathbf{Alg}_\tau \to \mathbf{Set}$. They are spanning arrows in the sense of the lemma, provided a subobject of a is redefined to be a morphism $u: s \to a$ for which Gu is injective in **Set**.

Another example of the use of this lemma with the adjoint functor theorem is the proof of the existence of tensor products of modules. Given modules A and B over a commutative ring K, a tensor product is a universal element of the set Bilin $(A, B; C)$ of bilinear functions $\beta: A \times B \to C$ to some third K-module C. This set is (the object function of) a functor of C. To get a solution set for given A and B, it suffices to consider only those bilinear β which span C (do not factor through a proper submodule of C). Then C consists of all finite sums $\Sigma \beta(a_i, b_i)$, so the solution set condition holds; since K-**Mod** is small-complete and Bilin: K-**Mod** $\to$ **Set** is continuous, a tensor product $\otimes: A \times B \to A \otimes B$ exists. The usual (more explicit) construction is wholly needless, since all the properties of the tensor product follow directly from the universality.

Exercises

1. Use the adjoint functor theorem to construct the coproduct in **Grp** (the coproduct $G \amalg H$ in **Grp** is usually called the *free product*). Using the product $G \times H$, show also that the injections $G \to G \amalg H$ and $H \to G \amalg H$ of the coproduct are both monic, and that their images intersect in the identity subgroup.
2. Make a similar construction for the coproduct of rings.
3. If R is a ring, A a right R-module and B a left R-module, use the adjoint functor theorem to construct $A \otimes_R B$ (this tensor product is an abelian group, with a function $\langle a, b \rangle \mapsto a \otimes b \in A \otimes_R B$ which is biadditive, has $ar \otimes b = a \otimes rb$ for all $a \in A$, $r \in R$, and $b \in B$, and is universal with these properties). Prove that $A \otimes_R B$ is spanned (as an abelian group) by the elements $a \otimes b$. If $S \to R$ is a morphism of rings, examine the relation of $A \otimes_S B$ to $A \otimes_R B$.
4. Construct coequalizers in $\mathbf{Alg}_\tau$ by the adjoint functor theorem.

8. The Special Adjoint Functor Theorem

We now consider another existence theorem for adjoints which avoids the solution set condition by assuming a small set of objects which cogenerates.

Theorem 1 *(Special Initial-Object Theorem). If the category D is small-complete, has small hom-sets, and a small cogenerating set Q, then D*

has an initial object provided every set of subobjects of each $d \in D$ has an intersection.

Proof. Form the product $q_0 = \Pi_{q \in Q} q$ of all the objects in the small cogenerating set Q and take the intersection r of all subobjects of q_0. For any object $d \in D$, there is at most one arrow $r \to d$, for if there were two different arrows, their equalizer would be a proper monic to r, hence a subobject of q_0 smaller than the intersection r.

To show r initial in D, we thus need only construct an arrow $r \to d$ for each d. So consider the set H of all arrows $h: d \to q \in Q$ and the (small) product $\Pi_{h \in H} q$. Take the arrow $j: d \to \Pi_{h \in H} q$ with components h (i.e., with $p_h \circ j = h$ for each projection p_h). Since the set Q cogenerates, j is monic. Form the pullback

$$\begin{array}{ccc} c & \xrightarrow{j'} & \Pi_{q \in Q} q = q_0 \\ \downarrow & & \downarrow k \\ d & \xrightarrow{j} & \Pi_{h \in H} q, \end{array}$$

where k is the arrow with components $p_h \circ k = p_q$ for each $h: d \to q$. Then j', as pullback of a monic j, is monic, so c is a subobject of q_0. But r was the intersection of all subobjects of q_0, so there is an arrow $r \to c$. The composite $r \to c \to d$ is the desired arrow.

Theorem 2 *(The Special Adjoint Functor Theorem). Let the category A be small-complete, with small hom-sets, and a small cogenerating set Q, while every set of subobjects of an object $a \in A$ has a pullback (and hence has an intersection). Let the category X have small hom-sets. Then a functor $G: A \to X$ has a left-adjoint if and only if G preserves all small limits and all pullbacks of families of monics.*

Proof. The conditions are necessary, since any right adjoint functor must indeed preserve all limits (in particular, all pullbacks). Conversely, it suffices as usual to construct for each $x \in X$ an initial object in the comma category $D = (x \downarrow G)$. We shall show that this category satisfies the hypotheses of the previous theorem for the construction of an initial object. First we verify that subobjects in $(x \downarrow G)$ have the expected form.

Lemma. *An arrow $h: \langle f: x \to Ga, a \rangle \to \langle f': x \to Ga', a' \rangle$ in the comma category $(x \downarrow G)$ is monic if and only if $h: a \to a'$ is monic in A.*

Proof. Trivially, $h: a \to a'$ monic implies $h: f \to f'$ monic. For the converse, observe that h monic means exactly that its kernel pair (the pullback of h with h) is $1_a, 1_a: a \rightrightarrows a$. On the other hand, by the lemma of § 6 the projection

$$(x \downarrow G) \to A, \qquad \langle f: x \to Ga, a \rangle \mapsto a$$

of the comma category creates all limits, and in particular, creates kernel pairs. Moreover, A has all kernel pairs. Therefore (Theorem 4.2), the projection of the comma category preserves all kernel pairs, in particular, the kernel pair $1_a, 1_a$, and consequently carries monics (in $(x\downarrow G)$) to monics in A, as desired.

Now return to the theorem. We are given a small cogenerating set Q in A. Since X has small hom-sets, the set Q' of all objects $k : x \to Gq$ with $q \in Q$ is small. It is, moreover, cogenerating in $(x\downarrow G)$. Given $s \neq t$: $\langle f : x \to Ga, a\rangle \to \langle f' : x \to Ga', a'\rangle$ in $(x\downarrow G)$, there is a $q_0 \in Q$ and an arrow $h : a' \to q_0$ with $hs \neq ht$, and this h can be regarded as an arrow

$$h : \langle f' : x \to Ga', a'\rangle \to \langle f_0 : x \to Gq_0, q_0\rangle\,,$$

where $f_0 = Gh \circ f'$, with $hs \neq ht$ in $(x\downarrow G)$. Therefore Q' cogenerates $(x\downarrow G)$.

Since A small-complete and G continuous imply $(x\downarrow G)$ small-complete it remains only to construct an intersection in $(x\downarrow G)$ for every set of subobjects $h_i : \langle f_i : x \to Ga_i, a_i\rangle \to \langle f : x \to Ga, a\rangle$, where $i \in J$. By the lemma, the corresponding arrows $h_i : a_i \to a$ are monics in A. By hypothesis, they then have a pullback $h : b \to a$ in A

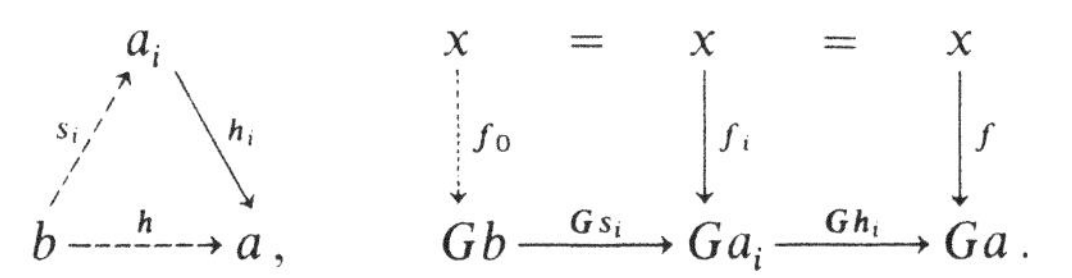

The functor G preserves pullbacks, so $Gh : Gb \to Ga$ with $Gh = Gh_i \circ Gs_i$ is a pullback of the Gh_i in X. Since also $Gh_i \circ f_i = f$ for all $i \in J$, there is a unique $f_0 : x \to Gb$ with $f_i = Gs_i \circ f_0$; the resulting arrow $h : \langle f_0, b\rangle \to \langle f, a\rangle$ is then a pullback in $(x\downarrow G)$ of the given h_i (again, because the projection of the comma category creates pullbacks). This pullback is the required intersection of the h_i.

There is another form of this theorem. Define a category to be *well-powered* when the subobjects of each object $a \in A$ can be indexed by a small set; that is, when there is to each a a small set J_a and a bijection from J_a to the set of all subobjects of a. Many familiar large categories – **Top**, **Grp**, R-**Mod**, etc. – are well powered; the dual notion is called *co-well-powered*. If A is well-powered and small-complete, then any set of subobjects of an $a \in A$ has an intersection, formed by the usual pullback. Therefore the special adjoint functor theorem specializes as follows:

Corollary. *If A is small complete, well-powered, with small hom-sets, and a small cogenerating set, while X has small hom-sets, then a functor $G : A \to X$ has a left adjoint if and only if it is continuous. In particular, any continuous $K : A \to$* **Set** *is representable.*

This classical form of the special adjoint functor theorem (sometimes called $SAFT$) often appears without an explicit "small hom-set" hypothesis – in sources which consider only categories with small hom-sets. Some authors use "locally small category" to mean "well-powered"; others use it to mean "has small hom-sets", so we avoid this term!

The classical form of $SAFT$ can be deduced directly from the adjoint functor theorem by constructing a solution set (as in Freyd [1964, p. 89], or Schubert [1970, p. 88]).

A typical example is the inclusion functor

$$G: \mathbf{Comp\ Haus} \subset \mathbf{Top} \qquad (1)$$

of the full subcategory of compact Hausdorff spaces in **Top**. As already noted, **Comp Haus** is small complete; it also has small hom-sets. The Urysohn lemma states that to any two points $x \neq y$ in a compact Hausdorff space X there is a continuous function $f: X \to I$ to the unit interval I with $fx = 0, fy = 1$. It follows that I is a cogenerator for **Comp Haus**. Hence the special adjoint functor theorem gives a left adjoint for the inclusion G above. This left-adjoint (or sometimes, its restriction to the full subcategory of completely regular spaces) is called the *Stone-Čech compactification*. This includes the case of a discrete space, as done in § 6.

Watt's Theorem [1960] is another example. Any ring R is a generator in the category R-**Mod**, hence a cogenerator in $(R\text{-}\mathbf{Mod})^{\mathrm{op}}$. It follows that any contravariant additive functor T on R-**Mod** to **Ab** which takes small colimits to limits is representable by a group isomorphism $T \cong \hom_R(-, C)$ for some R-module C. Indeed, by the special adjoint functor theorem $T: (R\text{-}\mathbf{Mod})^{\mathrm{op}} \to \mathbf{Ab}$ has a left adjoint F; since T is additive, the adjunction

$$\mathbf{Ab}(G, TA) \cong \hom_R(A, FG), \qquad G \in \mathbf{Ab}, \qquad A \in R\text{-}\mathbf{Mod},$$

is an isomorphism of additive groups; set $G = \mathbf{Z}$ to get

$$TA \cong \mathbf{Ab}(\mathbf{Z}, TA) \cong \hom_R(A, F\mathbf{Z}).$$

Exercises

1. Let $K: A \to \mathbf{Set}$ be any functor. If K has a left adjoint, prove that it is representable. Conversely, if A has all small copowers and K is representable as $K \cong A(a, -)$ for some $a \in A$ prove that K has a left adjoint (which assigns to each set X the small copower $X \cdot a$).
2. For A a left R-module, B a right R-module and G an abelian group, establish adjunctions
   (a) $\hom_R(A, \hom_{\mathbf{Z}}(B, G)) \cong \hom_{\mathbf{Z}}(B \otimes_R A, G) \cong \hom_R(B, \hom_{\mathbf{Z}}(A, G))$, where $\hom_{\mathbf{Z}}(B, G)$ has a suitable (left or right) R-module structure, and where $\hom_R$ denotes the hom-set in R-**Mod**, $\hom_{\mathbf{Z}}$ that in **Ab**.
   (b) The additive group $\mathbf{Q}/\mathbf{Z}$ of rational numbers modulo 1 is known to be an injective cogenerator of **Ab**. Use (a) to prove that $\hom_{\mathbf{Z}}(R, \mathbf{Q}/\mathbf{Z})$ is an injective cogenerator of R-**Mod** ("injective" object as defined in § 4).

3. Use Exercise 2(b) and the special adjoint functor theorem to prove that any continuous additive functor $T: R\text{-}\mathbf{Mod} \to \mathbf{Ab}$ is representable. (Watt's theorem).
4. (Stone-Čech compactification.) If X is a completely regular topological space, show that the universal arrow $X \to GFX$ for the left adjoint to (1) is an injection. (Use the Urysohn lemma: For $x \neq y$ in X completely regular there exists a continuous $f: X \to I$ with $fx \neq fy$ and I the unit interval.) Classical sources describe this compactification only when X is completely regular. This restriction is needless; it arose from the idea of considering just universal injections, not universal arrows.

9. Adjoints in Topology

Top is the category with objects all (small) topological spaces $X, Y, \ldots$ and arrows all continuous maps $f: X \to Y$. The standard forgetful functor (usually a nameless orphan!)

$$G: \mathbf{Top} \to \mathbf{Set},$$

sends X to GX, the set of points in X, is faithful, and has a left adjoint D which assigns to each set S the discrete topology on S (i.e., all subsets of S are open). Therefore G preserves all limits which may exist in **Top** (this is *why* the underlying set of the product of spaces is the cartesian product of their underlying sets). The forgetful functor G also has a right adjoint D', which assigns to each set S the indiscrete topology on S (with only S and $\emptyset$ open). Therefore G preserves all colimits which may exist in **Top** – and this is why the coproduct of two spaces is formed by putting a topology on the disjoint union of the underlying sets.

Next consider the subspace topology on a set $S \subset GX$.

If X is a fixed topological space, G induces a functor

$$G{\downarrow}X : (\mathbf{Top}{\downarrow}X) \to (\mathbf{Set}{\downarrow}GX)$$

$$\begin{array}{ccc} Y & \xrightarrow{f} & X \\ {\scriptstyle h}\downarrow & & \| \\ Y' & \xrightarrow{f'} & X \end{array} \quad \mapsto \quad \begin{array}{ccc} GY & \xrightarrow{Gf} & GX \\ {\scriptstyle Gh}\downarrow & & \| \\ GY' & \xrightarrow{Gf'} & GX, \end{array} \tag{1}$$

here f and f' are objects and h an arrow of the comma category $(\mathbf{Top}{\downarrow}X)$. This functor $G \downarrow X$ has a right adjoint L. Indeed, an object $t: S \to GX$ in $(\mathbf{Set}{\downarrow}GX)$ is a set S and a function t on S to GX. Put on S the topology with open sets all $t^{-1}U$ for U open in X, and call the resulting space LS; then t is a continuous map $Lt: LS \to X$. (For example, if S is a subset of GX, then LS is just S with the usual "subspace topology.") This topology on LS has the familiar universal property: Any continuous map $f: Y \to X$ which factors through t as $Gf = t \circ s$, in **Set**,

$$\begin{array}{ccc} GY & \xrightarrow{Gf} & GX \\ \downarrow s & & \| \\ S & \xrightarrow{t} & GX, \end{array} \qquad Gf = t \circ s,$$

has $s: Y \to LS$ continuous. This property just restates the desired adjunction: $\hom(Gf, t) \cong \hom(f, Lt)$. Observe that $(G{\downarrow}X) \circ L = \mathrm{Id}$; L is a "right-adjoint-right-inverse" to $(G \downarrow X)$.

Note especially that the universal property of the subspace topology on a subset $S \subset GX$ refers not only to the other subspaces of X, but to other spaces Y and *any* continuous $f: Y \to X$ which factors through the inclusion $t: S \to GX$ (i.e., has image contained in the subset S).

This adjoint may be used to construct (the usual) equalizers in **Top** by the following general process:

Proposition 1. *If $G: C \to D$ is a faithful functor, if D has equalizers, and if, for each $x \in C$, $(G \downarrow x): (C \downarrow x) \to (D \downarrow Gx)$ has a right-adjoint-right-inverse L, then C has equalizers.*

Proof. To get the equalizer of a parallel pair $f, f': x \to y$, apply G, take the equalizer $t: s \to Gx$ of Gf, Gf' in D and apply L; the universal property of the adjunction shows $Lt: Ls \to x$ an equalizer in C.

This argument is just an element-free version of the usual definition of the equalizer: Given two continuous maps $f, f': X \to Y$, take the set S of points x of X with $fx = f'x$ and impose the subspace topology. The adjunction explains *why* the subspace topology.

Now **Top** is well known to be complete: To prove this one needs only equalizers (of parallel pairs) and products. The product of any family X_i, $i \in J$, of spaces is constructed by taking the product ΠGX_i of the underlying sets and putting on it the (universal) topology in which all projections $p_i: \Pi GX_i \to GX_i$, $i \in J$, are continuous. The general fact that to spaces X_i, a set S, and functions $t_i: S \to GX_i$ there is a "universal" topology with exactly those open sets on S required to make all t_i continuous can be expressed categorically (Exercise 3).

Colimits may be treated in dual fashion. For any space X the functor

$$(X{\downarrow}G): (X{\downarrow}\mathbf{Top}) \to (GX{\downarrow}\mathbf{Set})$$

has a left adjoint M. Indeed, an object of $(GX{\downarrow}\mathbf{Set})$ is a function $t: GX \to S$ to a set S. Put on S the topology with open sets all subsets $V \subset S$ with $t^{-1}V$ open in X and call the resulting space MS. (If $t: GX \to S$ is a surjection, this is the familiar "quotient topology" or "identification topology" on S.) Then the function t is a continuous map $Mt: X \to MS$.

Moreover, $f : X \to Y$ continuous and $Gf = k \circ t$ for some function k,

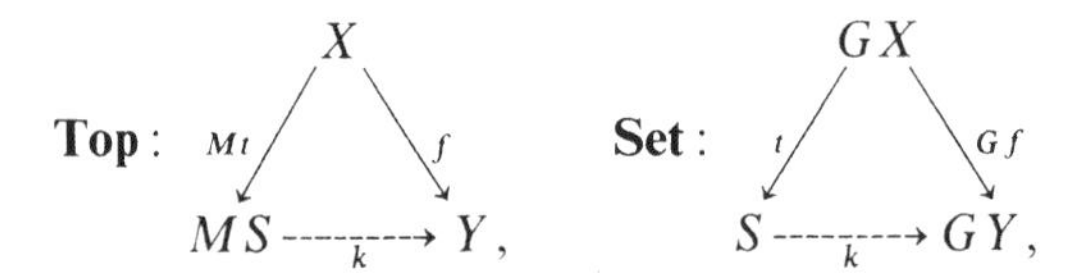

implies that $k : MS \to Y$ is continuous. Thus $k \mapsto k$ is an adjunction

$$(X \downarrow \mathbf{Top})(Mt, f) \cong (GX \downarrow \mathbf{Set})(t, Gf)$$

with unit the identity map, so M is left-adjoint-right-inverse to $X \downarrow G$.

Now Proposition 1 was proved just from the axioms for a category, so its dual is also true. This dual proposition and the above adjunction prove that **Top** has coequalizers.

Similar constructions yield coproducts (= disjoint unions) and general colimits in **Top**. Such colimits appear often, usually under other names, as for instance in the basic process of constructing spaces by gluing pieces together. For example, let $\{U_i \mid i \in J\}$ be an open cover of a space X. Each continuous $f : X \to Y$ determines a J-indexed family of restrictions $f \mid U_i : U_i \to Y$; conversely, a familiar result states that a J-indexed family of continuous maps $f_i : U_i \to Y$ determines a map f continuous on all of X if and only if $f_i|(U_i \cap U_j) = f_j|(U_i \cap U_j)$ for all i and j. This result may be expressed by the statement that the following diagram is an equalizer

$$\mathbf{Top}(X, Y) \to \Pi_i \mathbf{Top}(U_i, Y) \rightrightarrows \Pi_{i,j} \mathbf{Top}(U_i \cap U_j, Y) ,$$

where the arrows are given by restriction, as above. This result may equally well be expressed by the statement that X is the colimit in **Top**, with colimiting cone the inclusion maps $U_i \to X$, of the functor $U : J' \to \mathbf{Top}$, where J' is the category with objects the pairs of indices $\langle i, j \rangle$, the single indices $\langle i \rangle$, and the (non-identity) maps $\langle i,j \rangle \to \langle i \rangle$, $\langle i,j \rangle \to \langle j \rangle$, while U is the functor with $U\langle i, j \rangle = U_i \cap U_j$, $U\langle i \rangle = U_i$, with U on (non-identity) arrows the inclusion maps.

Another coequalizer is the space X/A obtained from the space X by *collapsing the subset* A to a point. It is the coequalizer

$$* \ \begin{array}{c} \xrightarrow{a} \\ \vdots \\ \xrightarrow[a']{} \end{array} \ X \to X/A$$

of the set of all the arrows sending the one point space $*$ to one of the points $a \in A$. It is used in homotopy theory. If we consider the category $\mathbf{Top}^{(2)}$ whose objects are pairs $\langle X, A \rangle$ (a space X with a subset A) and whose arrows $\langle X, A \rangle \to \langle X', A' \rangle$ are continuous maps $X \to X'$ sending

A to A', then the definition of X/A, for Y a pointed topological space, reads:

$$\mathbf{Top}_*(X/A, Y) = \mathbf{Top}^{(2)}(\langle X, A\rangle, \langle Y, *\rangle).$$

Thus $\langle X, A\rangle \mapsto X/A$ is left adjoint to the functor $Y \mapsto \langle Y, *\rangle$ which sends each pointed space to the pair $\langle Y, *\rangle$.

There are many familiar subcategories of **Top**.

Proposition 2. Haus, *the full subcategory of all Hausdorff spaces in* **Top**, *is complete and cocomplete. The inclusion functor* **Haus**$\rightarrow$**Top** *has a left adjoint H, as does the forgetful functor* **Haus**$\rightarrow$**Set**.

Proof. The left adjoint H will be obtained by the adjoint functor theorem. First, any product of Hausdorff spaces or subspace of a Hausdorff space is also Hausdorff, hence **Haus** is complete and the inclusion functor is continuous (i.e., it preserves small limits). It remains only to verify the solution set condition for every topological space X. But any continuous map of X to a Hausdorff space Y factors through the image, a subspace of Y, hence Hausdorff. This image is a quotient set of X with some topology, so there is at most a small set of (non-isomorphic) surjections $X \rightarrow Y$ to a Hausdorff Y. This is the solution set condition. The resulting left adjoint H assigns to each space X a Hausdorff space HX and a continuous map $\eta : X \rightarrow HX$, universal from X to a Hausdorff space. Now η universal implies that η is a surjection, so HX may be described as the "largest Hausdorff quotient" of X. If X is already Hausdorff, we may take $HX = X$ and $\eta = 1$, so H is a left-adjoint-left-inverse to the inclusion.

Since H is a left adjoint, it preserves colimits. It follows that **Haus** has all small colimits (is cocomplete). In particular, the coproduct in **Haus** is the coproduct in **Top** (because a coproduct of Hausdorff spaces is Hausdorff), while a coequalizer in **Haus** is the largest Hausdorff quotient of the coequalizer in **Top**.

The full subcategory of compactly generated Hausdorff spaces is especially convenient because it is cartesian closed (§ VII.8).

Exercises

1. For the full subcategory **L conn** of locally connected spaces in **Top**, prove that D : **Set**$\rightarrow$**L conn** has a left adjoint C, assigning to each space X the set of its connected components, but show that this functor C can have no left adjoint (because of misbehavior on equalizers).
2. Show that the right adjoint D' : **Set** $\rightarrow$ **Top** to the forgetful functor has no right adjoint (misbehavior on coproducts).
3. (Categorical construction of the usual products in **Top**.)
   (a) For diagonal functors $\Delta : C \rightarrow C^J$, $\Delta' : D \rightarrow D^J$, and $T \in C^J$, each $G : C \rightarrow D$ defines $G_* : (\Delta \downarrow T) \rightarrow (\Delta' \downarrow GT)$ by $\langle \tau : c \dot\rightarrow T\rangle \mapsto \langle G\tau : Gc \dot\rightarrow GT\rangle$. If G_* has a left adjoint and GT a limit in D, prove that T has a limit in C.

(b) For G the forgetful functor **Top**$\to$**Set** and J discrete, construct a left adjoint for G_*, showing that it constructs on a set S the weakest topology making a given J-indexed family of functions $f_j : S \to GX_j$ continuous.

(c) Conclude that **Top** has all (the usual) products.

4. Construct left adjoints for each of the inclusion functors $\mathbf{Top}_{n+1} \to \mathbf{Top}_n$, $n = 0, 1, 2, 3$, where $\mathbf{Top}_n$ denotes the full subcategory of all T_n-spaces in **Top**, with $T_4 =$ Normal, $T_3 =$ Regular, $T_2 =$ Hausdorff, etc.
5. Show that the inclusion **Haus**$\to$**Top** has no *right* adjoint, by showing that a coequalizer in **Top** of Hausdorff spaces need not be Hausdorff. Conclude that the forgetful functor **Haus**$\to$**Set** has no right adjoint.

Notes.

Instances and special cases of the adjoint functor theorem abound; there have been many partial discoveries or rediscoveries. One version is Bourbaki's condition [1957] for the existence of universal arrows; this version clearly formulated a solution set condition, but was cumbersome because Bourbaki's notion of "structures" did not make use of categorical ideas. The present version of the adjoint functor theorem was formulated and popularized by Freyd [1964], who also formulated *SAFT*. Our version of the special initial-object theorem is due to G.M. Kelly (private communication).

VI. Monads and Algebras

In this chapter we will examine more closely the relation between universal algebra and adjoint functors. For each type τ of algebras (§ V.6), we have the category $\mathbf{Alg}_\tau$ of all algebras of the given type, the forgetful functor $G : \mathbf{Alg}_\tau \to \mathbf{Set}$, and its left adjoint F, which assigns to each set S the free algebra FS of type τ generated by elements of S. A trace of this adjunction $\langle F, G, \varphi \rangle : \mathbf{Set} \rightharpoonup \mathbf{Alg}_\tau$ resides in the category **Set**; indeed, the composite $T = GF$ is a functor $\mathbf{Set} \to \mathbf{Set}$, which assigns to each set S the set of all elements of its corresponding free algebra. Moreover, this functor T is equipped with certain natural transformations which give it a monoid-like structure, called a "monad". The remarkable part is then that the whole category $\mathbf{Alg}_\tau$ can be reconstructed from this monad in **Set**. Another principal result is a theorem due to Beck, which describes exactly those categories A with adjunctions $\langle F, G, \varphi \rangle : X \rightharpoonup A$ which can be so reconstructed from a monad T in the base category X. It then turns out that algebras in this last sense are so general as to include the compact Hausdorff spaces (§ 9).

1. Monads in a Category

Any endofunctor $T: X \to X$ has composites $T^2 = T \circ T: X \to X$ and $T^3 = T^2 \circ T: X \to X$. If $\mu : T^2 \dot{\to} T$ is a natural transformation, with components $\mu_x : T^2 x \to Tx$ for each $x \in X$, then $T\mu : T^3 \dot{\to} T^2$ denotes the natural transformation with components $(T\mu)_x = T(\mu_x) : T^3 x \to T^2 x$ while $\mu T : T^3 \dot{\to} T^2$ has components $(\mu T)_x = \mu_{Tx}$. Indeed, $T\mu$ and μT are "horizontal" composites in the sense of § II.5.

Definition. *A monad $T = \langle T, \eta, \mu \rangle$ in a category X consists of a functor $T: X \to X$ and two natural transformations*

$$\eta : I_X \dot{\to} T, \qquad \mu : T^2 \dot{\to} T \tag{1}$$

which make the following diagrams commute

$$\begin{array}{ccc} T^3 & \xrightarrow{T\mu} & T^2 \\ {\scriptstyle \mu T}\downarrow & & \downarrow{\scriptstyle \mu} \\ T^2 & \xrightarrow{\mu} & T, \end{array} \qquad \begin{array}{ccccc} IT & \xrightarrow{\eta T} & T^2 & \xleftarrow{T\eta} & TI \\ \| & & \downarrow{\scriptstyle \mu} & & \| \\ T & = & T & = & T. \end{array} \tag{2}$$

Formally, the definition of a monad is like that of a monoid M in sets, as described in the introduction. The set M of elements of the monoid is replaced by the endofunctor $T: X \to X$, while the cartesian product $\times$ of two sets is replaced by composite of two functors, the binary operation $\mu: M \times M \to M$ of multiplication by the transformation $\mu: T^2 \dot{\to} T$ and the unit (identity) element $\eta: 1 \to M$ by $\eta: I_X \dot{\to} T$. We shall thus call η the *unit* and μ the *multiplication* of the monad T; the first commutative diagram of (2) is then the *associative* law for the monad, while the second and third diagrams express the left and right *unit laws*, respectively. All told, a monad in X is just a monoid in the category of endofunctors of X, with product $\times$ replaced by composition of endofunctors and unit set by the identity endofunctor.

Terminology. These objects $\langle X, T, \eta, \mu \rangle$ have been variously called "dual standard construction", "triple", "monoid", and "triad". The frequent but unfortunate use of the word "triple" in this sense has achieved a maximum of needless confusion, what with the conflict with ordered triple, plus the use of associated terms such as "triple derived functors" for functors which are not three times derived from anything in the world. Hence the term *monad*.

Every adjunction $\langle F, G, \eta, \varepsilon \rangle : X \rightharpoonup A$ gives rise to a monad in the category X. Specifically, the two functors $F: X \to A$ and $G: A \to X$ have composite $T = GF$ an endofunctor, the unit η of the adjunction is a natural transformation $\eta: I \dot{\to} T$ and the counit $\varepsilon: FG \dot{\to} I_A$ of the adjunction yields by horizontal composition a natural transformation $\mu = G\varepsilon F: GFGF \dot{\to} GF = T$. The associative law of (2) above for this μ becomes the commutativity of the first diagram below

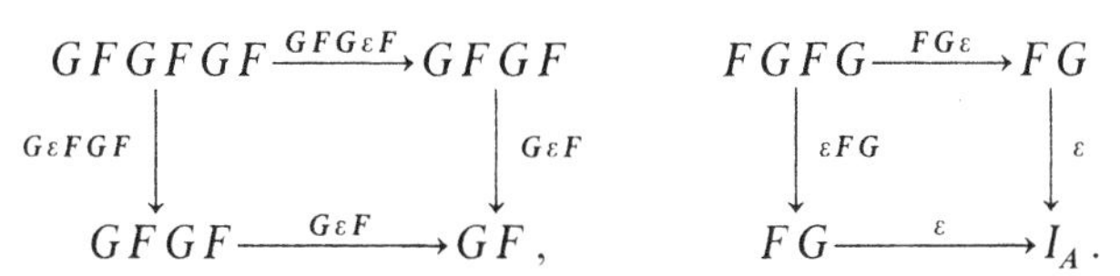

Dropping G in front and F behind, this amounts to the commutativity of the second diagram, which holds by the very definition (§ II.4) of the (horizontal) composite $\varepsilon\varepsilon = \varepsilon \cdot (FG\varepsilon) = \varepsilon \cdot (\varepsilon FG)$ (i.e., by the "interchange law" for functors and natural transformations). Similarly, the left and right unit laws of (2) reduce to the diagrams

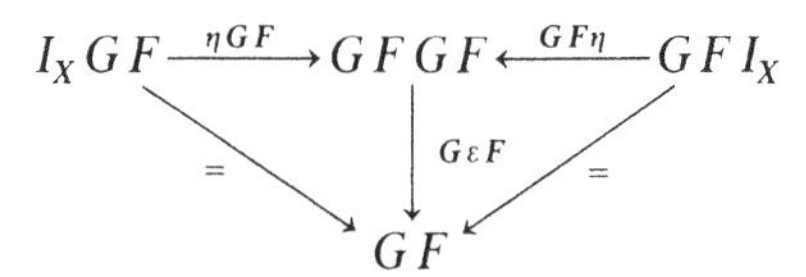

which are essentially just the two triangular identities

$$1 = G\varepsilon \cdot \eta G : G \dot{\to} G \qquad 1 = \varepsilon F \cdot F\eta : F \dot{\to} F$$

for an adjunction. Therefore $\langle GF, \eta, G\varepsilon F\rangle$ is indeed a monad in X. Call it the *monad defined by the adjunction* $\langle F, G, \eta, \varepsilon\rangle$.

For example, the *free group monad* in **Set** is the monad defined by the adjunction $\langle F, G, \varphi\rangle : \mathbf{Set} \rightharpoonup \mathbf{Grp}$, with $G : \mathbf{Grp} \to \mathbf{Set}$ the usual forgetful functor.

Dually, a comonad in a category consists of a functor L and transformations

$$L : A \to A\,, \qquad \varepsilon : L \dot{\to} I\,, \qquad \delta : L \dot{\to} L^2 \tag{1^{op}}$$

which render commutative the diagrams

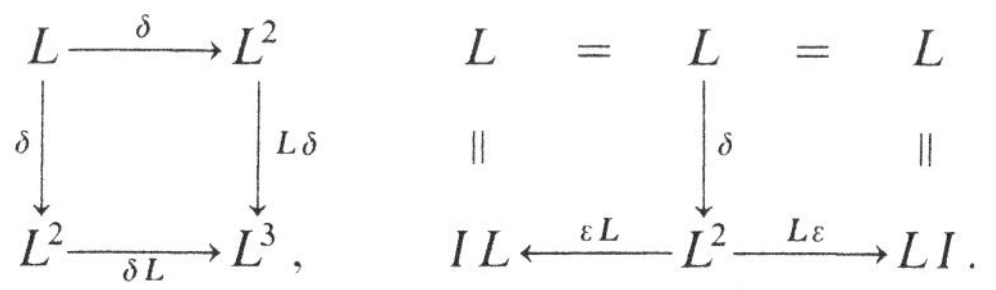

$$\begin{array}{ccc} L & \xrightarrow{\delta} & L^2 \\ {\scriptstyle \delta}\downarrow & & \downarrow{\scriptstyle L\delta} \\ L^2 & \xrightarrow[\delta L]{} & L^3\,, \end{array} \qquad \begin{array}{ccccc} L & = & L & = & L \\ \| & & \downarrow{\scriptstyle \delta} & & \| \\ IL & \xleftarrow{\varepsilon L} & L^2 & \xrightarrow{L\varepsilon} & LI\,. \end{array}$$

Each adjunction $\langle F, G, \eta, \varepsilon\rangle : X \rightharpoonup A$ defines a comonad $\langle FG, \varepsilon, F\eta G\rangle$ in A.

What is a monad in a preorder P? A functor $T: P \to P$ is just a function $T: P \to P$ which is monotonic ($x \leqq y$ in P implies $Tx \leqq Ty$); there are natural transformations η and μ as in (1) precisely when

$$x \leqq Tx\,, \qquad T(Tx) \leqq Tx \tag{3}$$

for all $x \in P$; the diagrams (2) then necessarily commute because in a preorder there is at most one arrow from here to yonder. The first equation of (3) gives $Tx \leqq T(Tx)$. Now suppose that the preorder P is a partial order ($x \leqq y \leqq x$ implies $x = y$). Then the Eqs. (3) imply that $T(Tx) = Tx$. Hence a monad T in a partial order P is just a *closure operation* t in P; that is, a monotonic function $t : P \to P$ with $x \leqq tx$ and $t(tx) = tx$ for all $x \in P$.

We leave the reader to describe a morphism $\langle T, \mu, \eta\rangle \to \langle T', \mu', \eta'\rangle$ of monads (a suitable natural transformation $T \dot{\to} T'$) and the category of all monads in a given category X.

2. Algebras for a Monad

The natural question, "Can every monad be defined by a suitable pair of adjoint functors?" has a positive answer, in fact there are two positive answers provided by two suitable pairs of adjoint functors. The first answer (due to Eilenberg-Moore [1965]) constructs from a monad $\langle T, \eta, \mu\rangle$ in X a category of X^T of "T-algebras" and an adjunction $X \rightharpoonup X^T$ which defines $\langle T, \eta, \mu\rangle$ in X. Formally, the definition of a T-algebra is that of a set on which the "monoid" T acts (cf. the introduction).

Definition. *If $T=\langle T,\eta,\mu\rangle$ is a monad in X, a T-algebra $\langle x,h\rangle$ is a pair consisting of an object $x\in X$ (the underlying object of the algebra) and an arrow $h: Tx\to x$ of X (called the structure map of the algebra) which makes both the diagrams*

$$\begin{array}{ccc} T^2x & \xrightarrow{Th} & Tx \\ {\scriptstyle \mu_x}\downarrow & & \downarrow{\scriptstyle h} \\ Tx & \xrightarrow{h} & x \end{array} \qquad \begin{array}{ccc} x & \xrightarrow{\eta_x} & Tx \\ & {\scriptstyle 1}\searrow & \downarrow{\scriptstyle h} \\ & & x \end{array} \tag{1}$$

commute. (The first diagram is the associative law, the second the unit law.) A morphism $f:\langle x,h\rangle\to\langle x',h'\rangle$ of T-algebras is an arrow $f: x\to x'$ of X which renders commutative the diagram

$$\begin{array}{ccc} x & \xleftarrow{h} & Tx \\ {\scriptstyle f}\downarrow & & \downarrow{\scriptstyle Tf} \\ x' & \xleftarrow{h'} & Tx' \end{array} . \tag{2}$$

Theorem 1 *(Every monad is defined by its T-algebras). If $\langle T,\eta,\mu\rangle$ is a monad in X, then the set of all T-algebras and their morphisms form a category X^T. There is an adjunction*

$$\langle F^T, G^T;\ \eta^T, \varepsilon^T\rangle : X \rightharpoonup X^T$$

in which the functors G^T and F^T are given by the respective assignments

$$G^T:\ \begin{array}{ccc} \langle x,h\rangle & \longmapsto & x \\ \downarrow{\scriptstyle f} & & \downarrow{\scriptstyle f} \\ \langle x',h'\rangle & \longmapsto & x', \end{array} \qquad F^T:\ \begin{array}{ccc} x & \longmapsto & \langle Tx,\mu_x\rangle \\ \downarrow{\scriptstyle f} & & \downarrow{\scriptstyle Tf} \\ x' & \longmapsto & \langle Tx',\mu_{x'}\rangle, \end{array} \tag{3}$$

while $\eta^T=\eta$ and $\varepsilon^T\langle x,h\rangle=h$ for each T-algebra $\langle x,h\rangle$. The monad defined in X by this adjunction is the given monad $\langle T,\eta,\mu\rangle$.

The proof is straightforward verification. If $f:\langle x,h\rangle\to\langle x',h'\rangle$ and $g:\langle x',h'\rangle\to\langle x'',h''\rangle$ are morphisms of T-algebras, so is their composite gf; with this composition of arrows, the T-algebras evidently form a category X^T, as asserted. The functor $G^T: X^T\to X$ is the evident functor which simply forgets the structure map of each T-algebra. On the other hand, for each $x\in X$ the pair $\langle Tx,\mu_x: T(Tx)\to Tx\rangle$ is a T-algebra (the *free T-algebra* on x), in view of the associative and (left) unit laws for the monad T. Hence $x\mapsto\langle Tx,\mu_x\rangle$ does indeed define a functor $F^T: X\to X^T$, as asserted. Then $G^TF^Tx=G^T\langle Tx,\mu_x\rangle=Tx$, so the unit η of the given monad is a natural transformation $\eta=\eta^T: I_X \dot\to G^TF^T$. On the other hand, $F^TG^T\langle x,h\rangle=\langle Tx,\mu_x\rangle$, while the first square in the definition (1) of a T-algebra $\langle x,h\rangle$ states that the structure map $h: Tx\to x$

is a morphism $\langle Tx, \mu_x \rangle \to \langle x, h \rangle$ of T-algebras. The resulting transformation

$$\varepsilon^T_{\langle x,h \rangle} = h : F^T G^T \langle x, h \rangle \to \langle x, h \rangle$$

is natural, by the definition (above) of a morphism of T-algebras. The triangular identities for an adjunction read

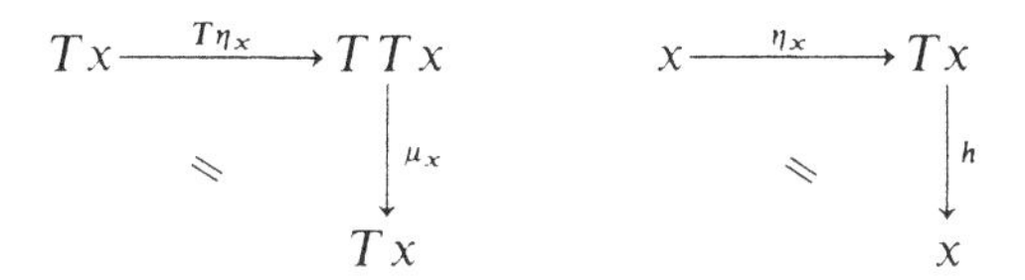

The first holds by the (right) unit law for T, the second by the unit law (see (1)) for a T-algebra. Therefore η^T and ε^T define an adjunction, as stated.

This adjunction thus determines a monad in X. The endofunctor $G^T F^T$ is the original T, its unit η^T is the original unit, and its multiplication $\mu^T = G^T \varepsilon^T F^T$ has $\mu^T x = G^T \varepsilon^T \langle Tx, \mu_x \rangle = G^T \mu_x = \mu_x$, so is the original multiplication of T. The proof is complete.

We now give several examples which show that the T-algebras for familiar monads are the familiar algebras.

Closure. A closure operation T on a preorder P is a monad in P (see § 1); a T-algebra is then an $x \in P$ with $Tx \leqq x$ (the structure map). Since $x \leqq Tx$ for all x, a T-algebra is simply an element $x \in P$ with $x \leqq Tx \leqq x$. If P is a partial order, this means that $x = Tx$, so that a T-algebra is simply an element x of the partial order which is *closed*, in the usual sense.

Group actions. If G is a (small) group, then for every (small) set X the definitions

$$TX = G \times X\,, \qquad X \xrightarrow{\eta_X} G \times X\,, \qquad G \times (G \times X) \longrightarrow G \times X\,,$$
$$x \longmapsto \langle u, x \rangle\,, \qquad \langle g_1, \langle g_2, x \rangle \rangle \longmapsto \langle g_1 g_2, x \rangle$$

for $x \in X$, $g_1, g_2 \in G$ and u the unit element of G, define a monad $\langle T, \eta, \mu \rangle$ on **Set**. A T-algebra is then a set X together with a function $h: G \times X \to X$ (the structure map) such that always

$$h(g_1 g_2, x) = h(g_1, h(g_2, x)), \qquad h(u, x) = x\,.$$

If we write $g \cdot x$ for $h(g, x)$, these are just the usual conditions that $\langle g, x \rangle \mapsto g \cdot x$ defines an action of the group G on the set X. That T-algebras for the monad T are just the group actions is not a surprise, since our definition of T-algebras was constructed on the model of group actions.

Modules. If R is a (small) ring, then for each (small) abelian group A the definitions

$$TA = R \otimes A\,, \qquad A \to R \otimes A\,, \qquad R \otimes (R \otimes A) \to R \otimes A\,,$$
$$a \mapsto 1 \otimes a\,, \qquad r_1 \otimes (r_2 \otimes a) \mapsto r_1 r_2 \otimes a\,,$$

for $a \in A$, $r_1, r_2 \in R$, define a monad on **Ab**. Much as in the previous case, the T-algebras are exactly the left R-modules.

Exercises

1. Complete semi-lattices (E. Manes; thesis). Recall that a complete semi-lattice is a partial order Q in which every subset $S \subset Q$ has a supremum (least upper bound) in Q. Let $\mathcal{P}$ be the *covariant power set functor* on **Set** so that $\mathcal{P}X$ is the set of all subsets $S \subset X$, while for each function $f: X \to Y$, $(\mathcal{P}f)S$ is the direct image of S under f. For each set X, let $\eta_X : X \to \mathcal{P}X$ send each $x \in X$ to the one point set $\{x\}$, while $\mu_X : \mathcal{P}\mathcal{P}X \to \mathcal{P}X$ sends each set of sets into its union.
   (a) Prove that $\langle \mathcal{P}, \eta, \mu \rangle$ is a monad $\mathcal{P}$ on **Set**.
   (b) Prove that each $\mathcal{P}$-algebra $\langle X, h \rangle$ is a complete semi-lattice when $x \leqq y$ is defined by $h\{x, y\} = y$, and $\sup S = hS$ for each $S \subset X$.
   (c) Prove conversely that every (small) complete semi-lattice is a $\mathcal{P}$-algebra in this way.
   (d) Conclude that the category of $\mathcal{P}$-algebras is the category of all (small) complete semi-lattices, with morphisms the order and sup-preserving functions.
2. Show that $G^T : X^T \to X$ creates limits.
3. (a) For monads $\langle T, \eta, \mu \rangle$ and $\langle T', \eta', \mu' \rangle$ on X, define a morphism θ of monads as a suitable natural transformation $\theta : T \dot{\to} T'$, and construct the category of all monads in X.
   (b) From θ construct a functor $\theta^* : X^{T'} \to X^T$ such that $G^T \circ \theta^* = G^{T'}$ and a natural transformation $F^T \dot{\to} \theta^* \circ F^{T'}$.

3. The Comparison with Algebras

Suppose we start with an adjunction $X \rightharpoonup A$, construct the monad T defined in X by the adjunction and then the category X^T of T-algebras; we then ask: How is this related to the original category A? A full answer will relate not only the categories, but the adjunctions, and is provided by the following comparison theorem.

Theorem 1 *(Comparison of adjunctions with algebras). Let*

$$\langle F, G, \eta, \varepsilon \rangle : X \rightharpoonup A$$

be an adjunction, $T = \langle GF, \eta, G\varepsilon F \rangle$ the monad it defines in X. Then there is a unique functor $K : A \to X^T$ with $G^T K = G$ and $KF = F^T$.

Proof. The conclusion asserts that we can fill in the arrow K in the following diagram so that both the F-square and the G-square commute

$$\begin{array}{ccc} A & \overset{K}{\dashrightarrow} & X^T \\ F\uparrow\downarrow G & & F^T\uparrow\downarrow G^T \\ X & = & X. \end{array} \qquad (1)$$

Now the counit ε of the given adjunction defines for each $a \in A$ an arrow $G\varepsilon_a : GFGa \to Ga$. This arrow may be considered as a structure map h for a T-algebra structure on the object $Ga = x$, for the requisite diagrams (cases of (2.1)) are

$$\begin{array}{ccc} GFGFGa & \xrightarrow{GFG\varepsilon_a} & GFGa \\ {\scriptstyle \mu_{Ga} = G\varepsilon FGa}\downarrow & & \downarrow{\scriptstyle G\varepsilon_a} \\ GFGa & \xrightarrow[G\varepsilon_a]{} & Ga, \end{array} \qquad \begin{array}{ccc} Ga & \xrightarrow{\eta_{Ga}} & GFGa \\ & {\scriptstyle 1}\searrow & \downarrow{\scriptstyle G\varepsilon_a} \\ & & Ga. \end{array}$$

They commute (the first is the definition of $G\varepsilon\varepsilon$, the second is one of the triangular identities for the given adjunction). Therefore for any $f : a \to a'$ in A we define K by

$$Ka = \langle Ga, G\varepsilon_a \rangle, \qquad Kf = Gf : \langle Ga, G\varepsilon_a \rangle \to \langle Ga', G\varepsilon_{a'} \rangle; \qquad (2)$$

since ε is natural, the proposed arrow Kf commutes with $G\varepsilon$ and so is a morphism of T-algebras. It is routine to verify that K is a functor with

$$KF = F^T, \qquad G^T K = G. \qquad (3)$$

It remains to show K unique. First, each Ka must be a T-algebra, and the commutativity requirement $G^T K = G$ means that the underlying X-object of this T-algebra Ka is Ga. Therefore Ka must have the form $Ka = \langle Ga, h \rangle$ for some structure map h; moreover $G^T K = G$ means that the value of K on an arrow f in A must be $Kf = Gf$, exactly as in (2) above. It remains only to determine the structure map h. Now (1) commutes, and the two adjunctions $\langle F, G, \ldots \rangle$ and $\langle F^T, G^T, \ldots \rangle$ have the *same* unit η, so the two functors $K : A \to X^T$ and the identity $I : X \to X$ define a map of the first adjunction to the second, in the sense considered in § IV.7. Proposition IV.7.1 for this map then states that $K\varepsilon = \varepsilon^T K$. But K on arrows is G, so $K\varepsilon_a = G\varepsilon_a$ for each $a \in A$, while the definition of the counit ε^T of an algebra gives $\varepsilon^T Ka = \varepsilon^T \langle Ga, h \rangle = h$. Thus $K\varepsilon = \varepsilon^T K$ implies $G\varepsilon_a = h$, so the structure map h is determined and K is unique.

For many familiar adjunctions $\langle F, G, \ldots \rangle$ this comparison functor K will be an isomorphism; we then say that G is *monadic* (tripleable). For other authors (Barr-Wells [1985]), "triplable" means only that K be an equivalence of categories. However, here is an easy example when K is not an isomorphism, and not even an equivalence. The forgetful functor $G : \mathbf{Top} \to \mathbf{Set}$ has a left adjoint D which assigns to each set X the discrete topological space (all subsets open in X), for the identity arrow $\eta_X : X \to GDX$ is trivially universal from the object X to the functor G. This adjunction $\langle D, G, \eta, \dots \rangle : \mathbf{Set} \rightharpoonup \mathbf{Top}$ defines on **Set** the monad $I = \langle I, 1, 1 \rangle$ which is the identity (identity functor, identity as unit and as multiplication). The I-algebras in **Set** are just the sets, so the comparison functor $\mathbf{Top} \to \mathbf{Top}^I = \mathbf{Set}$ is in this case the given forgetful functor G.

4. Words and Free Semigroups

The comparison functor can be illustrated explicitly in the case of semigroups. A semigroup is a set S equipped with an associative binary operation $\nu : S \times S \to S$. The free semigroup WX on a set X is like the free monoid on X (§ II.7). It consists of all words $\langle x_1 \rangle \dots \langle x_n \rangle$ of positive length n spelled in letters $x_i \in X$, where we write $\langle x \rangle$ to distinguish the word $\langle x \rangle$ in WX from the element $x \in X$. Words are multiplied by juxtaposition,

$$(\langle x_1 \rangle \dots \langle x_n \rangle)(\langle y_1 \rangle \dots \langle y_m \rangle) = \langle x_1 \rangle \dots \langle x_n \rangle \langle y_1 \rangle \dots \langle y_m \rangle ;$$

this multiplication ν is associative, so makes $FX = \langle WX, \nu \rangle$ a semigroup, with the set WX the disjoint union $\amalg X^n$, $n = 1, 2, \dots$. If $G : \mathbf{Smgrp} \to \mathbf{Set}$ is the forgetful functor from the category of all small semigroups (forget the multiplication), then the arrow $\eta_X : X \to GFX$ defined by $x \mapsto \langle x \rangle$ (send each x to the one-letter word in x) is universal from X to G. Therefore F is a functor, left adjoint to G, and η defines an adjunction

$$\langle F, G, \eta, \varepsilon \rangle : \mathbf{Set} \rightharpoonup \mathbf{Smgrp} .$$

If S is any semigroup (set S with an associative binary operation $S \times S \to S$, written as multiplication) the counit ε_S of this adjunction is by definition that morphism $\varepsilon_S : FGS \to S$ of semigroups for which the composite $G\varepsilon_S \cdot \eta_{GS} : GS \to GFGS \to GS$ is the identity; in other words, ε_S is the unique morphism of semi-groups which sends each generator $\langle s \rangle$ to s. This means that

$$\varepsilon_S(\langle s_1 \rangle \dots \langle s_n \rangle) = s_1 \dots s_n \quad \text{(product in } S\text{)} \tag{1}$$

for all $s_i \in S$: The counit ε removes the "pointy bracket" $\langle\ \rangle$.

Proposition 1. *The monad on* **Set** *determined by the adjunction* **Set**$\rightharpoonup$**Smgrp** *is*

$$W = \langle W : \mathbf{Set} \to \mathbf{Set},\ \eta : I \dot{\to} W,\ \mu : W^2 \dot{\to} W \rangle$$

where $WX = \coprod_{n=1}^{\infty} X^n$, $\eta_X x = \langle x \rangle$ *for each* $x \in X$, *while* μ_X *is*

$$\mu_X(\langle\langle x_{11}\rangle \dots \langle x_{1n_1}\rangle\rangle \dots \langle\langle x_{k1}\rangle \dots \langle x_{kn_k}\rangle\rangle)$$
$$= \langle x_{11}\rangle \dots \langle x_{1n_1}\rangle \dots \langle x_{k1}\rangle \dots \langle x_{kn_k}\rangle$$

for all positive integers k, *all* k*-tuples* $n_1, \dots, n_k$ *of positive integers, and all* $x_{ij} \in X$.

Proof. By definition, $\eta x = \langle x \rangle$, while $\mu = G\varepsilon F : W^2 \dot{\to} W$ is determined by the formula above for ε_S, where we have written each element of $W^2 X$ as a word (of length k) in k words of the respective lengths $n_1, \dots, n_k$. More briefly, μ_X applied to a word of words removes the *outer* pointy brackets.

Note that this description allows direct verification of the unit and associative laws for the monad W, without overt reference to the notion of a semi-group. For example, the associative law for μ amounts to an observation on *three* layers of pointy brackets, that removing first the middle brackets and then the outer brackets gives the same result as removing first the outer brackets and then the (newly) outer brackets.

Proposition 2. *For the above word-monad* W *in* **Set**, *the* W*-algebras have the form* $\langle S, \nu_1, \nu_2, \dots \rangle$: *A set* S *equipped with one* n*-ary operation* $\nu_n : S^n \to S$ *for each positive integer* n, *such that* $\nu_1 = 1$ *while for every positive* k *and every* k*-tuple of positive integers* $n_1, \dots, n_k$ *one has the identity*

$$\nu_k(\nu_{n_1} \times \cdots \times \nu_{n_k}) = \nu_{n_1 + \cdots + n_k} : S^{n_1 + \cdots + n_k} \to S . \tag{2}$$

A morphism $f : \langle S, \nu_1, \dots \rangle \to \langle S', \nu'_1, \dots \rangle$ *of* W*-algebras is a function* $f : S \to S'$ *which commutes with each* ν_n, *so that* $f\nu_n = \nu'_n f^n : S^n \to S'$.

Proof. Consider a W-algebra $\langle S, h : WS \to S \rangle$. Since $WS = \coprod S^n$, the structure map h is a list of n*-ary* operations $\nu_n : S^n \to S$, one for each n. The unit law for the algebra requires that $h\eta_X = 1$, hence that ν_1 be the identity. On the other hand, since the product of sets is distributive over the coproducts of sets,

$$W(WX) = \coprod_k \Big(\coprod_n X^n\Big)^k \cong \coprod_k \coprod_n (X^{n_1} \times \cdots \times X^{n_k}) \cong \coprod_k \coprod_n X^{n_1 + \cdots + n_k},$$

where n at the middle and the right runs over all k-tuples $\langle n_1, \dots, n_k \rangle$. With this notation, the associative law for the structure map h takes the stated form (2).

The simplest case of this identity (2), for $3 = 2 + 1 = 1 + 2$ and ν_1 the identity, is

$$\nu_2(\nu_2 \times 1) = \nu_3 = \nu_2(1 \times \nu_2) : S \times S \times S \to S .$$

If we write the binary operation ν_2 as multiplication, this states that the ternary operation ν_3 satisfies, for all elements $x, y, z \in S$,

$$(xy)z = \nu_3(x, y, z) = x(yz) .$$

Similarly, v_n must be the n-fold product. An easy induction proves

Corollary. *The system $\langle S, v_1, v_2, \ldots\rangle$ is a W-algebra, as above, if and only if $v_1 = 1$, $v_2 : S \times S \to S$ is an associative binary operation on S, and for all $n \geqq 2$, $v_{n+1} = v_n(v_2 \times 1) : S^{n+1} \to S$.*

Thus, if we start with semigroups, regarded as sets $\langle S, v\rangle$ with one associative binary operation, define the resulting monad W on **Set**, and then construct the category of W-algebras, we get the same semigroups, now regarded as algebraic system $\langle S, v_1, v_2, \ldots\rangle$, where $v_1 = 1$, $v_2 = v$, and v_{n+1} is v_2 iterated. The comparison functor $K : \mathbf{Smgrp} \to \mathbf{Set}^W$ is the evident map $\langle S, v\rangle \mapsto \langle S, 1, v_2, \ldots, v_n, \ldots\rangle$ where v_n is the iterate of the binary v. In other words, K is an isomorphism, but it replaces the algebraic system $\langle S, v\rangle$ with one associative binary operation by the same set with all the iterated operations derived from this binary operation.

A similar description applies to algebras over other familiar monads (Exercises 1, 2).

Exercises

1. Let W_0 be the monad in **Set** defined by the forgetful functor $\mathbf{Mon} \to \mathbf{Set}$. Show that a W_0-algebra is a set M with a string $v_0, v_1, \ldots$ of n-ary operations v_n, where $v_0 : * \to M$ is the unit of the monoid M and v_n is the n-fold product.
2. For any ring R with identity, the forgetful functor $G : R\text{-}\mathbf{Mod} \to \mathbf{Set}$ from the category of left R-modules has a left adjoint and so defines a monad $\langle T_R, \eta, \mu\rangle$ in **Set**.
   (a) Prove that this monad may be described as follows: For each set X, $T_R X$ is the set of all those functions $f : X \to R$ with only a finite number of non-zero values; for each function $t : X \to Y$ and each $y \in Y$, $[(T_R t)f]_y = \Sigma' f_x$, with sum taken over all $x \in X$ with $tx = y$; for each $x \in X$, $\eta_x x : X \to R$ is defined by $(\eta_x x)x = 1$, $(\eta_x x)x' = 0$; for each $k \in T_R(T_R X)$, $\mu_x k : X \to R$ is defined for $x \in X$ by $(\mu_x k)_x = \Sigma_f k_f f_x$, the sum taken over all $f \in T_R X$.
   (b) From this description, verify directly that $\langle T_R, \eta, \mu\rangle$ is a monad.
   (c) Show that the $\langle T_R, \eta, \mu\rangle$-algebras are the usual R-modules, described not via addition and scalar multiple, but via all operations of linear combination (The structure map h assigns to each f the "linear combination with coefficients f_x for each $x \in X$".)
3. Give a similar complete description of the adjunction defined by the forgetful functor $\mathbf{CRng} \to \mathbf{Set}$, noting that TX is the ring of all polynomials with integral coefficients in letters (i.e., indeterminates) $x \in X$.
4. The adjunction $\langle F, G, \varphi\rangle : \mathbf{Ab} \to \mathbf{Rng}$ with G the functor "forget the multiplication in a ring" defines a monad T in **Ab**.
   (a) Give a direct description of this monad, like that in the text for W, with X^n replaced by the n-fold tensor power and coproduct $\amalg$ by the (infinite) direct sum of abelian groups.
   (b) Give the corresponding description of T-algebras and show that the comparison functor from rings to T-algebras is an isomorphism.

5. Free Algebras for a Monad

Given an adjunction

$$\langle F, G, \varphi \rangle : X \rightharpoonup A \, ,$$

any full subcategory $B \subset A$ which contains all the objects Fx for $x \in X$ leads to another adjunction

$$\langle F_B, G_B, \varphi_B \rangle : X \rightharpoonup B$$

where the functor F_B is just F with its codomain restricted from A to B, G_B is G with domain restricted to B, while for $x \in X$ and $b \in B$ the given adjunction leads to a bijection φ_B

$$\hom_B(F_B x, b) = \hom_A(Fx, b) \cong \hom_X(x, Gb) = \hom_X(x, G_B b) \, ,$$

which is manifestly natural in x and b. Moreover, this second adjunction φ_B defines in X the same monad as did the first. This observation shows that one and the same monad in X can usually be defined by many different adjunctions. The "smallest" such adjunction will be the one where B is FX, the full subcategory of A with objects all the "free" objects $Fx \in A$. The familiar properties of arrows $Fx \to Fy$ between such free objects do suggest a way of constructing this subcategory FX and the adjunction φ_B directly from the monad. Here is the suggested construction, which really gives this category directly and not as a subcategory (cf. Exercise 3).

Theorem 1 *(The Kleisli category of a monad,* [1965].) *Given a monad $\langle T, \eta, \mu \rangle$ in a category X, consider to each object $x \in X$ a new object x_T and to each arrow $f : x \to Ty$ in X a new arrow $f^\flat : x_T \to y_T$. These new objects and arrows constitute a category when the composite of $f^\flat$ with $g^\flat : y_T \to z_T$ is defined by*

$$g^\flat \circ f^\flat = (\mu_z \circ Tg \circ f)^\flat \, . \tag{1}$$

Moreover, functors $F_T : X \to X_T$ and $G_T : X_T \to X$ are defined by

$$F_T : \; k : x \to y \mapsto (\eta_y \circ k)^\flat : x_T \to y_T \, , \tag{2}$$

$$G_T : \; f^\flat : x_T \to y_T \mapsto \mu_y \circ Tf : Tx \to T^2 y \to Ty \tag{3}$$

respectively, so that $G_T x_T = Tx$ on objects. Then the bijection $f^\flat \mapsto f$ gives an adjunction $\langle F_T, G_T, \varphi_T \rangle : X \rightharpoonup X_T$ which defines in X precisely the given monad $\langle T, \eta, \mu \rangle$.

Sketch of proof. The definition of the arrows $f^\flat$ amounts to a bijection $X_T(x_T, y_T) \cong X(x, Ty)$ on hom-sets, while the definition of the composite in X_T refers to the composite

$$x \xrightarrow{f} Ty \xrightarrow{Tg} T^2 z \xrightarrow{\mu_z} Tz$$

in X. A suitable large diagram shows the new composition associative: Other diagrams prove that $(\eta_x)^{\flat}: x_T \to x_T$ is a left and right unit for this composition. Another calculation shows that F_T and G_T as described are indeed functors. By construction, $f^{\flat} \mapsto f$ is a bijection

$$X_T(F_T x, y_T) = X_T(x_T, y_T) \cong X(x, Ty) = X(x, G_T y_T);$$

it is natural in x and y_T, so yields the desired adjunction φ_T. Its unit is η, and its counit ε_T is given by $(\varepsilon_T)y_T = (1_{Ty})^{\flat}: (Ty)_T \to y_T$. The resulting multiplication in X is $G_T \varepsilon_T F_T$, which by the definition of G_T is exactly the given multiplication μ. Therefore the adjunction does define the original monad T.

Theorem 2 *(The comparison theorem for the Kleisli construction). Let $\langle F, G, \eta, \varepsilon\rangle : X \rightharpoonup A$ be an adjunction and $T = \langle GF, \eta, G\varepsilon F\rangle$ the monad it defines in X. Then there is a unique functor $L: X_T \to A$ with $GL = G_T$ and $LF_T = F$.*

We leave the proof to the reader, noting that the uniqueness of L requires another (and somewhat different) application of Proposition IV.7.1 on maps of adjunctions.

The two comparison theorems may be summarized as follows:

Theorem 3. *Given a monad $\langle T, \eta, \mu\rangle$ in X, consider the category with objects all those adjunctions $\langle F, G, \eta, \varepsilon\rangle : X \rightharpoonup A$ which define $\langle T, \eta, \mu\rangle$ in X, and with arrows those maps of adjunctions (§ IV.7) which are the identity on X. This category has an initial object – the Kleisli construction – and a terminal object $\langle F^T, G^T, \eta, \varepsilon^T\rangle : X \rightharpoonup X^T$ with the comparison functor:*

$$X_T \overset{L}{\dashrightarrow} A \overset{K}{\dashrightarrow} X^T.$$

Exercises

1. Construct the Kleisli comparison functor L, prove its uniqueness, and show that the image of X_T under L is the full subcategory FX of A with objects all Fx, $x \in X$.
2. Show that the restriction of L gives an equivalence of categories $X_T \to FX$.
3. Construct an example of an adjunction where F is not a bijection on objects. Deduce that the equivalence $X_T \to FX$ in Exercise 2 need not be an isomorphism. (Suggestion: $S \mapsto T(S) =$ the one-point-set defines a monad in **Set**.)
4. In the summary comparison Theorem 3, does the category of all adjunctions really exist?
5. If $\langle F, G, \eta, \varepsilon\rangle : X \rightharpoonup B$ defines the monad $\langle T, \eta, \mu\rangle$ in X, while a second adjunction $\langle L, R, \eta', \varepsilon'\rangle : B \rightharpoonup A$ defines the identity monad in B (i.e., $RL = I_B$, $\eta' = 1$, and $R\varepsilon' L = 1$), prove that the composite adjunction $X \rightharpoonup A$ defines in X the same monad $\langle T, \eta, \mu\rangle$.

6. Split Coequalizers

We need certain special types of coequalizers. By a *fork* in a category C we mean a diagram

$$a \overset{\partial_0}{\underset{\partial_1}{\rightrightarrows}} b \xrightarrow{e} c \tag{1}$$

in C with $e\partial_0 = e\partial_1$. A fork is thus just a cone from the diagram $a \rightrightarrows b$ to the vertex c. Recall that an arrow e is a *coequalizer* of the parallel pair of arrows ∂_0 and ∂_1 if it is a fork and if any $f: b \to d$ with $f\partial_0 = f\partial_1$ has the form $f = f'e$ for a unique $f': c \to d$. An arrow e is called an *absolute coequalizer* of ∂_0 and ∂_1 in C if for any functor $T: C \to X$ (to any category X whatever) the resulting fork

$$Ta \overset{T\partial_0}{\underset{T\partial_1}{\rightrightarrows}} Tb \xrightarrow{Te} Tc$$

still has Te a coequalizer (of $T\partial_0$ and $T\partial_1$). In particular, an absolute coequalizer is automatically a coequalizer. In the same way one can define absolute colimits (or absolute limits) of any other type (Paré [1971a]).

A *split fork* in C is a fork (1) with two more arrows

$$a \xleftarrow{t} b \xleftarrow{s} c \tag{2}$$

which satisfy with the arrows (1) the conditions

$$e\partial_0 = e\partial_1, \qquad es = 1, \qquad \partial_0 t = 1, \qquad \partial_1 t = se. \tag{3}$$

We say that s and t *split* the fork (1). These conditions imply that e is a split epi, with right inverse s. A split fork can also be represented as a pair of commutative squares

$$\begin{array}{ccccc} b & \xrightarrow{t} & a & \xrightarrow{\partial_0} & b \\ \downarrow{\scriptstyle e} & & \downarrow{\scriptstyle \partial_1} & & \downarrow{\scriptstyle e} \\ c & \xrightarrow[s]{} & b & \xrightarrow[e]{} & c \end{array}$$

such that both horizontal composites are the identity. Put differently: The arrows ∂_1 and e are objects in the functor category $C^{\mathbf{2}}$ and $\langle \partial_0, e \rangle : \partial_1 \to e$ is an arrow between them which has $\langle t, s \rangle : e \to \partial_1$ as its right inverse: $\langle \partial_0, e \rangle \langle t, s \rangle = \langle 1, 1 \rangle$.

Lemma. *In every split fork, e is the coequalizer of ∂_0 and ∂_1.*

Proof. For any arrow $f: b \to d$ with $f\partial_0 = f\partial_1$, take $f' = fs: c \to d$. Then, using the Eqs. (3) defining a split fork,

$$f'e = fse = f\partial_1 t = f\partial_0 t = f,$$

so f factors through e. On the other hand, $f = ke$ for some $k: c \to d$ implies $fs = kes = k$, so k is necessarily $f' = fs$, and f' is unique.

By a *split coequalizer* of ∂_0 and ∂_1 we shall mean the arrow e of such a split fork on ∂_0 and ∂_1. It is possible to characterize those parallel pairs ∂_0, ∂_1 for which any (and hence every) coequalizer is split (Exercise 2).

Since a split fork is defined by equations involving only composites and identities, it remains a split fork under the application of any functor. Hence,

Corollary. *In every split fork, e is an absolute coequalizer of ∂_0 and ∂_1.*

Here is an example of a fork in **Cat**, for C any category:

$$C^2 \underset{\partial_1}{\overset{\partial_0}{\rightrightarrows}} C \xrightarrow{e} \mathbf{1}.$$

C^2 is the category whose objects are the arrows of C; ∂_0 and ∂_1 are the functors assigning to each arrow its domain and its codomain, respectively, while e is the functor which sends every object of C to the unique object of $\mathbf{1}$. If C has a terminal object a_0, this fork is a split by the functor s which sends the unique object of $\mathbf{1}$ to a_0, and the functor t which sends each $c \in C$ to the unique arrow $c \to a_0$.

Here is an example of a fork in **Grp**. Let $N \lhd G$ be any normal subgroup of G and form the semidirect product $G \times {}_0N$, which has elements the pairs $\langle x, n \rangle$ for $x \in G$, $n \in N$ with the (evidently associative) multiplication $\langle x, n \rangle \langle y, m \rangle = \langle xy, (y^{-1}ny)m \rangle$. Then

$$G \times {}_0N \underset{\partial_1}{\overset{\partial_0}{\rightrightarrows}} G \xrightarrow{p} G/N$$

is a fork, where p is the usual projection to the quotient group G/N, while $\partial_0 \langle x, n \rangle = x$, $\partial_1 \langle x, n \rangle = xn$. Moreover, in this fork p is clearly the coequalizer of ∂_0 and ∂_1. This fork is not in general split, but if we apply the standard forgetful functor $U : \mathbf{Grp} \to \mathbf{Set}$, the resulting fork in **Set** is split. Take s to be a function sending each coset (element of G/N) to a representative element in G, while $tx = \langle x, x^{-1}(spx) \rangle$. This example, incidentally, gives one way in which any quotient group can be regarded as a coequalizer in the category of groups.

Exercises

1. In **Rng** give a similar construction to show that every quotient R/A of a ring R by an ideal A can be represented as a coequalizer, and show that the resulting fork is split after the application of the forgetful functors to sets.
2. A parallel pair $\partial_0, \partial_1 : a \rightrightarrows b$ is said to be *contractible* (Beck) if there is an arrow $t : b \to a$ with $\partial_0 t = 1$ and $\partial_1 t \partial_0 = \partial_1 t \partial_1$.
   (a) In any split fork (1), prove ∂_0, ∂_1 contractible;
   (b) If a contractible pair has a coequalizer, prove that this coequalizer is split.

7. Beck's Theorem

A basic construction in familiar categories of algebras is the formation of coequalizers – in **Grp**, via factor groups, in R-**Mod** via quotient modules, and the like. Beck's theorem will characterize the category of T-algebras for any monad T as a category with an adjunction in which the "forgetful" functor creates *suitable* coequalizers. We recall (§ V.1) that a functor $G: A \to X$ *creates coequalizers* for a parallel pair $f, g: a \rightrightarrows b$ in A when to each coequalizer $u: Gb \to z$ of Gf, Gg in X there is a unique object c and a unique arrow $e: b \to c$ with $Gc = z$ and $Ge = u$ and when moreover this unique arrow is a coequalizer of f and g.

Theorem 1 *(Beck's theorem characterizing algebras). Let*

$$\langle F, G, \eta, \varepsilon \rangle : X \rightharpoonup A \tag{1}$$

be an adjunction, $\langle T, \eta, \mu \rangle$ the monad which it defines in X, X^T the category of T-algebras for this monad, and

$$\langle F^T, G^T, \eta^T, \varepsilon^T \rangle : X \rightharpoonup X^T \tag{2}$$

the corresponding adjunction. Then the following conditions are equivalent:

(i) *The (unique) comparison functor $K: A \to X^T$ is an isomorphism;*

(ii) *The functor $G: A \to X$ creates coequalizers for those parallel pairs f, g in A for which Gf, Gg has an absolute coequalizer in X;*

(iii) *The functor $G: A \to X$ creates coequalizers for those parallel pairs f, g in A for which Gf, Gg has a split coequalizer in X.*

Proof. We first show that (i) implies (ii). Consider two maps

$$\langle x, h \rangle \underset{d_1}{\overset{d_0}{\rightrightarrows}} \langle y, k \rangle$$

of T-algebras for which the corresponding arrows in X have an absolute coequalizer

$$x \underset{d_1}{\overset{d_0}{\rightrightarrows}} y \xrightarrow{e} z .$$

To create a coequalizer for this parallel pair we must first find a unique T-algebra structure $m: Tz \to z$ on z such that e becomes a map of T-algebras, and then prove that this e is, in fact, a coequalizer of d_0, d_1 in the category X^T of T-algebras. But on the left side of the diagram

$$\begin{array}{ccccc} Tx & \underset{Td_1}{\overset{Td_0}{\rightrightarrows}} & Ty & \xrightarrow{Te} & Tz \\ \downarrow h & & \downarrow k & & \vdots\, m \\ x & \underset{d_1}{\overset{d_0}{\rightrightarrows}} & y & \xrightarrow{e} & z \end{array}$$

both the upper square (with d_0) and the lower square (with d_1) commute, because d_0 and d_1 are maps of algebras; it follows that ek has equal composites with Td_0 and Td_1. But e is an absolute coequalizer, so Te is still a coequalizer: Therefore there is a unique vertical map m, as shown, which makes the right square commute.

We now wish to show that this m is a structure map for z. The associative law for m (outer square below) may be compared with the associative law for the structure map k (inner square below) by the diagram

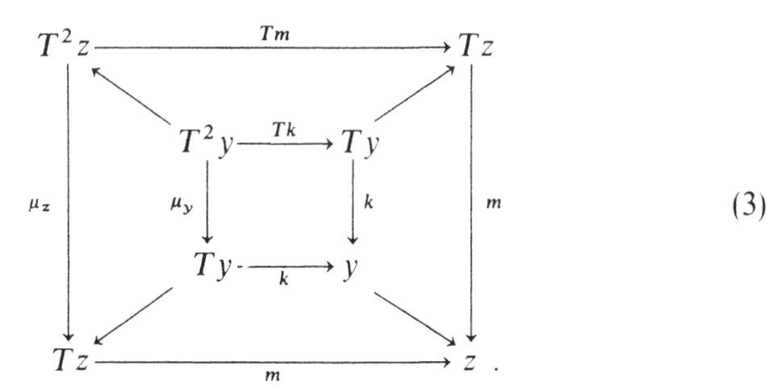

(3)

The left hand trapezoid commutes since μ is natural, and the other three trapezoids commute by the definition of m above in terms of k and e. Therefore

$$m \circ Tm \circ T^2 e = m \circ \mu_z \circ T^2 e .$$

But e is an absolute coequalizer, so $T^2 e$ is a coequalizer and thus is epi; cancelling $T^2 e$ gives the associative law for m. The same style of argument will prove that m satisfies the unit law $m \circ \eta_z = 1 : z \to z$.

We have found the desired unique T-algebra structure m on z, with e a map of T-algebras by the construction of m. To show that e is a coequalizer in X^T, consider any other map $f : \langle y, k \rangle \to \langle w, n \rangle$ of T-algebras with $fd_0 = fd_1$. Then $f : y \to w$ is an arrow in X with $fd_0 = fd_1$, while e is an (absolute) coequalizer of $d_0, d_1 : x \rightrightarrows y$. Therefore there is a map $f' : z \to w$ with $f = f'e$. An argument just like that for the diagram (3) shows that f' is in fact a map of T-algebras. Since it is unique with $f = f'e$, this completes the proof that e is a coequalizer in X^T, and hence that (i) implies (ii).

Next, every split coequalizer is an absolute coequalizer, hence condition (ii) of the theorem requires more creativity of G than does condition (iii). Therefore (ii) implies (iii).

It remains to prove that (iii) implies (i). As a preliminary, consider a T-algebra $\langle x, h \rangle$; the conditions that $h : Tx \to x$ be a structure map of an algebra are exactly the conditions that

$$T^2 x \underset{Th}{\overset{\mu_x}{\rightrightarrows}} Tx \xrightarrow{h} x \tag{4}$$

be a fork in X split by $T^2 x \xleftarrow{\eta_{Tx}} Tx \xleftarrow{\eta_x} x$. Indeed, the fork condition $h \circ \mu_x = h \circ Th$ for (4) is just the associative law for h, the composite

$h \circ \eta_x$ is 1 because of the unit law for $\langle x, h \rangle$, while the equations

$$\mu_x \circ \eta_{Tx} = 1, \qquad Th \circ \eta_{Tx} = \eta_x \circ h$$

hold by the unit law for the monad T and the naturality of η.

For each object $a \in A$, the adjunction $\langle F, G, \varepsilon, \eta \rangle : X \rightharpoonup A$ provides a fork

$$FGFGa \underset{FG\varepsilon_a}{\overset{\varepsilon_{FGa}}{\rightrightarrows}} FGa \xrightarrow{\varepsilon_a} a \tag{5}$$

in A which we call the "canonical presentation" of a. It does correspond to a familiar presentation if $A = \mathbf{Grp}$; then ε_a is just the projection on the group a of the free group generated by all the elements of a. If the functor G is applied to the fork (5) we get a split fork in X; indeed, that special case of the split fork (4) when $\langle x, h \rangle$ is the T-algebra $\langle Ga, G\varepsilon_a \rangle$ used in the comparison theorem.

Now consider any other adjunction $\langle F', G', \eta', \varepsilon' \rangle : X \rightharpoonup A'$ which defines the same monad in X. By a *comparison* (of F' to F) we mean a functor $M : A' \to A$ with $MF' = F$ and $GM = G'$; as already noted, such a comparison is a morphism of adjunctions and hence satisfies $M\varepsilon' = \varepsilon M$.

Lemma. *If G satisfies hypothesis* (iii) *of the theorem on the creation of coequalizers, then there is a unique comparison $M : A' \to A$.*

Since G^T is now known to satisfy this hypotheses, this lemma will incidentally provide a new proof of the comparison theorem (§ 3).

Proof. If M exists, then $FGM = MF'G'$ and $M\varepsilon' = \varepsilon M$, so M must carry the canonical presentation of a' to the canonical presentation of Ma'. In other words, the object Ma' must fit in a fork

$$FGFG'a' = FG'F'G'a' \underset{FG'\varepsilon'_{a'}}{\overset{\varepsilon_{FG'a'}}{\rightrightarrows}} FG'a' \overset{k}{\dashrightarrow} Ma'$$

in A, and moreover k must be $M\varepsilon'_{a'} = \varepsilon_{Ma'}$. Map this fork to the category X by the functor G. The result is the fork

$$GFGFG'a' \underset{TG'\varepsilon'_{a'}}{\overset{G\varepsilon_{FG'a'}}{\rightrightarrows}} GFG'a' \xrightarrow{G'\varepsilon'_{a'}} G'a'$$

in X which is split – since $T = GF$, it is a case of the fork (4) above, for $x = G'a'$. But the hypothesis (iii) ensures that G creates coequalizers in this case. Therefore there is exactly one possible choice for k and Ma' above; (moreover, once Ma' is chosen, $\varepsilon_{Ma'}$ has the property required of k, so must be k.) This shows that the comparison M is unique if it exists.

Now choose k and Ma' in this way. To show M a functor consider any $f : a' \to b'$ in A'. In the diagram

$$\begin{array}{ccccc}
FG'F'G'a' & \rightrightarrows & FG'a' & \xrightarrow{k} & Ma' \\
\downarrow{\scriptstyle FG'F'G'f} & & \downarrow{\scriptstyle FG'f} & & \vdots{\scriptstyle Mf} \\
FG'F'G'b' & \rightrightarrows & FG'b' & \xrightarrow{k_b} & Mb'
\end{array}$$

both left-hand squares commute, so $k_b \circ FG'f$ must factor though the first coequalizer k by a unique arrow $Ma' \to Mb'$ as shown. Taking this arrow to be Mf clearly makes M a functor $A' \to A$, just as required for the lemma.

By this lemma we construct both the original comparison functor $K: A \to X^T$ and a comparison functor $M: X^T \to A$. The composite $MK: A \to A$ is then a comparison (of the adjunction $F\ldots$ to itself), hence must be the identity, again by the lemma. Similarly, $KM: X^T \to X^T$ is a comparison of F^T to F^T, hence must be the identity. Now $MK = 1$ and $KM = 1$ prove K an isomorphism, as required for (i).

The construction of M in this theorem may be further analyzed, using for parallel pairs the following notion of "reflection" of colimits:

Definition. *A functor $G: A \to X$ reflects colimits of $T: J \to A$ when every cone $\lambda: T \dot{\to} a$ from T to $a \in A$ for which $G\lambda: GT \dot{\to} Ga$ is a colimiting cone in X is already a colimiting cone in A.*

In particular, G reflects coequalizers when every fork in A which becomes a coequalizer in X is already a coequalizer in A. Similarly, G *reflects isomorphisms* when, for all arrows t of A, Gt an isomorphism implies t an isomorphism.

Beck's theorem has an acronym PTT for "precise tripleability theorem". There are many other versions: A "weak" version, easier to prove, where there are hypotheses on the coequalizers of more pairs (Exercises 2, 3), an "equivalence" version, which gives conditions that the comparison functor $K: A \to X^T$ be not an isomorphism but an equivalence of categories (Exercises 2, 6), a "constructive" version which analyses the hypotheses (certain hypotheses suffice to give a left adjoint for K; others make this adjunction an equivalence: Exercises 2, 5), a "crude" version (CTT or VTT) with strong hypotheses which apply well to the composite of several "forgetful" functors (Exercises 9–11). However, note that there are more authoritative definitions of VTT and CTT in Barr-Wells [1985].

Exercises

(Throughout, "coequalizers" means "coequalizers of parallel pairs".)

1. If G creates coequalizers, prove that it also reflects coequalizers.
2. Weak Tripleability Theorem (Beck's thesis). Given the adjunction (1) and the corresponding comparison functor K, give a direct proof of the following:
   (a) If A has all coequalizers, then K has a left adjoint L.
   (b) If, in addition, G preserves all coequalizers, then the unit of this adjunction is an isomorphism $I \cong KL$.
   (c) If, in addition, G reflects all coequalizers, then the counit of this adjunction is an isomorphism $LK \cong I$.

3. (Alternative hypothesis for Exercise 2.) If A has all coequalizers, G preserves all coequalizers, and G reflects isomorphisms, prove that G reflects all coequalizers.
4. (a) Show that the canonical presentation of a T-algebra $\langle x, h\rangle$ is

$$\langle T^2x, \mu_{Tx}\rangle \underset{Th}{\overset{\mu_x}{\rightrightarrows}} \langle Tx, \mu_x\rangle \xrightarrow{h} \langle x, h\rangle .$$

   (b) Show that the comparison functor $M : X^T \to A$ in Beck's theorem appears as a coequalizer diagram

$$FGFx \underset{Fh}{\overset{\varepsilon_{Fx}}{\rightrightarrows}} Fx \dashrightarrow M(x, h) .$$

5. Given the data (1), (2), and the comparison functor K, let P be the set of all those parallel pairs $f, g : a \rightrightarrows b$ in A such that Gf, Gg has a split coequalizer. Using Exercise 4 (b), prove
   (a) If A has coequalizers of all pairs in P, K has a left adjoint M.
   (b) If, in addition, G preserves all coequalizers of pairs in P, then the unit $\eta : I \dot{\to} KM$ of this adjunction is an isomorphism.
   (c) If, in addition to (a), G reflects coequalizers for all pairs in P, then the counit $MK \dot{\to} I$ of this adjunction is an isomorphism.
6. Use the results of Exercise 5 and Theorem IV.4.1 to prove the following version of Beck's theorem, characterizing the category of T-algebras up to equivalence: Given the data (1) and (2), the following assertions are equivalent:
   (i) The comparison functor $K : A \to X^T$ is an equivalence of categories.
   (ii) If f, g is any parallel pair in A for which Gf, Gg has an absolute coequalizer, then A has a coequalizer for f, g, and G preserves and reflects coequalizers for these pairs.
   (iii) The same, with "absolute coequalizer" replaced by "split coequalizer".

   The next exercises use certain definitions of properties CTT, VTT, PTT for a functor $G : A \to X$. Let C_G (respectively S_G) be the set of all those parallel pairs $\langle f, g\rangle$ in A such that $\langle Gf, Gg\rangle$ has a coequalizer in X (respectively, a split coequalizer). Then G has CTT when G has a left adjoint, preserves and reflects all coequalizers which exist, and when A has coequalizers of all pairs in C_G. Next, G has VTT when G has a left adjoint, reflects coequalizers of all pairs in S_G, and when A has split coequalizers of all pairs in S_G. Finally, G is PTT when G has a left adjoint, preserves and reflects coequalizers for all pairs in S_G, and when A has coequalizers of all pairs in S_G. Clearly, CTT and VTT imply PTT.
7. CTT (Crude Tripleability Theorem; Barr-Beck). If G is CTT, prove that the comparison functor K is an equivalence of categories.
8. VTT (Vulgar tripleability theorem). If G is VTT, prove that the comparison functor is an equivalence of categories.
9. Given functors $G_1 : A \to X$, $G_2 : X \to Y$, $G_3 : Y \to Z$ with G_1 CTT, G_2 PTT, and G_3 VTT, prove that the composite functor $G_3 G_2 G_1$ is PTT.
10. Prove that the composite of two VTT functors is VTT.
11. Prove that the composite of two CTT functors is CTT.

8. Algebras Are T-Algebras

For semi-groups, monoids, and rings, we already know (§ 4) that the comparison functor is an isomorphism. This result holds more generally for any variety, as defined in § V.6):

Theorem 1. *Let Ω be a set of operators, E a set of identities (on the operators derived from Ω), G the forgetful functor from the category $\langle \Omega, E\rangle$-**Alg** of all small $\langle \Omega, E\rangle$-algebras to* **Set**, *and T the resulting monad in* **Set**. *Then the comparison functor $K: \langle \Omega, E\rangle\text{-}\mathbf{Alg} \to \mathbf{Set}^T$ is an isomorphism.*

The proof will use Beck's theorem. Consider any parallel pair $f, g: A \rightrightarrows B$ of morphisms of $\langle \Omega, E\rangle$-algebras for which the underlying functions have an absolute coequalizer e:

$$GA \underset{Gg}{\overset{Gf}{\rightrightarrows}} GB \xrightarrow{e} X. \tag{1}$$

To "create coequalizers" we must show that the set-map e lifts to a unique morphism $B \to ?$ of algebras, and then that this map is a coequalizer of the algebra maps f, g. So consider any n-ary operator $\omega \in \Omega$ with its given actions ω_A and ω_B on the sets A and B (as usual, we confuse the algebra A with its underlying set $|A|$). In the diagram below (ignore the right hand square)

$$\begin{array}{ccccccc} A^n & \underset{g^n}{\overset{f^n}{\rightrightarrows}} & B^n & \xrightarrow{e^n} & X^n & \overset{h'^n}{\dashrightarrow} & C^n \\ \downarrow{\scriptstyle \omega_A} & & \downarrow{\scriptstyle \omega_B} & & \vdots{\scriptstyle \omega_X} & & \downarrow{\scriptstyle \omega_C} \\ A & \underset{g}{\overset{f}{\rightrightarrows}} & B & \xrightarrow{e} & X & \overset{h'}{\dashrightarrow} & C \end{array} \tag{2}$$

the two left hand squares (with f and g, respectively) commute because f and g are morphisms of Ω-algebras. The function e is an absolute coequalizer in **Set** and therefore its n-th power e^n is still a coequalizer (of f^n and g^n). But

$$e\omega_B f^n = ef\omega_A = eg\omega_A = e\omega_B g^n,$$

so $e\omega_B$ must factor uniquely through this coequalizer as $e\omega_B = \omega_X e^n$. This defines the operation ω_X on X so that the square (2) on e commutes; that is, so that e is a morphism of Ω-algebras. The same diagram applies to all the derived operators λ and defines λ_X uniquely; it follows that any identity $\lambda_B = \mu_B$ valid in B is also valid in X, so X is a $\langle \Omega, E\rangle$-algebra.

It remains to show e a coequalizer for algebras. So consider any morphism $h: B \to C$ of algebras with $hf = hg$. Then $hf = hg$ in **Set** (apply the forgetful functor G), so h factors as $h = h'e$ for a unique function h'. We must show that the right hand square in (2) above commutes for every operator ω. But h is a morphism of algebras, so

$$h'\omega_X e^n = h'e\omega_B = h\omega_B = \omega_C h^n = \omega_C h'^n e^n$$

and e^n a coequalizer means e^n epi, hence gives $h'\omega_X = \omega_C h'^n$, as required.

Exercises

1. Prove Theorem 1, using split coequalizers rather than absolute coequalizers, noting that each ω_X must be defined in terms of a splitting $\langle s, t\rangle$ of the fork (1) as

$$\omega_X(x_1, \ldots, x_n) = e\,\omega_B(sx_1, \ldots, sx_n), \qquad x_i \in X .$$

(For $n=2$, observe that this is like the usual definition of the product of cosets of a normal subgroup.)

2. If K is a commutative ring, show that Beck's theorem applies to the forgetful functor K-**Alg**$\to K$-**Mod**.

9. Compact Hausdorff Spaces

Theorem 1. *The standard forgetful functor*

$$G : \textbf{Cmpt Haus} \to \textbf{Set},$$

which assigns to each (small) compact Hausdorff space its underlying set, is monadic.

Proof. We already know that G has a left adjoint F; indeed, we may take each FX to be the Stone-Čech compactification (V.6.2) of the set X with the discrete topology.

For the remainder of the proof (given in a form due to R. Paré [1971]) it is convenient to regard a topological space as a pair $(X, (^-)_X)$ consisting of a set X and a closure operation $S \mapsto \bar{S}$ defined for all subsets $S, T \subset X$ with the standard properties

$$\bar{\emptyset} = \emptyset, \qquad S \subset \bar{S}, \qquad \bar{\bar{S}} = \bar{S}, \qquad \overline{S \cup T} = \bar{S} \cup \bar{T},$$

with $\emptyset$ the empty subset. A *continuous* map $f : (X, (^-)_X) \to (Y, (^-)_Y)$ is then a function $f : X \to Y$ such that $f\bar{S} \subset \overline{fS}$ for all $S \subset X$. Also a function $f : X \to Y$ is closed if $f\bar{S} \supset \overline{fS}$ for all $S \subset X$. We recall the well-known

Lemma. *If X is a compact space and Y a Hausdorff space, then every continuous $f : (X, (^-)_X) \to (Y, (^-)_Y)$ is closed.*

We must verify that the forgetful functor G,

$$(X, (^-)_X) \mapsto X,$$

creates coequalizers for suitable pairs. So let $f, g : (X, (^-)_X) \rightrightarrows (Y, (^-)_Y)$ be a pair of continuous maps such that there is a set W and an absolute coequalizer e,

$$X \overset{f}{\underset{g}{\rightrightarrows}} Y \xrightarrow{e} W,$$

in **Set**. Let P denote the covariant power set functor **Set**$\to$**Set**; thus for each subset $S \subset Y$, $(Pe)S \subset W$ is the usual direct image of S under e.

Since e is an absolute coequalizer, Pe is still a coequalizer, in the diagram (of sets)

$$\begin{array}{ccccc} PX & \overset{Pf}{\underset{Pg}{\rightrightarrows}} & PY & \xrightarrow{Pe} & PW \\ \downarrow{\scriptstyle (^{-})_X} & & \downarrow{\scriptstyle (^{-})_Y} & & \vdots{\scriptstyle (^{-})_W} \\ PX & \overset{Pf}{\underset{Pg}{\rightrightarrows}} & PY & \xrightarrow{Pe} & PW \end{array} \tag{1}$$

Since f and g are both continuous maps, both squares on the left (the square with f, and that with g) are commutative. It follows that

$$Pe \circ (^{-})_Y \circ Pf = Pe \circ (^{-})_Y \circ Pg .$$

But Pe is a coequalizer, so $Pe \circ (^{-})_Y$ factors through Pe. This gives a unique function $(^{-})_W$ – the dotted arrow in (1) – which makes the right hand square in the diagram commute. This function may thus be described as follows: Given a subset $T \subset W$, choose any subset $S \subset Y$ with $(Pe)S = T$; then $\overline{T} = (Pe)\overline{S}$, independent of the choice of S. In particular, if $e^{-1}T \subset Y$ is the usual inverse image of T, then $\overline{T} = Pe(\overline{e^{-1}T})$. It is now routine to verify that this is a closure operation on W, hence that W is a topological space.

By the commutativity of the diagram, e is then continuous and closed. Since Y is compact and $e: Y \to W$ is surjective, W is also compact. Since Y is Hausdorff, each point in Y is a closed set there; since e is a closed map and is surjective, the points of W are closed. To show W Hausdorff, consider two points $w_1 \neq w_2 \in W$. They are closed in W, so $e^{-1}w_1$ and $e^{-1}w_2$ are disjoint closed sets in Y. By a familiar property of the compact Hausdorff space Y, disjoint closed sets can be separated by disjoint open sets (every compact Hausdorff space is normal), so there are disjoint open sets $U_1, U_2 \subset Y$ with $e^{-1}w_i \in U_i$. Their complements U_1' and U_2' in Y are then closed sets with $U_1' \cup U_2' = Y$. Since e is a closed map, $e(U_1')$ and $e(U_2')$ are closed sets in W with

$$(Pe)(U_1') \cup (Pe)(U_2') = W, \qquad w_i \notin (Pe)(U_i').$$

So take complements again, this time in W: $[(Pe)U_1']'$ and $[(Pe)U_2']'$ are disjoint open neighborhoods of w_1 and w_2, respectively, in W. Therefore W is a Hausdorff space.

We have produced from the absolute coequalizer e in **Set** a unique topology on its codomain W such that e is continuous; moreover, this topology is compact Hausdorff. It remains to show that the continuous map $e: (Y, (^{-})_Y) \to (W, (^{-})_W)$ is a coequalizer in **Cmpt Haus**. So consider any compact Hausdorff $(Z, (^{-})_Z)$ and a continuous map $h: Y \to Z$, such that both composites in

$$X \overset{f}{\underset{g}{\rightrightarrows}} Y \xrightarrow{h} Z$$

are equal. Since e is a coequalizer in **Set**, there is a unique function $h': W \to Z$ with $h = h'e: Y \to Z$; it remains to show h' continuous. Take $T \subset W$ and $S \subset Y$ with $(Pe)S = T$. Then $\bar{T} = (Pe)\bar{S}$, so

$$\begin{aligned}(Ph')\,\bar{T} &= (Ph')\,(Pe)\,\bar{S} = P(h)\,\bar{S} = \overline{P(h)S} \\ &= \overline{(Ph')\,(Pe)S} = \overline{(Ph')\,T}.\end{aligned}$$

Therefore h' is continuous (and closed). The proof is complete.

Exercises

1. Show that the topology on W introduced in the proof above is the "quotient topology" on W defined by $e: Y \to W$ (i.e., that a set is open in W if and only if its inverse image is open in Y).

Notes.

The recognition of the power and simplicity of the use of monads and comonads came quite slowly, and started from their use in homological algebra (see § VII.6). Mac Lane [1956] mentioned in passing (his § 3) that all the standard resolutions could be obtained from universal arrows (i.e., from adjunctions). Then Godement [1958] systematized these resolutions by using standard constructions (comonads). P. J. Huber [1961], starting from "homotopy theory" in the Eckmann-Hilton sense, explored the examples of derived functors which can be defined by comonads and then in [1962] studied the resulting functorial simplicial resolutions for more general abelian categories. Then Hilton (and others) raised the question as to whether any monad arises from an adjunction. Two independent answers appeared: Kleisli's constructions in [1965] of the "free algebra" realization and the decisive construction by Eilenberg-Moore [1965] of the category of algebras for a monad. Stimulated by this description of the algebras, Barr-Beck in [1966] showed how the resolutions derived from monads and comonads can be used even in non-abelian categories – obtaining the surprising result that the free group monad in **Set** does lead to the standard cohomology of groups. Subsequent developments in this direction are sumarized in their paper [1969].

Thus, about 1965, it became urgent to decide how to characterize the category of algebras over a monad. Linton [1966] treated the case for monads in **Set**, and then Beck established his theorem (unpublished, but presented at a conference in 1966). The absolute "coequalizer" form of the theorem, due to Paré [1971], made possible Paré's elegant proof (§ 9) that compact Hausdorff spaces are monadic. Many other developments in this direction are summarized in Manes' thesis (cf. [1969]).

The description of algebras by monads is closely related to another description by algebraic theories (Lawvere [1963], described in Pareigis [1970]).

VII. Monoids

This chapter will explore the general notion of a monoid in a category. As we have already seen in the introduction, an ordinary monoid in **Set** is defined by the usual diagrams relative to the cartesian product $\times$ in **Set**, while a ring is a monoid in **Ab**, relative to the tensor product $\otimes$ there. Thus we shall begin with categories B equipped with a suitable bifunctor such as $\times$ or $\otimes$, more generally denoted by $\Box$. These categories will themselves be called "monoidal" categories because the bifunctor $\Box : B \times B \to B$ is required to be associative. Usually it is associative only "up to" an isomorphism; for example, for the tensor product of vector spaces there is an isomorphism $U \otimes (V \otimes W) \cong (U \otimes V) \otimes W$. Ordinarily we simply "identify" these two iterated product spaces by this isomorphism. Closer analysis shows that more care is requisite in this identification – one must use the *right* isomorphism, and one must verify that the resulting identification of multiple products can be made in a "coherent" way.

Once the coherence question for monoidal categories is settled, we proceed to define monoids in such categories, the actions of monoids on objects of the category, and the construction of free monoids. Next, we introduce the simplicial category Δ, which turns out to be the basic monoidal category because it contains a "universal" monoid and because of its role in simplicial resolutions and simplicial topology. Finally, compactly generated spaces are used to illustrate closed monoidal categories.

1. Monoidal Categories

A category is monoidal when it comes equipped with a "product" like the direct product $\times$, the direct sum $\oplus$, or the tensor product $\otimes$. We write this product as $\Box$ (many authors write $\otimes$) to cover all cases impartially. We consider first categories equipped with a multiplication $\Box$ which is strictly associative and has a strict two-sided identity object e. In detail, *a strict monoidal category* $\langle B, \Box, e \rangle$ is a category B with a bifunctor $\Box : B \times B \to B$ which is associative,

$$\Box(\Box \times 1) = \Box(1 \times \Box) : B \times B \times B \to B\,, \tag{1}$$

and with an object e which is a left and right unit for $\Box$,

$$\Box(e \times 1) = \mathrm{id}_B = \Box(1 \times e)\,. \tag{2}$$

In writing the associative law (1), we have identified $(B \times B) \times B$ with $B \times (B \times B)$; in writing the unit law (2), we mean $e \times 1$ to be the functor $c \mapsto \langle e, c\rangle : B \to B \times B$. The bifunctor $\Box$ assigns to each pair of objects $a, b \in B$ an object $a \Box b$ of B and to each pair of arrows $f : a \to a'$, $g : b \to b'$ an arrow $f \Box g : a \Box b \to a' \Box b'$. Thus $\Box$ a bifunctor means that the interchange law

$$1_a \Box 1_b = 1_{a \Box b}, \qquad (f' \Box g')(f \Box g) = (f'f) \Box (g'g)\,, \tag{3}$$

holds whenever the composites $f'f$ and $g'g$ are defined. The associative law (1) states that the binary operation $\Box$ is associative *both* for objects and for arrows; similarly, the unit law (2) means that $e \Box c = c = c \Box e$ for objects c *and* that $1_e \Box f = f = f \Box 1_e$ for arrows f.

Any monoid M (in the usual sense, in **Set**), regarded as a discrete category, is a strict monoidal one with $\Box$ the multiplication of elements of M. If X is any category, the category $\mathrm{End}(X)$ with objects all endofunctors $S : X \to X$ and arrows all natural transformations $\theta : S \dot{\to} T$ is strict monoidal, with $\Box$ the composition of functors.

A (relaxed) monoidal category is a category B with a bifunctor $\Box$, its *multiplication*, which is associative "up to" a natural isomorphism α, and which has an object e which is a left unit for $\Box$ up to a natural isomorphism λ and a right unit up to ϱ. Moreover, "all" diagrams involving α, λ, and ϱ must commute.

Formally, a *monoidal* category $B = \langle B, \Box, e, \alpha, \lambda, \varrho\rangle$ is a category B, a bifunctor $\Box : B \times B \to B$, an object $e \in B$, and three natural isomorphisms α, λ, ϱ. Explicitly,

$$\alpha = \alpha_{a,b,c} : a \Box (b \Box c) \cong (a \Box b) \Box c \tag{4}$$

is natural for all $a, b, c \in B$, and the pentagonal diagram

$$\begin{array}{ccccc} a \Box (b \Box (c \Box d)) & \xrightarrow{\alpha} & (a \Box b) \Box (c \Box d) & \xrightarrow{\alpha} & ((a \Box b) \Box c) \Box d \\ \downarrow{\scriptstyle 1 \Box \alpha} & & & & \uparrow{\scriptstyle \alpha \Box 1} \\ a \Box ((b \Box c) \Box d) & & \xrightarrow{\qquad\alpha\qquad} & & (a \Box (b \Box c)) \Box d \end{array} \tag{5}$$

commutes for all $a, b, c, d \in B$. Again, λ and ϱ are natural

$$\lambda_a : e \Box a \cong a, \qquad \varrho_a : a \Box e \cong a \tag{6}$$

for all objects $a \in B$, the triangular diagram

$$\begin{array}{ccc} a\square(e\square c) & \xrightarrow{\alpha} & (a\square e)\square c \\ \downarrow{\scriptstyle 1\square\lambda} & & \downarrow{\scriptstyle \varrho\square 1} \\ a\square c & = & a\square c \end{array} \tag{7}$$

commutes for all $a, c \in B$, and also

$$\lambda_e = \varrho_e : e\square e \to e \,. \tag{8}$$

Soon we shall see that these three diagrams imply that all such diagrams commute. For the moment, we observe (Exercise 1) that they imply commutativity in the diagrams

$$\begin{array}{ccc} e\square(b\square c) & \xrightarrow{\alpha} & (e\square b)\square c \\ \downarrow{\scriptstyle \lambda} & & \downarrow{\scriptstyle \lambda\square 1} \\ b\square c & = & b\square c, \end{array} \qquad \begin{array}{ccc} a\square(b\square e) & \xrightarrow{\alpha} & (a\square b)\square e \\ {\scriptstyle 1\square\varrho}\downarrow & & \downarrow{\scriptstyle \varrho} \\ a\square b & = & a\square b. \end{array} \tag{9}$$

Any category with finite products is monoidal, if we take $a\square b$ to be (any chosen) product of the objects a, b and e to be a terminal object, while α, λ, and ϱ are the unique isomorphisms (Prop. III. 5.1) which commute with the respective projections. Then the pentagon (5) commutes (both legs commute with the projections of the four fold products), and so does the triangle (7). Similarly, any category with finite coproducts is monoidal, with $\square$ the coproduct and e an initial object.

The usual "tensor products" give monoidal categories. For example, the tensor product of two abelian groups A and B is defined by the condition that there is a function $A \times B \to A \otimes B$, $a, b \mapsto a \otimes b$, universal among bilinear functions on $A \times B$ to abelian groups. By iteration, there is a universal trilinear $A \times (B \times C) \to A \otimes (B \otimes C)$; by the uniqueness of a universal, there is then a unique isomorphism $\alpha : A \otimes (B \otimes C) \to (A \otimes B) \otimes C$ which is natural (because of its uniqueness); the corresponding pentagon (5) commutes because both legs are the unique comparisons of universal quadrilinear functions. The isomorphisms $\lambda : \mathbf{Z} \otimes A \cong A$, $\varrho : A \otimes \mathbf{Z} \cong A$ are well known (and used to identitfy $\mathbf{Z} \otimes A$ with A). All told, $\langle \mathbf{Ab}, \otimes, \mathbf{Z}, \alpha, \lambda, \varrho \rangle$ is a monoidal category.

The pentagonal condition (5) for $\otimes$ in **Ab** may also be verified directly on elements $a \in A$, $b \in B$, and $c \in C$, by noting that $\alpha[a \otimes (b \otimes c)] = (a \otimes b) \otimes c$. This suggests one role of this condition: It avoids the possible use of the "wrong" associativity; for example, of the isomorphism $\alpha' : A \otimes (B \otimes C) \to (A \otimes B) \otimes C$ defined on elements of these abelian groups as $\alpha'[a \otimes (b \otimes c)] = -(a \otimes b) \otimes c$. For this α', (5) fails to commute by a sign.

There are many other examples. A discussion like that for **Ab** shows for each commutative ring K that $\langle K\text{-}\mathbf{Mod}, \otimes_K, K\rangle$ is monoidal. The same holds for graded K-modules and for differential graded K-modules (= chain complexes of K-modules) under the customary definition of the tensor product for such objects (Mac Lane [1963]). Similarly, the category of all K-algebras (or, all differential graded K-algebras) is monoidal, under the familiar tensor product of algebras. For any ring R, the category of all $R-R$ bimodules is monoidal under $\otimes_R$.

A (strict) *morphism* of monoidal categories.

$$T:(B, \Box, e, \alpha, \lambda, \varrho) \to (B', \Box', e', \alpha', \lambda', \varrho'),$$

is a functor $T: B \to B'$ such that, for all a, b, c, f, and g

$$T(a \Box b) = Ta \Box' Tb, \qquad T(f \Box g) = Tf \Box' Tg, \qquad Te = e', \tag{10}$$

$$T\alpha_{a,b,c} = \alpha'_{Ta,Tb,Tc}, \qquad T\lambda_a = \lambda'_{Ta}, \qquad T\varrho_a = \varrho'_{Ta}. \tag{11}$$

With these morphisms as arrows, we can form **Moncat**, the category of all small monoidal categories. This category has (the obvious) finite products; in particular **1** with the evident (strict) monoidal structure is terminal in **Moncat**. There is also a full subcategory consisting of all *strict* monoidal categories; naturally, the definition of morphisms T for these can omit the conditions (11) on α, λ, and ϱ.

Many useful morphisms between monoidal categories are, however, not strict in the sense of (10) and (11). For example, the forgetful functor $U: \langle K\text{-}\mathbf{Mod}, \otimes_K, -\rangle \to \langle \mathbf{Ab}, \otimes, \ldots\rangle$ is not strict; indeed, for K-modules A and B, we have not an equality $U(A \otimes_K B) = UA \otimes UB$ nor even an isomorphism, but just a natural morphism $UA \otimes UB \to U(A \otimes_K B)$, expressing the fact that $A \otimes_K B$ is a quotient of $A \otimes_{\mathbf{Z}} B$. A similar situation arises for the forgetful functor $\langle \mathbf{Ab}, \otimes, \ldots\rangle \to \langle \mathbf{Set}, \times, \ldots\rangle$. We shall not formulate here the properties of these "relaxed" morphisms between monoidal categories (for this, see §XII.5).

One might be tempted to avoid all this fuss with α, λ, and ϱ by simply identifying all isomorphic objects in B. This will not do, by the following argument due to Isbell. Let $\mathbf{Set}_0$ be the skeleton of the category of sets; it has a product $X \times Y$ with projections p_1 and p_2 as usual. If D is a (the) denumerable set, then $D = D \times D$, and both projections of this product are epis $p_1, p_2: D \to D$. Now suppose that the isomorphism $\alpha: X \times (Y \times Z) = (X \times Y) \times Z$, defined as usual to commute with the three projections, were always the identity; it is then the identity for $X = Y = Z = D$; since α is natural, $f \times (g \times h) = (f \times g) \times h$ for any three $f, g, h: D \to D$. But $\times$ on functions is defined in terms of the projections p_1 and p_2 above, so

$$fp_1 = p_1(f \times (g \times h)) = p_1((f \times g) \times h) = (f \times g) p_1 : D \to D,$$

and p_1 is epi, so $f = f \times g$. The corresponding argument with p_2 gives $f \times g = g$, hence $f = g$ for any $f, g: D \to D$, an absurdity. A similar argument applies to the skeleton of $\langle \mathbf{Ab}, \otimes, \ldots\rangle$.

Exercises

1. Prove that (5) and (7) imply (9). Hint: Take the pentagon (5) with $a=b=e$ and fill in the inside, adding ϱ in two places, the basic identity (7) twice, and suitable naturalities to get $(\lambda\Box 1)\alpha\lambda=\lambda\lambda: e\Box(e\Box(c\Box d))\rightarrow c\Box d$, and hence ($\lambda$ an isomorphism) $(\lambda\Box 1)\alpha=\lambda$.
2. Construct the product in **Moncat** of two monoidal categories.
3. For B monoidal, show that B^{op} has the (evident) monoidal structure.
4. For B monoidal and C any category, show that the functor category B^C is monoidal, with multiplication $S\Box T$ defined by $(S\Box T)c=Sc\Box Tc$ and $e: C\rightarrow B$ the constant functor e. Show that the adjunction $B^{C\times D}\cong(B^C)^D$ is an isomorphism of monoidal categories.
5. Prove: A strict monoidal category with one object is a set (the set of arrows) with two binary operations $\circ$, $\Box$ which satisfy the interchange law and have a common (left and right) unit id_e. Apply Ex. 5 of § II.5.
6. Show by examples that the axioms (5) and (7) are independent.

2. Coherence

A coherence theorem asserts: "Every diagram commutes"; more modestly, that every diagram of a certain class commutes. The class of diagrams at issue now are the diagrams in a monoidal category which, like the pentagon (1.5), are built up from instances of α, λ, and ϱ by multiplications $\Box$. However, two apparently or formally different vertices of such a diagram might become equal in a particular monoidal category, in such a way as to spoil the commutativity. Hence we prove only that every "formal" diagram commutes, where a formal diagram is one in which the vertices are iterated formal $\Box$-products of "variables". We call these formal products "binary words"; they are exactly like the well-formed formulas and terms used in logical syntax in proof theory.

The precise definition is by recursion. A *binary word* of length 0 is the symbol e_0 (the empty word); a binary word of length 1 is the symbol $(-)$ (the variable or the place holder); if v and w are binary words of lengths m and n, respectively, then the symbol $v\Box w=(v)\Box(w)$ is a binary word of *length* $m+n$. For example, $((-\Box-)\Box e_0)\Box-$ is a binary word of length 3 – an iterated 4-fold product, with chosen arrangement of parentheses, and a specified argument set equal to e_0. For any two binary words v and w of the same length, introduce one arrow $v\rightarrow w$. These words with these arrows form a category W (a preorder with every arrow invertible). It is a monoidal category under multiplication $v, w\mapsto v\Box w$, with unit e_0, and with α, λ, and ϱ the appropriate (and necessarily unique) arrows.

By its very construction (unique arrows $v\rightarrow w$) every diagram in W will commute. Morphisms from W to B then give the desired diagrams which commute in other monoidal categories B. These morphisms are given by the following theorem, which states in effect that W is the free monoidal category on one generator $(-)$:

Theorem 1. *For any monoidal category B and any object $b \in B$, there is a unique morphism $W \to B$ of monoidal categories with $(-) \mapsto b$.*

Proof. We write the desired morphism as $w \mapsto w_b$, to suggest that it means "Substitute b in all the blanks of the word w". On objects w we must set

$$(e_0)_b = e, \qquad (-)_b = b, \qquad (v \Box w)_b = v_b \Box w_b\,; \tag{1}$$

by induction, these formulas uniquely determine all w_b.

For words of fixed length n we now construct a certain "basic" graph $G_n = G_{n,b}$. Its vertices are all words w of length n which do not involve e_0 while its edges $v \to w$ are to be identical with certain arrows $v_b \to w_b$ in B. Call them the "basic" arrows. Here each instance

$$\alpha : u_b \Box (v_b \Box w_b) \to (u_b \Box v_b) \Box w_b$$

of associativity and each instance of α^{-1} is basic, as are all arrows $\beta \Box 1$ or $1 \Box \beta$ with $1 : v_b \to v_b$ an identity and β already recognized as basic. Intuitively, each basic arrow is an arrow such as $(1 \Box \alpha) \Box (1 \Box 1)$ – one instance of α, boxed with identities. Observe then that each basic arrow is either "directed" (it involves α) or "antidirected" (with α^{-1}). In the graph G_n the paths from u to w are thus the composable sequences of basic arrows from u_b to w_b; by composition each path yields an arrow $u_b \to w_b$ in B. The crux of our proof will be to show that any two paths from u to w yield by composition the *same* arrow $u_b \to w_b$ in B – i.e., that the graph G_n is a commutative diagram in B.

First, take $w^{(n)}$ to be the unique word of length n which has all pairs of parentheses starting in front. There is a directed path in G_n from any w of length n to $w^{(n)}$; indeed, we may choose such a path in a canonical way, successively moving outermost parentheses to the front by instances of α. For any two words v and w of length n the two canonical paths combine to give a path $v \to w^{(n)} \to w$; this observation is really just the known proof of the "general associative law" for a product ab, given the usual associative law $a(bc) = (ab)c$.

Define the *rank* ϱ of a word w by recursion, setting $\varrho e_0 = 0$, $\varrho(-) = 0$, and

$$\varrho(v \Box w) = \varrho(v) + \varrho(w) + \text{length}(w) - 1\,;$$

observe that $\varrho w = 0$ means that all pairs of parentheses in w start at the front.

Now we show that G_n commutes. Along any path from v to w, join each vertex to the "bottom" vertex $w^{(n)}$ by the canonical directed path.

A glance at the diagram

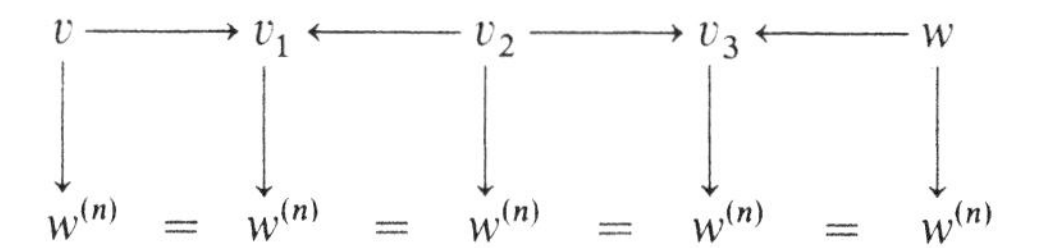

indicates that it will suffice to show that any two *directed* paths (all α's, no α^{-1}) from a v_i to $w^{(n)}$ are equal. This will be proved by induction on the rank of $v_i = v$. Suppose it true for all v of smaller rank, and consider two different directed paths starting at v with (directed) basic arrows β and γ, as in the figure

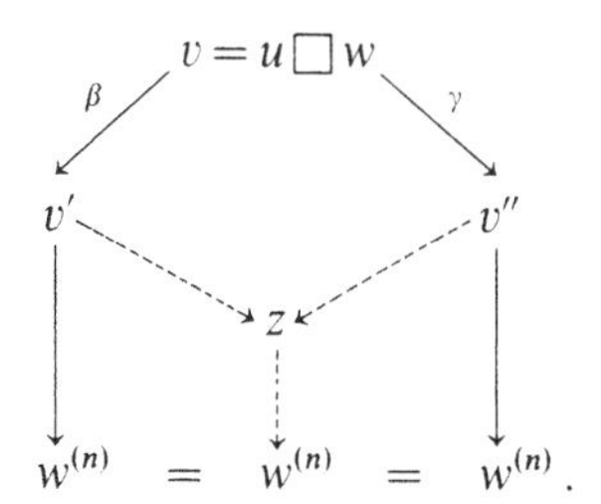

Both β and γ decrease the rank. Hence it will suffice to show that one can "rejoin" their codomains v' and v'' by directed paths to some common vertex z in such a way that the diamond from v to z is commutative. This is done by a case subdivison. If $\beta = \gamma$, take $z = v' = v''$. If $\beta \neq \gamma$, write v as $v = u \square w$ and observe that β has one of the following three forms:

$$
\begin{aligned}
&\beta = \beta' \square 1_w; && \beta \text{ acts "inside" the first factor } u,\\
&\beta = 1_u \square \beta''; && \beta \text{ acts inside the second factor,}\\
&\beta = \alpha_{u,s,t}, && \text{where } v = u \square w = u \square (s \square t).
\end{aligned}
$$

For γ there are three corresponding cases.

Now compare the cases for β and γ. If both act inside the same factor u, we can use induction on the length n. If β acts inside u and γ inside w, use the diamond

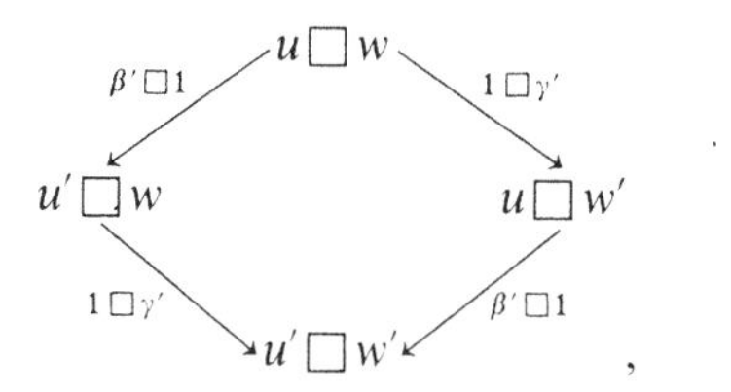

which commutes because $\square$ is a bifunctor. There remains the case when one of β or γ, say β, is $\beta = \alpha = \alpha_{u,s,t}$ as in the third case above. Since

$\gamma \neq \beta$, γ must act inside u or inside w. If γ acts inside u, we use a diamond from $u \Box (s \Box t)$ to $(u' \Box s) \Box t$, which commutes because α is natural. If γ is inside $w = s \Box t$ and actually inside s or inside t, naturality of α gives a similar diamond. There remains only the case where γ is inside $s \Box t$ but not inside s or t. Then γ must be an instance of α, t must be a product $t = p \Box q$, and our diamond must then start with

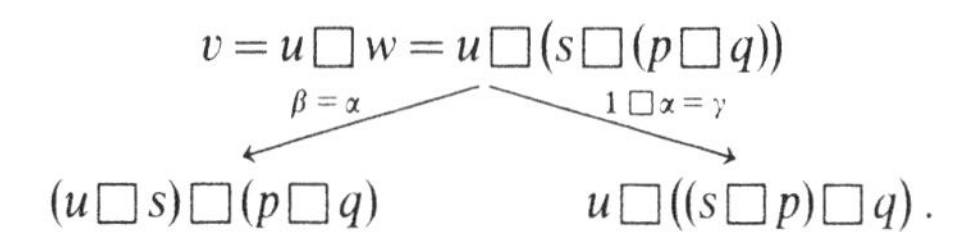

This we can complete to a "diamond" by taking that diamond to be the pentagon of (5). This shows that the graph G_n is commutative in B; it completes the coherence proof as far as associativity alone is concerned.

It is trivial to "fold in" to this proof the applications of λ and ϱ. Formally, consider the graph G'_n with vertices all words of length n, including words involving e_0, and with edges all basic arrows constructed, just as above, by boxing instances of α, λ, and ϱ (and their inverses) with identities. This graph G'_n is infinite, but contains the previous (finite) graph G_n built from α alone. It remains to show G'_n commutative in B. For each word w, there is still at least one path $w \to w^{(n)}$. But the composite arrow obtained from any such path is equal to that for a different path which first removes all e's, then applies α. Indeed, if some e is removed by $\lambda : e \Box b \cong b$ after some application of α, then that e can be removed before – either by naturality of α, or by (7), or by (9). Moreover by (8) it does not matter in $e \Box e$ whether e is removed by λ or by ϱ. Finally, this reduced path has composite equal to that for a canonical path in which all the e's are removed in some specified order (say, starting with the left-most occurence of e). This process reduces G'_n to G_n and proves that G'_n is commutative in B, since G_n is.

We can now define the morphism $W \to B$ required in the theorem. The category W was constructed with exactly one arrow $v \to w$ between words v and w of the same length n; the morphism will send this arrow to the composite arrow for any path $v_b \to w_b$ in G'_n, since we now know the composite to be unique (independent of the choice of the path). In virtue of this same uniqueness, this construction does define a functor $W \to B$. Moreover this functor is a morphism of monoidal categories because

$$f \Box g = f \circ 1 \Box 1 \circ g = (f \Box 1) \circ (1 \Box g)$$

for any arrows f and g.

The coherence result can be formulated in terms of graphs whose edges are the natural transformations α, λ, and ϱ. To state this, note first that each word w of length n (in one variable) determines for each monoidal category B a functor $w_B: B^n = B \times \cdots \times B \to B$ of n variables, obtained by replacing each blank $(-)$ in the word w by the identity functor of B. The explicit definition of this functor, like (1), is by recursion; $(e_0)_B: \mathbf{1} \to B$ is the constant functor $e \in B$ and $(-)_B$ is the identity functor $B \to B$, while if w_B and w'_B are already determined for words w and w' of the respective lengths n and n', then $(w \square w')_B$ is the composite functor

$$(w \square w')_B : B^{n+n'} = B^n \times B^{n'} \xrightarrow{w_B \times w'_B} B \times B \xrightarrow{\quad\square\quad} B\,. \tag{2}$$

With this formulation, the coherence result is as follows:

Corollary. *Let B be a monoidal category. There is a function which assigns to each pair of words v, w of the same length n a (unique) natural isomorphism*

$$\mathrm{can}_B(v, w) : v_B \dot{\to} w_B : B^n \to B\,, \tag{3}$$

called the canonical map from v_B to w_B, in such a way that the identity arrow $e \to e$ is canonical (between functors of 0 variables), the identity transformation $\mathrm{id}_B : I_B \dot{\to} I_B$ is canonical, $\alpha, \alpha^{-1}, \lambda, \lambda^{-1}, \varrho$ and ϱ^{-1} are canonical, and the composite as well as the $\square$-product of two canonical maps is canonical.

This sort of formulation, as will appear from the proof, applies also to the case considered in the theorem itself: For each $b \in B$ there is a function which assigns to each pair of words v, w of the same length a canonical arrow $\mathrm{can}_b(v, w) : v_b \to w_b$, with properties like those stated for can_B.

Proof. From the given monoidal category B we construct a category $It(B)$ with objects all pairs $\langle n, T \rangle$, T any functor $T : B^n \to B$, and with arrows $f : \langle n, T \rangle \to \langle n, T' \rangle$ all natural transformations $f : T \dot{\to} T'$. In this category we define a multiplication by $\langle m, S \rangle \square \langle n, T \rangle = \langle m+n, S \square T \rangle$, where $S \square T$ is the composite

$$S \square T : B^{m+n} \cong B^m \times B^n \xrightarrow{S \times T} B \times B \xrightarrow{\quad\square\quad} B\,,$$

we take the unit e to be the functor $\mathbf{1} \to B$ constant at e and define $\lambda : e \square T \dot{\to} T$ for each T and then for each $a \in B^n$ as the arrow $\lambda_{Ta}: e \square Ta \to Ta$ of B. This λ is natural in T. Similar pointwise definitions give ϱ and α; it is routine to verify that $It(B)$ is a (relaxed) monoidal category.

The identity functor $I : B \to B$ is an object of $It(B)$. Hence the theorem above stating that W is free monoidal on $(-)$ gives (for $b = I$) a unique morphism $W \to It(B)$ of monoidal categories with $(-) \mapsto I$. In particular, this morphism sends each word w to the functor w_B described in (2) above, while the unique arrow $v \to w$, for v and w of the same length, is sent to a natural transformation $v_B \dot{\to} w_B$ which we call $\mathrm{can}_B(v, w)$, as in (3). Since the functor is a morphism, it must preserve α, λ, and ϱ. Thus, using our notation for words,

$$\mathrm{can}_B(e_0, e_0) = 1_e : e \dashrightarrow e, \qquad \mathrm{can}_B((-), (-)) = \mathrm{id}_B : B \dashrightarrow B\,,$$

$$\mathrm{can}_B(-\Box(-\Box-), (-\Box-)\Box-) = \alpha : B\Box(B\Box B) \dashrightarrow (B\Box B)\Box B\,,$$

$$\mathrm{can}_B(e_0\Box-, (-)) = \lambda, \qquad \mathrm{can}_B(-\Box e_0, (-)) = \varrho\,,$$

$$\mathrm{can}_B(v\Box v', w\Box w') = \mathrm{can}_B(v, w)\Box \mathrm{can}_B(v', w')\,.$$

This corollary states that every diagram of the following sort is commutative:

Vertices. Words w of length n representing functors $w_B : B^n \to B$.

Edges. Natural transformations 1_e, id_B, α, λ, ϱ, and their $\Box$ products. Moreover, the functors in question are e, I, $-\Box-$ and their composites, and each edge is a natural transformation between the functors represented by the vertices at its ends.

Exercises

1. Draw a diagram showing all canonical maps between binary words of length 5. (It can be regarded as a polyhedral subdivision of the surface of the sphere into 19 regions – 16 pentagons (instances of α) and 3 squares (which commute by naturality).)
2. (Stasheff [1963].) Show that the diagram giving all canonical maps between words of length $n+3$ can be regarded as a polyhedral subdivision of the surface of the n-sphere.
3. Construct the free monoidal category on any set X, and prove for it the appropriate universal property. (Hint: Its objects are words, with any $x \in X$ a word of length 1, and there is a surjection $W_X \to M_X$ from the set W_X of words to the free monoid on X. There is a (unique) arrow $v \to w$ if and only if v and w are words with the same image in M_X.)

3. Monoids

Following the ideas suggested in the introduction, we can now define the notion of a monoid in an arbitrary monoidal category $\langle B, \Box, e\rangle$.

A *monoid* c in B is an object $c \in B$ together with two arrows $\mu : c\Box c \to c$, $\eta : e \to c$ such that the diagrams

$$\begin{array}{ccccc} c\Box(c\Box c) & \xrightarrow{\alpha} & (c\Box c)\Box c & \xrightarrow{\mu\Box 1} & c\Box c \\ \downarrow{\scriptstyle 1\Box\mu} & & & & \downarrow{\scriptstyle \mu} \\ c\Box c & & \xrightarrow{\quad\mu\quad} & & c, \end{array} \tag{1}$$

$$\begin{array}{ccccc} e\Box c & \xrightarrow{\eta\Box 1} & c\Box c & \xleftarrow{1\Box\eta} & c\Box e \\ & \searrow{\scriptstyle\lambda} & \downarrow{\scriptstyle\mu} & \swarrow{\scriptstyle\varrho} & \\ & & c & & \end{array} \tag{2}$$

are commutative. A morphism $f:\langle c,\mu,\eta\rangle\to\langle c',\mu',\eta'\rangle$ of monoids is an arrow $f:c\to c'$ such that

$$f\mu=\mu'(f\Box f):c\Box c\to c', \qquad f\eta=\eta':e\to c'.$$

With these arrows, the monoids in B constitute a category $\mathbf{Mon}_B$, and $\langle c,\mu,\eta\rangle\mapsto c$ defines a forgetful functor $U:\mathbf{Mon}_B\to B$.

This definition includes a variety of cases; some already noted in our introduction:

Monoidal Category	*Monoids Therein*
$\langle\mathbf{Set},\times,1\rangle$	(ordinary) Monoids
$\langle\mathbf{Top},\times,*\rangle$	Topological monoids
$\langle C^C,\circ,\mathrm{Id}\rangle$	Monads (cf. Chapter VI!)
$\langle\mathbf{Ab},\otimes,\mathbf{Z}\rangle$	Rings
$\langle K\text{-}\mathbf{Mod},\otimes_K,K\rangle$	K-algebras
$\langle$Graded modules, ...$\rangle$	Graded algebras
$\langle DG\text{-}K\text{-}\mathbf{Mod},\otimes_K,K\rangle$	Differential graded K-algebras
$\langle B^{\mathrm{op}},\Box^{\mathrm{op}},e\rangle$	Comonoids in B
$\langle K\text{-}\mathbf{Mod}^{\mathrm{op}},\otimes_K^{\mathrm{op}},K\rangle$	K-coalgebras
$\langle\mathbf{Cat},\times,\mathbf{1}\rangle$	Strict Monoidal categories
$\langle O\text{-}\mathbf{Grph},\times_O,O\to O\rangle$	Categories (cf. (II. 7. 3)).

There is a "general associative law" which states that in a monoid $\langle c,\mu,\eta\rangle$ any two n-fold products are equal. Specifically, if w is any binary word and $w_c\in B$ the corresponding object of B, as defined in Theorem 2.1, the w-fold product μ_w is an arrow $\mu_w:w_c\to c$ defined by the following recursion: If $w=e_0$, $\mu_{e_0}:e\to c$ is η; if $w=(-)$, $\mu_{(-)}:c\to c$ is the identity; if $w=(-)\Box(-)$, μ_w is μ, and in general if $w=u\Box v$, $\mu_{u\Box v}$ is the evident composite

$$(u\Box v)_c=u_c\Box v_c\xrightarrow{\mu_u\Box\mu_v}c\Box c\xrightarrow{\ \Box\ }c. \tag{3}$$

Proposition 1 *(General Associative Law). For $\langle c,\mu,\eta\rangle$ a monoid in B, the iterated products μ_v and μ_w for any two words v and w of the same length n satisfy*

$$\mu_w\circ\mathrm{can}_c(v,w)=\mu_v:v_c\to c, \tag{4}$$

where $\mathrm{can}_c(v,w):v_c\to w_c$ is the canonical arrow of Theorem 2.1.

Proof. The axioms (1) and (2) for a monoid are exactly those cases of (4) where the canonical arrow in question is α, λ, or ϱ. From these cases, (4) may be verified by induction, since all canonical arrows are composites of α's, λ's, and ϱ's.

For example, one may define the n-th $\square$-power of every $b \in B$ to be

$$b^n = (b \square b) \square \cdots \square b \tag{5}$$

with "all parentheses in front"; thus $b^0 = e$, $b^1 = b$, $b^{n+1} = b^n \square b$. For the monoid $\langle c, \mu, \eta \rangle$ the n-fold product $\mu^{(n)} : c^n \to c$ is then defined by recursion as

$$\mu^{(0)} = \eta, \quad \mu^{(1)} = \mathrm{id}_c, \quad \mu^{(2)} = \mu, \quad \text{and} \quad \mu^{(n+1)} = \mu(\mu^{(n)} \square 1). \tag{6}$$

Then (4) includes the more familiar equation ("general associative law")

$$\mu^{(n)}(\mu^{(k_1)} \square \cdots \square \mu^{(k_n)}) = \mu^{(k_1 + \cdots + k_n)} \tag{7}$$

valid for all natural numbers n and $k_1, \ldots, k_n$.

Theorem 2 *(Construction of free monoids). If the monoidal category B has denumerable coproducts, and if for each $a \in B$ the functors $a \square -$ and $- \square a : B \to B$ preserve these coproducts, then the forgetful functor $U : \mathbf{Mon}_B \to B$ has a left adjoint.*

Note: In many cases ($B = \mathbf{Set}$, $B = \mathbf{Ab}$, ...) the functors $a \square -$ and $- \square a$ themselves have right adjoints, hence automatically preserve coproducts.

Proof. The distributive law $\theta : \coprod_n (a \square b_n) \cong a \square \coprod_n b_n$ holds for each denumerable coproduct $\coprod_n b_n$ of objects $b_n \in B$ because $a \square -$ preserves coproducts. Indeed, the definition of the coproduct injections $i_n : b_n \to \coprod_n b_n$ and j_n shows that there is a unique arrow θ which makes the diagram

$$\begin{array}{ccc} a \square b_n & = & a \square b_n \\ \downarrow{\scriptstyle 1 \square i_n} & & \downarrow{\scriptstyle j_n} \\ a \square \coprod_n b_n & \xleftarrow{\ \theta\ } & \coprod_n (a \square b_n) \end{array}$$

commute, and "preserves coproducts" means exactly that θ is an isomorphism. Its inverse is constructed similarly.

For given a, take $b_n = a^n$ to be the n-th power defined as in (5) and define a multiplication μ on $\coprod_n a^n$ by "juxtaposition" $a^m \square a^n \cong a^{m+n}$. Formally, μ is the unique arrow defined by the commutative diagram

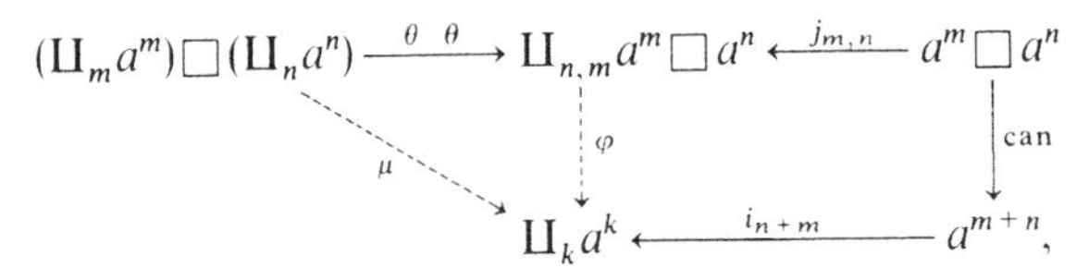

where the vertical map "can" is the canonical map (iterated associativity) given by the coherence theorem for B, φ is that unique map on the coproduct $\coprod_{n,m}$ which makes the square with the coproduct injections $j_{m,n}$ and i_{n+m} commute for all the natural numbers m and n, the map $\theta \circ \theta$ is the composite of two canonical isomorphisms θ above (because $\square$ is distributive over $\coprod_m$ and $\coprod_n$), and the multiplication μ is $\mu = \varphi(\theta\ \theta)$.

A large but routine diagram (exercise!) shows this μ to be associative, in the sense (1). A corresponding unit $\eta_a : e \to \amalg_n a^n$ is defined to be the injection $i_0 : e = a^0 \to \amalg_n a^n$ of the coproduct. All told, $\langle \amalg_n a^n, \mu, \eta_a \rangle$ is a monoid in B. The injection $\varrho_a = i_1 : a = a^1 \to \amalg_n a^n$ of the coproduct is an arrow

$$\varrho_a : a \to U\langle \amalg_n a^n, \mu, \eta_a \rangle$$

to the forgetful functor $U : \mathbf{Mon}_B \to B$.

This arrow is universal from a to U. For let $\langle c, \mu_c, \eta_c \rangle$ be any monoid in B and $f : a \to c = U(c, \mu_c, \eta_c)$ an arrow in B. Then we define an arrow $f' : \amalg_n a^n \to c$ as the composite on the bottom of the commutative diagram

$$\begin{array}{ccccc} a^n & \xrightarrow{f^n} & c^n & \xrightarrow{\mu_w} & c \\ \downarrow{\scriptstyle i_n} & & \downarrow{\scriptstyle j_n} & & \| \\ \amalg_n a^n & \dashrightarrow & \amalg_n c^n & \dashrightarrow & c \end{array}$$

constructed as follows. First, take w to be the word of length n with all parentheses in front, so that $w_b = b^n$, by our definition of b^n; then $\mu_w : c^n \to c$ is the n-fold product defined in the general associative law (6), i_n and j_n are coproduct injections, and the dotted arrows on the bottom are constructed, by universality of the coproducts, so as to make the indicated squares commute (for all n). A routine large diagram will prove that f' is a morphism of monoids; by construction $f' \circ \varrho_a = f$, so ϱ_a is indeed universal and therefore $\amalg_n a^n$ is a free monoid on a, as asserted in the theorem.

The point of this quite formal proof is that it contains many separate instances of the same sort of formality. If $B = \langle \mathbf{Set}, \times, 1, \ldots \rangle$, this is the standard construction (Corollary II. 7. 2) of the free monoid on the set a; in this case a^n is the set of words of length n spelled in letters of a, and the free monoid is the disjoint union $\amalg_n a^n$, with product given by composition. If $B = \langle K\text{-}\mathbf{Mod}, \otimes_K, K, \ldots \rangle$, this is the standard construction (e.g., Mac Lane [1963b], p. 179) of the tensor algebra $\oplus_n A^n$ on the K-module A. The same construction also gives "differential graded" tensor algebras, free topological monoids, etc.

Exercises

1. Prove: if B has finite products, so does $\mathbf{Mon}_B$.
2. (Coherence for monoids.) Interpret the proposition about the canonical maps μ_w for a monoid $\langle c, \mu, \eta \rangle$ as the following coherence theorem. Consider a graph with vertices the binary words w and with arrows $v \to w$ those arrows $v_c \to w_c$ which are $1, \mu, \eta$, instances $\alpha(u'_c, v'_c, w'_c)$ of α, instances of λ and of ϱ, and all $\Box$-products of such arrows. Prove that any two paths $w \to (-)$ in this graph have equal composites, but show that this would not hold when the ending is not $(-)$ as above but the word $(-)\Box(-)$ of length 2.

3. (a) (Substitution of words in a word.) Each word u of length n determines a functor $u_W : W^n \to W$. If $v_1, \ldots, v_n$ are n words, show that the word $u_W(v_1, \ldots, v_n)$ has length the sum of the lengths of the v_i, and that it corresponds (intutively) to substituting $v_1, \ldots, v_n$, in order, for the n blanks $(-)$ in the word u.
   (b) If $w = u_W(v_1, \ldots, v_n)$, show that the canonical maps μ_w of Proposition 1 have the property that the composite
$$w_c = u_w(v_{1c}, \ldots, v_{nc}) \xrightarrow{\mu_W(\mu_{v_1} \cdots \mu_{v_n})} u_c \xrightarrow{\mu_u} c$$
   is equal to $\mu_w : w_c \to c$. Show that this result includes Proposition 1.

4. Actions

Again, we work in a fixed monoidal category B. A left *action* of a monoid $\langle c, \mu, \eta \rangle$ on an object $a \in B$ is an arrow $v : c \Box a \to a$ of B such that the diagram

$$\begin{array}{ccccccc} c\Box(c\Box a) & \xrightarrow{\alpha} & (c\Box c)\Box a & \xrightarrow{\mu\Box 1} & c\Box a & \xleftarrow{\eta\Box 1} & e\Box a \\ \downarrow{\scriptstyle 1\Box v} & & & & \downarrow{\scriptstyle v} & & \downarrow{\scriptstyle \lambda} \\ c\Box a & & \xrightarrow{\quad v \quad} & & a & = & a \end{array} \qquad (1)$$

commutes. For example, c acts on itself by the map $\mu : c \Box c \to c$; this is the "left regular representation" of c. A *morphism* $f : v \to v'$ of left actions of c is an arrow $f : a \to a'$ in B such that $v'(1 \Box f) = fv : c \Box a \to a'$. With these morphisms as arrows, the left actions v for a fixed monoid c form a category ${}_c\mathbf{Lact}$. These definitions clearly include familiar cases: an action of an ordinary monoid on a set, a left R-module regarded as an action of the ring R on an abelian group, and similarly with rings replaced by K-algebras, or $D\,G$-algebras ($D\,G =$ differential graded).

There is a forgetful functor ${}_c\mathbf{Lact} \to B$, defined by $\langle v : c \Box a \to a \rangle \mapsto a$; it has a left adjoint which sends each $b \in B$ to $c \Box b$, with action of c on $c \Box b$ defined by the composite
$$c\Box(c\Box b) \xrightarrow{\alpha} (c\Box c)\Box b \xrightarrow{\mu\Box 1} c\Box b\,.$$

Right actions $\sigma : b \Box c \to b$ of c are defined similarly, and commuting left-and-right actions of c on a may be defined to parallel the usual bimodules (left and right R-modules).

Exercises

1. (Dubuc [1970], Prop II.1.1.) Let $\langle T, \eta, \mu \rangle$ be a monad in a category X. Show that the monad T has an action on an endofunctor $S : X \to X$ if and only if S can be lifted to the category X^T of T-algebras as $S = G^T S'$, and show that these actions correspond one-one to the liftings $S' : X \to X^T$.
2. Let a small strict monoidal category B (as a monoid in $\langle \mathbf{Cat}, \times, \ldots \rangle$) act on a category C. Define then the action of a monoid in B on an object in C, and use this to extend the result of Exercise 1 to the case of functors $S : A \to X$ from any category A.

3. Describe the actions of a K-coalgebra.
4. If B has coproducts preserved by all functors $a\Box -$, show that ${}_c\mathbf{Lact}$ has coproducts preserved by the forgetful functor to B.
5. If the base category B has finite products, so does the category ${}_c\mathbf{Lact}$, in such a way that the projections $a \times a' \to a, a'$ of the product (in B) become morphisms of actions (in ${}_c\mathbf{Lact}$).
6. (Generalization of the tensor product of a right module by a left module.) If B has coequalizers, c is a monoid, $\sigma: b\Box c \to b$ a right action, and $v: c\Box a \to a$ a left action, construct a "tensor product" $b\Box_c a \in B$ as the coequalizer of two maps $b\Box(c\Box a) \to b\Box a$ given by the actions, and prove $\Box_c$ a functor $\Box_c: \mathbf{Ract}_c \times {}_c\mathbf{Lact} \to B$.
7. (Coherence result for an action.) Given a left action $v: c\Box a \to a$ of a monoid c, describe the properties of canonical maps $v_w: w_{c,a} \to a$, where w is any word of length $\geqq 1$ with "last argument" $(-)$ (define what this means), while $w_{c,a}$ results from substituting a for the last argument and c for all the other arguments in w.

5. The Simplicial Category

We now describe a particular strict monoidal category Δ which plays a central role in topology and also provides a "universal" monoid.

This category Δ has as objects all finite ordinal numbers $n = \{0, 1, \ldots, n-1\}$ and as arrows $f: n \to n'$ all (weakly) monotone functions; that is, all functions f such that $0 \leqq i \leqq j < n$ implies $f_i \leqq f_j$. In this category, the ordinal number 0 is initial, while the number 1 is terminal. *Ordinal addition* is a bifunctor $+ : \Delta \times \Delta \to \Delta$, defined on ordinals n, m as the usual (ordered) sum $n + m$ and on arrows $f: n \to n'$, $g: m \to m'$ as

$$\begin{aligned} (f+g)(i) &= f i, & i &= 0, \ldots, n-1 \\ &= n' + g(i-n), & i &= n, \ldots, n+m-1\,. \end{aligned}$$

(Thus the function $f+g$ is just f and g placed "side by side".) Moreover, $\langle \Delta, +, 0\rangle$ is a strict monoidal category. Since 1 is terminal in Δ, there are unique arrows $\mu: 2 \to 1$, $\eta: 0 \to 1$; for the same reason, these arrows form a monoid $\langle 1, \mu, \eta\rangle$ in Δ. It is "universal" in the following sense.

Proposition 1. *Given a monoid $\langle c, \mu', \eta'\rangle$ in a strict monoidal category $\langle B, \Box, e\rangle$, there is a unique morphism $F: \langle \Delta, +, 0\rangle \to \langle B, \Box, e\rangle$ such that $F1 = c$, $F\mu = \mu'$ and $F\eta = \eta'$, as in the figure*

$$\begin{array}{ccccc} 0 & \xrightarrow{\eta} & 1 & \xleftarrow{\mu} & 2 = 1+1 \\ \vdots & & \vdots & & \vdots \\ \downarrow & & \downarrow & & \downarrow \\ e & \xrightarrow{\eta'} & c & \xleftarrow{\mu'} & c\Box c, \end{array} \qquad \begin{array}{c} \langle \Delta, +, 0\rangle \\ \vdots\, F \\ \downarrow \\ \langle B, \Box, e\rangle\,. \end{array} \tag{1}$$

The proof depends on showing that the arrows of Δ are exactly the iterated formal products (for the binary product μ). In detail, write $\mu^{(k)}$ for the unique arrow $\mu^{(k)}: k \to 1$. Thus $\mu^{(0)} = \eta$, $\mu^{(1)}$ is the identity, $\mu^{(2)} = \mu : 2 \to 1$,

$$\mu^{(3)} = \mu(\mu + 1) = \mu(1 + \mu) : 3 \to 1 ,$$

and so on. Since 1 is terminal in Δ,

$$\mu^{(n)}(\mu^{(k_1)} + \cdots + \mu^{(k_n)}) = \mu^{(k_1 + \cdots + k_n)} . \tag{2}$$

(This is the "general associative law".) On the other hand, if $f : m \to n$ is any arrow of Δ, let m_i be the (ordinal) number of elements in the subset $f^{-1}i$ of m; then

$$f = \mu^{(m_0)} + \mu^{(m_1)} + \cdots + \mu^{(m_{n-1})}, \qquad \sum_{i=0}^{n-1} m_i = m \tag{3}$$

(note that some of the m_i may be zero). This shows that any f is a sum of iterated products constructed from μ and η.

Now consider the functor F required in the Proposition. Since $F(1) = c$ and F is to be a morphism of monoidal categories, F must have $Fn = c^{(n)}$; this determines the object function of F. Next, $F\mu = \mu'$ and $F\eta = \eta'$ imply that $F\mu^{(n)} = \mu'^{(n)}$; the representation (3) of any arrow f of Δ then determines the arrow function Ff of F. Thus F is unique. It remains only to show that the object and arrow functions so defined give a functor. But in Δ, composites are given by (2), which corresponds exactly to the general associative law valid in B. q.e.d.

This universal property gives a complete characterization of Δ. Its objects form the free monoid generated (under $+$) by 1; its arrows are generated by additions and compositions from $\mu : 2 \to 1$ and $\eta : 0 \to 1$, using the associative law for μ and the left and right unit laws for η and μ.

There is another description of the arrows of Δ, which starts by observing that a monotone function $f : n \to n'$ can be factored as $f = g \circ h$ where $h : n \to n''$ is surjective and monotone, $g : n'' \to n'$ is monotone and injective. Moreover, this injective function g will be determined just by giving the image of g, which is a subset of n'' ordinals in the set n'. In particular, there are exactly $n+1$ injective monotone functions $n \to n+1$; namely, for $i = 0, \ldots, n$, the injective monotone function $\delta_i^n : n \to n+1$ whose image omits i, thus

$$\delta_i^n : n \to n+1, \qquad \delta_i^n\{0, \ldots, n-1\} = \{0, \ldots, \hat{i}, \ldots, n\} , \tag{4}$$

where $\hat{i}$ on the right indicates that i is to be omitted. We display all these arrows (omitting the superscripts n) as

$$0 \xrightarrow{\delta_0} 1 \overset{\delta_0}{\underset{\delta_1}{\rightrightarrows}} 2 \rightrightarrows 3, \ldots, \qquad \delta_0, \ldots, \delta_n : n \to n+1 . \tag{5}$$

On the other hand, a monotone $h : n \to n''$ which is surjective is determined by the subset $\{j \mid hj = h(j+1), 0 \leqq j \leqq n-2\}$ of those $n - n''$ arguments j at which h does not increase. In particular, there are n such arrows $n+1 \to n$; for $i = 0, \ldots, n-1$ they are

$$\sigma_i^n : n+1 \to n, \qquad \sigma_i^n(i) = \sigma_i^n(i+1). \tag{6}$$

We display them (without superscripts) as

$$0 \longleftarrow 1 \xleftarrow{\sigma_0} 2 \overset{\sigma_0}{\underset{\sigma_1}{\leftleftarrows}} 3 \;\text{⭅}\; 4 \ldots, \qquad \sigma_0, \ldots, \sigma_{n-1} : n+1 \to n. \tag{7}$$

These arrows may also be expressed in terms of μ and η. Indeed $\delta_0 : 0 \to 1$ is η, $\sigma_0 : 2 \to 1$ is μ, and the definitions show that

$$\delta_i^n = 1_i + \eta + 1_{n-i} : n \to n+1, \qquad i = 0, \ldots, n, \tag{8}$$

$$\sigma_i^n = 1_i + \mu + 1_{n-i-1} : n+1 \to n, \qquad i = 0, \ldots, n-1. \tag{9}$$

Lemma. *In Δ, any arrow $f : n \to n'$ has a unique representation*

$$f = \delta_{i_1} \circ \cdots \circ \delta_{i_k} \circ \sigma_{j_1} \circ \cdots \circ \sigma_{j_h}, \tag{10}$$

where the ordinal numbers h and k satisfy $n - h + k = n'$, while the strings of subscripts i and j satisfy

$$n' > i_1 > \cdots > i_k \geqq 0, \qquad 0 \leqq j_1 < \cdots < j_h < n-1.$$

Proof. By induction on $i \in n$, any monotone f is determined by its image, a subset of n', and by the set of those $j \in n$ at which it does not increase $[f(j) = f(j+1)]$. Putting $i_1, \ldots, i_k$, in reverse order, for those elements of n' not in the image and $j_1, \ldots, j_h$, in order, for the elements j of n where f does not increase, it follows that the functions on both sides of (10) are equal.

In particular, the composite of any two δ's or σ's may be put into the canonical form (10). This yields the following list of three kinds of identities on these binary composites

$$\delta_i \delta_j = \delta_{j+1} \delta_i \qquad i \leqq j \tag{11}$$

$$\sigma_j \sigma_i = \sigma_i \sigma_{j+1} \qquad i \leqq j, \tag{12}$$

$$\left.\begin{aligned} \sigma_j \delta_i &= \delta_i \sigma_{j-1}, && i < j, \\ &= 1, && i = j,\, i = j+1 \\ &= \delta_{i-1} \sigma_j, && i > j+1. \end{aligned}\right\} \tag{13}$$

These identities may be verified directly. For example, (11) asserts that $\delta_i^{n+1} \delta_j^n = \delta_{j+1}^{n+1} \delta_i^n : n \to n+2$ for any $j \leqq n$; one checks that each side of this equation is a monotone injection, and that both sides have the same image.

Proposition 2. *The category* Δ, *with objects all finite ordinals, is generated by the arrows* $\delta_i^n : n \to n+1$ *and* $\sigma_j^n : n+1 \to n$ *subject to the relations* (11), (12), *and* (13).

Proof. These relations suffice to put any composite of δ's and σ's into the unique form (10) of the Lemma.

The category Δ has a direct geometric interpretation by affine simplices, which give a functor

$$\Delta : \Delta \to \mathbf{Top} \tag{14}$$

representing Δ as a subcategory of **Top**. On objects n of Δ, take Δ_0 to be the empty topological space, and Δ_{n+1} to be the "standard" n-dimensional affine simplex – the subspace of Euclidean $\mathbf{R}^{n+1}$ consisting of the following points

$$\Delta_{n+1} = \{p = (t_0, \ldots, t_n) \mid t_0 \geqq 0, \ldots, t_n \geqq 0, \Sigma\, t_i = 1\}\,;$$

here the non-negative real numbers $t_0, \ldots, t_n$ are the *barycentric coordinates* of the point $p \in \Delta_{n+1}$. On arrows $f: n+1 \to m+1$, $\Delta_f : \Delta_{n+1} \to \Delta_{m+1}$ is the (affine) map defined by

$$\Delta_f(t_0, \ldots, t_n) = (s_0, \ldots, s_m), \qquad s_j = \sum_{fi=j} t_i\,.$$

Note carefully that (in this notation) Δ_{n+1} has dimension n and $n+1$ vertices, while Δ_f is the (unique) affine map which sends the vertex i of Δ_{n+1} to the vertex fi of Δ_{m+1}; for example, $\Delta_{\delta_i} : \Delta_{n+1} \to \Delta_{n+2}$ is that affine map which sends the n-simplex Δ_{n+1} to that n-dimensional face of Δ_{n+2} which is opposite vertex number i. Geometrically, the "boundary" of a tetrahedron Δ_4 consists of the four triangular faces which are the images of Δ_3 under $\delta_0, \delta_1, \delta_2$, and δ_3. Using standard properties of affine geometry (Mac Lane-Birkhoff [1967], Chap. 12) one may verify (exercise) that Δ as defined is indeed a functor $\Delta \to \mathbf{Top}$.

Note that this functor Δ sends the ordinal number $n+1$ to the n-dimensional simplex: Δ is a subcategory of **Top**, but the geometric dimension is one less than the arithmetic one used in Δ.

By Δ^+ we denote the full subcategory of Δ with objects all the positive ordinals $\{1, 2, 3, \ldots\}$ (omit only 0). Topologists use this category, call it Δ, and rewrite its objects (using the geometric dimension) as $\{0, 1, 2, \ldots\}$. Here we stick to our Δ, which contains the real 0, an object which is necessary if all face and degeneracy operations are to be expressed, as in (3), in terms of binary product μ and unit η.

Contravariant functors on the category Δ^+ to **Set** are traditionally known as "simplicial sets".

Thus, a *simplicial object* S in a category X is defined to be a functor $S : (\Delta^+)^{\mathrm{op}} \to X$, and a *morphism* $S \to S'$ of simplicial objects is a natural transformation $\theta: S \dot{\to} S'$. If we write this functor S as

$$n+1 \mapsto S_n, \qquad \delta_i \mapsto d_i, \qquad \sigma_j \mapsto s_j,$$

so that S_n is in *geometric* dimension n, then a simplicial object in X may be described in the traditional (and more complicated) way as a list of $S_0, S_1, \ldots, S_n, \ldots$ of objects of X (S_n: the object of n-simplices) with arrows ("face operators") $d_i: S_n \to S_{n-1}$ for $i=0, \ldots, n$, and $n>0$, and arrows ("degeneracies") $s_i: S_n \to S_{n+1}$ for $i=0, \ldots, n, n \geqq 0$ which satisfy the identities dual to (11), (12), and (13).

$$d_i d_{j+1} = d_j d_i, \qquad i \leqq j \tag{11^{op}}$$

$$s_{j+1} s_i = s_i s_j \qquad i \leqq j \tag{12^{op}}$$

$$\left.\begin{aligned} d_i s_j &= s_{j-1} d_i, && i<j \\ &= 1, && i=j, j+1, \\ &= s_j d_{i-1} && i>j+1. \end{aligned}\right\} \tag{13^{op}}$$

For example, if Y is an affine simplex with its vertices linearly ordered, then $d_i Y$ is the "i-th face" obtained by omitting vertex i while $s_i Y$ is the degenerate simplex with vertex i doubled. The rules above then follow.

An *augmented simplicial* object in X is a functor $S': \Delta^{\text{op}} \to X$. A simplicial object S may be *augmented* (i.e., extended to a functor S') by finding one object $S_{-1} \in X$ and one arrow $\varepsilon: S_0 \to S_{-1}$ of X with $\varepsilon d_0 = \varepsilon d_1 : S_1 \to S_{-1}$; thus $S'(\delta_0) = \varepsilon$. Such an arrow ε is (traditionally) an *augmentation* of S.

A simplicial object S in an abelian category A (e.g. $A = \mathbf{Ab}$) gives homology, via a suitable "boundary" operation. Specifically construct from S the arrows

$$S_0 \xleftarrow{\ \partial\ } S_1 \xleftarrow{\ \partial\ } S_2 \xleftarrow{\ \partial\ } \cdots \tag{15}$$

where the *boundary homomorphism* $\partial: S_{n+1} \to S_n$ is the arrow defined as the alternating sum $\partial = d_0 - d_1 + \cdots + (-1)^{n+1} d_{n+1}$. The relations (11^{op}) on the faces d_i imply that $\partial\partial = 0$. (This means that the diagram (15) is a chain complex in A). Since $\partial\partial = 0$,

$$\operatorname{Im}\{\partial: S_{n+1} \to S_n\} \leqq \operatorname{Ker}\{\partial: S_n \to S_{n-1}\}$$

and we can take the quotient object (see Chap. VIII) to be the n-th *homology* of S:

$$H_0(S) = S_0/\operatorname{Im}\{\partial : S_1 \to S_0\}$$

and

$$H_n(S) = \operatorname{Ker}\{\partial : S_n \to S_{n-1}\}/\operatorname{Im}\{\partial : S_{n+1} \to S_n\}\,, \quad n > 0\,.$$

Each augmentation of the functor S yields an augmentation of this chain complex; that is, an object S_{-1} of A and an arrow $\varepsilon: S_0 \to S_{-1}$ with $\varepsilon\partial = 0$, hence an arrow $H_0(S) \to S_{-1}$.

The singular homology of a topological space is a classical example.

Consider the composite functor

$$\Delta^{\mathrm{op}} \times \mathbf{Top} \xrightarrow{\hom(\Delta -, \ldots)} \mathbf{Set} \xrightarrow{\mathbf{Z}} \mathbf{Ab}$$

where $\Delta : \Delta \to \mathbf{Top}$ is the functor described in (14), while $\mathbf{Z}$ assigns to each set the free abelian group generated by the elements of that set. This composite determines for each topological space X an augmented simplicial object $S = S(X)$ in $\mathbf{Ab}$. Each arrow $h \in \hom(\Delta_{n+1}, X)$ is a singular n-simplex in X, so S_{n+1} is the free abelian group generated by all such simplices (all finite linear combinations with integral coefficients of singular n-simplices). The associated chain complex is the *singular chain complex* of the space X, with its homology the *singular homology* (see e.g. Mac Lane [1963] Chap. II).

We may summarize the protean aspects of Δ thus:

(a) Δ is the category of finite ordinal numbers, hence a full subcategory of the category **Ord** of all (linearly) ordered sets.

(b) Δ is a full subcategory of **Cat**, if we interpret each ordinal n as a category (finite preorder); the objects of Δ are the categories $\mathbf{0}, \mathbf{1}, \mathbf{2}, \mathbf{3}, \ldots$.

(c) Δ is the strict monoidal category containing the universal monoid, its arrows are all "iterated multiplications" $\mu^{(m_0)} + \cdots + \mu^{(m_{n-1})}$.

(d) Δ is a subcategory of **Top**, consisting of the standard ordered simplices (one for each dimension), with order preserving affine mappings.

The simplicial objects defined via Δ provide a means of treating many questions in algebraic topology, especially those dealing with homology, CW-complexes, Eilenberg-Mac Lane spaces, and cohomology operations. This line of development is presented in May [1967], Lamotke [1968], and Gabriel-Zisman [1967], the last presentation making full use of categorical techniques.

Exercises

1. In Δ, show that an arrow $f : n \to n'$ is monic (or epi) if and only if the function f is injective (resp., surjective).
2. (a) Show that the subcategory $\Delta_{\mathrm{mon}} \subset \Delta$ of all monics in Δ is generated by the arrows δ_i, subject to the relations (11).

   (b) Show that every arrow in Δ_{mon} is uniquely an iterated sum of $\eta : 0 \to 1$ and $\mathrm{id} : 1 \to 1$.
3. (a) Show that the subcategory $\Delta_{epi} \subset \Delta$ of all epis in Δ is generated by the arrows σ_i subject to the relations (12). Show that Δ_{epi} is a strict monoidal category.

   (b) *A semigroup* $\langle c, \mu \rangle$ in a strict monoidal category $\langle C, \Box, e \rangle$ is an object c with an arrow $\mu : c \Box c \to c$ which is associative, in that $\mu(\mu \Box 1_c) = \mu(1_c \Box \mu)$. Show that $2 \to 1$ is a universal semigroup in Δ_{epi}.
4. Show that the category of simplicial objects in **Set** is small-complete.

6. Monads and Homology

Monads and their duals, the comonads, play via Δ a central role in homological algebra, as we may now briefly indicate. Let $L = \langle L, \varepsilon, \delta \rangle$

be a comonad in a category A; in other words $L: A \to A$ is an endofunctor, and the natural transformations $\varepsilon: L \dot{\to} \mathrm{Id}_A, \delta: L \dot{\to} L^2$ satisfy

$$\delta L \cdot \delta = L\delta \cdot \delta : L \dot{\to} L^3, \quad \varepsilon L \cdot \delta = 1_L = L\varepsilon \cdot \delta : L \dot{\to} L. \tag{1}$$

These are the duals to the definition of a monad in (2) of §VI.1. This amounts to saying that $\langle L, \varepsilon, \delta \rangle$ is a comonoid in the strict monoidal category A^A of endofunctors of A, where the functor $\Box$ (multiplication) is composition.

Now Δ contains the universal monoid $\langle 1, 0 \to 1, 1+1 \to 1 \rangle$, so Δ^{op} contains the universal comonoid $\langle 1, 1 \to 0, 1 \to 1+1 \rangle$. Thus, by the dual of Theorem 5.1, any comonoid in a strict monoidal category $\langle B, \Box, e \rangle$ determines a unique morphism $\Delta^{\mathrm{op}} \to B$ of monoidal categories, carrying the universal comonoid to the given one. This morphism $\Delta^{\mathrm{op}} \to B$ is an augmented simplicial object in B (and $(\Delta^+)^{\mathrm{op}} \to B$ is a simplicial object).

In particular, each comonad $\langle L, \varepsilon, \delta \rangle$ in A, as a comonoid in the functor category A^A, determines an augmented simplicial object (functor) $\Delta^{\mathrm{op}} \to A^A$, with

$$\begin{array}{cccccc} \langle 1, & 0 & \longleftarrow & 1 & \longrightarrow & 1+1 \rangle \\ \downarrow & \downarrow & & \downarrow & & \downarrow \\ \langle L, & \mathrm{Id} & \xleftarrow{\varepsilon} & L & \xrightarrow{\delta} & L \circ L \rangle . \end{array}$$

Thus $n \mapsto L^n = L \circ \cdots \circ L$, ε is the augmentation, $\delta = s_0 : L \to L^2$ is the degeneracy arrow, and the faces and degeneracies in higher dimensions are given by the duals of the equations (8) and (9) of § 5 (which express δ and σ in terms of μ and η):

$$d_i^n = L^i \varepsilon L^{n-i} : L^{n+1} \dot{\to} L^n, \qquad i = 0, \ldots, n, \tag{2}$$

$$s_i^n = L^i \delta L^{n-i-1} : L^n \dot{\to} L^{n+1}, \qquad i = 0, \ldots, n-1. \tag{3}$$

The whole simplicial object has the form

$$Smp\,L = \left\{ L \underset{d_1}{\overset{d_0}{\leftleftarrows}} L^2 \,\overleftarrow{\overleftarrow{\leftarrow}}\, L^3 \ldots; \quad L \longrightarrow L^2 \rightrightarrows L^3 \ldots \right\}.$$

Now suppose that A is an Ab-category (e.g., an abelian category, or that we have applied to $Smp\,L$ a functor to some Ab-category). The simplicial identities on the face operations d_i then show that the alternating sums

$$\partial = d_0 - d_1 + d_2 - \cdots + (-1)^n d_n : L^{n+1} a \to L^n a$$

satisfy $\partial\partial = 0$, so are the boundary morphisms of a chain complex called $L^* a$,

$$L^* a : La \xleftarrow{\partial} L^2 a \xleftarrow{\partial} L^3 a \xleftarrow{\partial} \cdots$$

with an augmentation $\varepsilon_a : La \to a$. This complex is a standard "resolution" of $a \in A$ in the sense of homological algebra, and so may be used to construct derived functors; in particular, various cohomology functors.

The cohomology of groups provides an example.

The forgetful functor $U : \mathbf{Rng} \to \mathbf{Mon}$ (forget the addition) has (by the adjoint functor theorem) a left adjoint $\mathbf{Z}$, sending each monoid M to the monoid ring $\mathbf{Z}M$. In particular, if $M = \Pi$ is a (multiplicative) group, $\mathbf{Z}\Pi$ is the group ring: Its additive group is the free abelian group generated by the elements $x \in \Pi$, and its multiplication is the unique bilinear map with $\langle x, y\rangle \to xy$, the product in Π, for all $x, y \in \Pi$. Let Π-**Mod** denote the category of left Π-modules A.

The forgetful functor $U : \Pi\text{-}\mathbf{Mod} \to \mathbf{Ab}$ has a left adjoint $\mathbf{Z}(\Pi) \otimes -$ which assigns to each (additive) abelian group B the left $\mathbf{Z}(\Pi)$-module $\mathbf{Z}(\Pi) \otimes B$. The unit and counit of this adjunction are the maps

$$\eta : B \to \mathbf{Z}(\Pi) \otimes B, \qquad b \mapsto 1 \otimes b, \qquad b \in B,$$
$$\varepsilon : \mathbf{Z}(\Pi) \otimes UA \to A, \qquad x \otimes a \mapsto xa, \qquad a \in A.$$

The composite $\Pi\text{-}\mathbf{Mod} \to \mathbf{Ab} \to \Pi\text{-}\mathbf{Mod}$ determines a comonad $\langle L, \varepsilon, \delta\rangle$ in the category Π-**Mod**, where $L : \Pi\text{-}\mathbf{Mod} \to \Pi\text{-}\mathbf{Mod}$ is the functor $L = \mathbf{Z}(\Pi) \otimes -$ (literally, $\mathbf{Z}(\Pi) \otimes U -$), $\varepsilon : L \dot{\to} \mathrm{Id}$ is as above, and $\delta : L \dot{\to} L^2$ is the natural transformation $\delta = \mathbf{Z}(\Pi) \otimes \eta U$ given explicitly for each Π-module A as

$$\delta_A : LA = \mathbf{Z}(\Pi) \otimes A \to \mathbf{Z}(\Pi) \otimes \mathbf{Z}(\Pi) \otimes A = L^2 A$$
$$x \otimes a \mapsto x \otimes 1 \otimes a, \qquad x \in \Pi, a \in A,$$

where 1 is the identity element of the group Π. Take the Π-module $A = \mathbf{Z} =$ the abelian group $\mathbf{Z}$ regarded as a trivial Π-module ($x \cdot m = m$ for all $x \in \Pi$ and all integers m). Then $\mathbf{Z}(\Pi) \otimes \mathbf{Z} \cong \mathbf{Z}(\Pi)$, and the simplicial object $(Smp\,L)\mathbf{Z}$ becomes

$$\mathbf{Z}(\Pi) \leftleftarrows \mathbf{Z}(\Pi)^{(2)} \overset{\leftarrow}{\leftleftarrows} \cdots \mathbf{Z}(\Pi)^{(n)} \leftleftarrows \mathbf{Z}(\Pi)^{(n+1)} \cdots$$

where $\mathbf{Z}(\Pi)^{(n)}$ denotes the n-fold tensor product $A_n = \mathbf{Z}(\Pi) \otimes \cdots \otimes \mathbf{Z}(\Pi)$. Explicitly, A_{n+1} is the free abelian group with generators all elements

$$x \otimes x_1 \otimes \cdots \otimes x_n = x \otimes [x_1 | \cdots | x_n]$$

(the alternative notation on the right is traditional) for all elements $x, x_i \in \Pi$. The Π-module structure is determined ($y \in \Pi$) by

$$\langle y, x[x_1 | \cdots | \cdots | x_n]\rangle \mapsto yx[x_1 | \cdots | x_n].$$

The face operators $d_i : \mathbf{Z}(\Pi)^{(n+1)} \to \mathbf{Z}(\Pi)^{(n)}$, as determined (2) by ε, are

$$\begin{aligned} d_i(x[x_1 | \cdots | x_n]) &= xx_1[x_2 | \cdots | x_n], & i &= 0, \\ &= x[x_1 | \cdots | x_i x_{i+1} | \cdots | x_n], & 0 &< i < n, \\ &= x[x_1 | \cdots | x_{n-1}], & i &= n. \end{aligned}$$

The degeneracy operators $s_i : \mathbf{Z}(\Pi)^{(n)} \to \mathbf{Z}(\Pi)^{(n+1)}$, as determined by δ according to (3), are the Π-module maps

$$s_i(x[x_1|\cdots|x_{n-1}]) = x[x_1|\cdots|x_{i-1}|1|x_i|\cdots|x_{n-1}] \qquad 0 \leqq i \leqq n-1 .$$

Since Π-**Mod** is already an abelian category, this (augmented) simplicial object determines an augmented chain complex in Π-**Mod** of the form

$$\mathbf{Z} \leftarrow \mathbf{Z}(\Pi) \leftarrow \mathbf{Z}(\Pi)^{(2)} \leftarrow \cdots \leftarrow \mathbf{Z}(\Pi)^{(n)} \leftarrow \cdots .$$

This is a "free resolution" of the trivial Π-module $\mathbf{Z}$; it is, in fact, the standard resolution used to define the homology and cohomology of the group Π. (Mac Lane [1963], Theorem IV. 5.1).

The cohomology of Π is obtained from the resolution as follows. Take a Π-module A and the corresponding functor $\hom_\Pi(-, A) : (\Pi\text{-}\mathbf{Mod})^{\mathrm{op}} \to \mathbf{Ab}$, where $\hom_\Pi(-, -)$ denotes the abelian group of Π-module morphisms. Apply this functor to the chain complex above (dropping the augmentation $\mathbf{Z}(\Pi) \to \mathbf{Z}$) to get a "cochain" complex

$$\hom_\Pi(\mathbf{Z}(\Pi), A) \xrightarrow{\ \delta\ } \hom_\Pi(\mathbf{Z}(\Pi)^{(2)}, A) \xrightarrow{\ \delta\ } \cdots$$

with coboundary $\delta = \hom_\Pi(\partial, A)$. The cohomology groups of this complex are exactly the cohomology groups $H^n(\Pi, A)$ of the group Π with coefficients in A. The formulas for d_i above give δ explicitly. Thus, for example, $H^0(\Pi, A) = \{a \mid a \in A \text{ and } xa = a \text{ for all } x\}$; $H_1(\Pi, A)$ is the group of "crossed homomorphisms" $\Pi \to A$ modulo the principal crossed homomorphisms, and $H^2(\Pi, A)$ is the group of all group extensions of the additive group A by the multiplicative group Π, with operations (conjugation) given by the Π-module structure of A (Mac Lane [1963a], IV.2, IV.3).

The higher cohomology groups of groups appear in obstruction problems (Mac Lane [1963a], IV.8), in the theory of the $K(\Pi, 1)$ spaces in topology (Mac Lane [1963a], IV.11), and class field theory (Cassels-Fröhlich [1967]).

The homology of Π with coefficients in a *right*-Π-module C is found in a similar way: To the standard resolutions apply not the functor $\hom_\Pi(-, A)$ but the (covariant, additive) functor $C \otimes_\Pi - : \Pi\text{-}\mathbf{Mod} \to \mathbf{Ab}$. The homology of the resulting chain complex in **Ab** is the homology $H_n(\Pi, C)$ of Π with coefficients in C. For example (Mac Lane [1963a], Prop. X.5.2)

$$H_0(\Pi, \mathbf{Z}) = \mathbf{Z}, \qquad H_1(\Pi, \mathbf{Z}) = \Pi/[\Pi, \Pi] ;$$

the latter is the factor commutator group of Π.

7. Closed Categories

The ideas broached in this chapter have extensive further developments which we shall indicate briefly. First, a monoidal category B is said to be *symmetric* when it is equipped with isomorphisms

$$\gamma_{a,b} : a \Box b \cong b \Box a , \tag{1}$$

natural in $a, b \in B$, such that the diagrams

$$\gamma_{a,b} \circ \gamma_{b,a} = 1, \qquad \varrho_b = \lambda_b \circ \gamma_{b,e} : b \Box e \cong b , \tag{2}$$

$$\begin{array}{ccccc} a\Box(b\Box c) & \xrightarrow{\alpha} & (a\Box b)\Box c & \xrightarrow{\gamma} & c\Box(a\Box b) \\ \downarrow{\scriptstyle 1\Box\gamma} & & & & \downarrow{\scriptstyle \alpha} \\ a\Box(c\Box b) & \xrightarrow{\alpha} & (a\Box c)\Box b & \xrightarrow{\gamma\Box 1} & (c\Box a)\Box b \end{array} \tag{3}$$

all commute. This selection of conditions suffices (Mac Lane [1963b]) to prove that "all" such diagrams commute, much as in the coherence theorem of § 2 above. Monoidal categories $\langle B, \Box, e, \ldots \rangle$, where $\Box$ is the categorical product or coproduct, are automatically symmetric when $\gamma : a \times b \cong b \times a$ is taken to be the (canonical) isomorphism which commutes with the projections. These ideas are elaborated in Chapter XI.

A *closed category* V is a symmetric, monoidal category in which each functor $-\Box b : V \to V$ has a specified right adjoint $(\ )^b : V \to V$. For example, $\langle \mathbf{Ab}, \otimes, - \rangle$ is closed; the adjoint is given for abelian groups A and B as $A^B = \hom(B, A)$, the abelian group of all morphisms $B \to A$. Similarly, $\langle K\text{-}\mathbf{Mod}, \otimes_K, \ldots \rangle$ is closed for any commutative ring K. The cartesian closed categories, such as **Set** and **Cat**, are also closed categories in this sense. In all these cases, the functor $(\ )^b : V \to V$ is a sort of "internal hom functor".

An Ab-category (and in particular, an abelian category) has already been described (§ I.8) as a category with "hom-sets" in **Ab**. Similarly, one can describe "categories" with "hom-sets" in any monoidal category B: A set R of "*objects*" r, s, t; to each pair of objects r, s an object $R(r, s) \in B$; to each ordered triple an arrow (composition!)

$$R(s, t) \Box R(r, s) \to R(r, t)$$

in B; to each object r, an arrow $e \to R(r, r)$ in B (unit!). These data are subject to the usual associativity and unit axioms on composition. The result is called a B-category, a *B-based* category, or a category *relative* to B – and often, replacing the letter B by V, a V-category. But observe that this structure R is not yet a category in the ordinary sense; it has only hom-*objects* $R(r, s)$ and not hom-*sets*. These can be obtained only applying to the hom-objects $R(r, s)$ a suitable functor $U : B \to \mathbf{Set}$, say $U = B(e, -)$, to get hom-sets $U R(r, s)$. When there are such hom-sets, one says that the ordinary category UR has been "*enriched*" by the objects $R(r, s) \in B$.

Practically all the basic theory of categories applies to enriched categories, provided that the basic category B is not just monoidal, but closed. This development (for a presentation, see Dubuc [1970] and Kelly [1982] and references there) may provide a powerful method of treating at one time the cases of ordinary categories, additive categories based on closed categories of chain complexes (for relative homological algebra), and categories based on a suitable cartesian closed variant of **Top**.

8. Compactly Generated Spaces

A convenient category of topological spaces should be cartesian closed. The familiar adjunction which makes **Set** cartesian closed,

$$\mathbf{Set}(X \times Y, Z) \cong \mathbf{Set}(X, Z^Y), \qquad Z^Y = \mathbf{Set}(Y, Z), \tag{1}$$

which sends each $f: X \times Y \to Z$ to $f^\sharp: X \to Z^Y$, with $(f^\sharp x)y = f(x, y)$, may be considered also for topological spaces X, Y, and Z. We obtain a topological space $\mathrm{Cop}(Y, Z)$ by imposing on the set $\mathbf{Top}(Y, Z)$ of all *continuous* maps $Y \to Z$ the *compact open* topology: A subbase for the open sets consists of the sets $N(C, U)$ where C is any compact subset of Y, U any open subset of Z, and $N(C, U)$ consists of all those continuous $h: Y \to Z$ for which $hC \subset U$. A standard argument (which we will not need) shows that the basic adjunction $f \mapsto f^\sharp$ of (1) restricts to give an adjunction

$$\mathbf{Top}(X \times Y, Z) \cong \mathbf{Top}(X, \mathrm{Cop}(Y, Z)), \tag{2}$$

provided Y is locally compact Hausdorff.

There have been many attempts to repair this situation for more general spaces Y by using a variety of other topologies on the function space or other topologies on the product space. The best device is to so restrict the category of topological spaces that the (categorical) product $X \mapsto X \times Y$ (with its intrinsic topology as a product) does always have a right adjoint (which will be a function space with a uniquely determined topology).

A topological space X is *compactly generated* when each subset $A \subset X$ which intersects every compact subset C of X in a closed set is itself closed. By **CGHaus** we denote the category with objects all compactly generated Hausdorff spaces (= Kelley spaces), with arrows all continuous functions $X \to X'$.

Proposition 1. CGHaus *is a full coreflective subcategory of* **Haus**.

It is a full subcategory by definition. To each Hausdorff space Y we construct a compactly generated space KY with the same points as Y (the "Kelleyfication" of Y) by requiring that $A \subset Y$ be closed in KY if and only if $A \cap C$ is closed in Y for all compact sets $C \subset Y$. Thus all closed sets of Y are closed in KY, KY is Hausdorff, and the identity function $\varepsilon_Y : KY \to Y$ is continuous. Any continuous map $f : X \to Y$ from a compactly generated Hausdorff space X factors as $f = \varepsilon f'$.

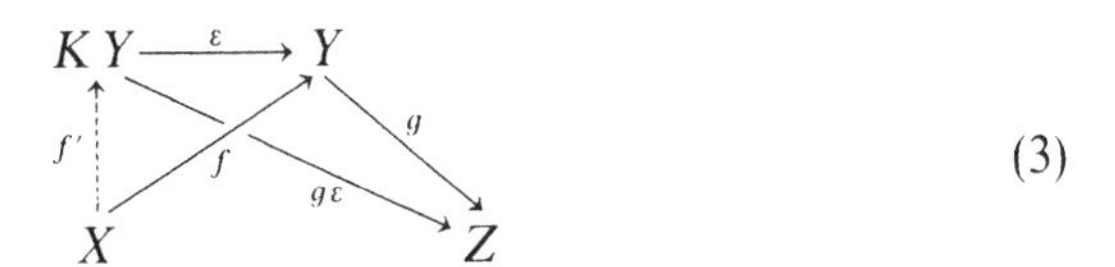

(3)

where $f' : X \to KY$ is the same function (as f) and is continuous because X is compactly generated. This shows that ε is universal from K to Y, so is the counit of an adjunction which makes **CGHaus** coreflective in **Haus**, as desired.

The description of KY means also (see Fig. (3)) that a function $g : Y \to Z$ to a topological space Z is continuous, on KY as $g\varepsilon : KY \to Z$, if and only if the original g is continuous on all compact subsets of Y. Observe also that metrizable spaces and locally compact Hausdorff spaces are compactly generated.

Proposition 2. CGHaus *is (small) complete and cocomplete.*

Proof. The category **Haus** is complete (Proposition V.9.2) and a right adjoint such as K preserves limits. Hence **CGHaus** is complete. In particular, the product (written $\square$) of two spaces X and Y in **CGHaus** is obtained from their "ordinary" product $X \times Y$ in **Haus** as

$$X \square Y = K(X \times Y)\,. \tag{4}$$

In other words, the $\square$-product of Kelley spaces is the product of the underlying sets, with the Kelleyfication of the usual product topology.

Cocompleteness follows readily. Since any coproduct in **Haus** (= disjoint union) of compactly generated spaces is also compactly generated, it will suffice to construct the coequalizer of a parallel pair f, $g : Y \rightrightarrows X$ in **CGHaus**. Take the coequalizer $p : X \to Q$ in **Haus** (Prop. V.9.2) and form KQ:

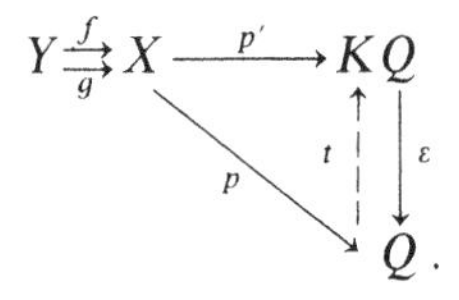

Since $\varepsilon : KQ \to Q$ is universal, there is a unique continuous $p' : X \to KQ$ with $\varepsilon p' = p$ and $p'f = p'g$. Since p' is also a map in **Haus**, and p is the coequalizer of f and g there, there is a continuous $t: Q \to KQ$ with $p' = tp$. Then $p = \varepsilon p' = \varepsilon tp$, so $\varepsilon t = 1$ and $\varepsilon t \varepsilon = \varepsilon$. But ε is monic (in **Haus**), so $t\varepsilon = 1$, and ε is an isomorphism: The coequalizer in **Haus** lies in **CGHaus**.

For example, if A is a subset of a compactly generated Hausdorff space X, then we get an identification space $X//A$ as a coequalizer in **CGHaus** (collapse all of A to a point in **CGHaus**). It is the largest Hausdorff quotient of the space X/A (collapsed in **Top**); its topology is automatically compactly generated.

Theorem 3. **CGHaus** *is a cartesian closed category.*

For two compactly generated Hausdorff spaces X and Y define

$$X^Y = K(\mathrm{Cop}(Y, X)), \tag{5}$$

the function space with the Kelleyfication of the compact-open topology. Define $e: X^Y \square Y \to X$ by evaluation; $\langle f, y\rangle \mapsto fy$. We claim that e is continuous; it suffices to prove that $e: X^Y \times Y \to X$ is continuous on compact sets. Since any compact subset of the product space is contained in the product of its projections, it suffices to show that e is continuous on any set of the form $D \times C$, where D is compact in $\mathrm{Cop}(Y, X)$ and C is compact in Y. Consider $\langle f, y\rangle \in D \times C$, and let U be an open set of X containing fy. Since $f: Y \to X$ is continuous, there exists a neighborhood M of y in C whose closure satisfies $f\,\overline{M} \subset U$. But $N(\overline{M}, U)$ as given before (2) is a set of the subbase for $\mathrm{Cop}(Y, X)$ and $[N(\overline{M}, U) \cap D] \times M$ is open in $D \times C$, contains $\langle f, y\rangle$, and is mapped by e into U. This proves e continuous.

It remains to show e universal from $- \square Y$ to X. So consider any map $h: Z \square Y \to X$ in **CGHaus**. Then we construct $k: Z \to \mathbf{Set}(Y, X)$ as $k = h^\sharp$; that is, so that $(kz)y = h(z, y)$ for all $z \in Z$ and $y \in Y$. A direct proof shows that $kz: Y \to X$ is continuous; thus $kz \in X^Y$. Next, we prove that $z \mapsto kz$ is continuous $Z \to X^Y$. Since Z is compactly generated, it is enough to show $Z \to \mathrm{Cop}(Y, X)$ continuous. So let $N(C, U)$ be one of the open sets for the subbase of the compact-open topology, and suppose that $kz \in N(C, U)$; thus $(h\{z\} \times C) \subset U$. Since U is open, C compact, and h continuous, there is a neighborhood V of z such that $h(V \times C) \subset U$. This implies that $kV \subset N(C, U)$. Therefore k is continuous.

We now have the commutative diagram

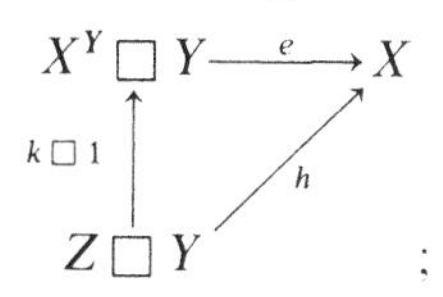

;

by the adjunction in **Set**, there is at most one k with $e\,(k \square 1) = h$, and we have just shown this k continuous. Therefore e is universal, and defines the desired adjunction

$$\mathbf{CGHaus}(Z \square Y, X) \cong \mathbf{CGHaus}(Z, X^Y). \tag{6}$$

Since $\square$ designates the product in **CGHaus**, this category is cartesian closed.

This adjunction (6) is a bijection of sets. One also wishes the corresponding homeomorphism

$$X^{Z\square Y} \cong (X^Y)^Z$$

of function spaces. This follows from the adjunction (6) for categorical reasons (Ex. IV.6.3).

This summarizes the basic properties of the category **CGHaus**. More extensive work (Steenrod [1967] and elsewhere) indicates that it is the convenient category for topological studies; Dubuc and Porta [1971] show that it is appropriate for topological algebra (extensions of the Gelfand duality). All told, this suggests that in **Top** we have been studying the wrong mathematical objects.

The right ones are the spaces in **CGHaus**.

Exercises

1. If Y is Hausdorff, show that KY is the colimit (in **Haus**) of the compact subspaces of Y, ordered by inclusion.
2. Prove that a closed (or open) subset of a space $X \in$ **CGHaus** with the usual subspace topology is itself in **CGHaus**.
3. Prove that the inclusion **CGHaus** $\to$ **Haus** creates colimits.
4. If Z is locally compact Hausdorff and $X \in$ **CGHaus**, prove that $Z \square X = Z \times X$.
5. Prove that **CGHaus** is equivalent to the following category: Objects, all Hausdorff spaces; arrows $f: X \to Y$ all functions continuous on compact sets.

9. Loops and Suspensions

For homotopy theory, we consider the category $\mathbf{CGHaus}_*$ of *pointed* compactly generated Hausdorff spaces – with objects the spaces $X \in$ **CGHaus** with a selected base point, $*_X$, and with arrows the continuous maps preserving the base point. Let $X^{(*)Y}$ be the subspace of X^Y consisting of all base-point preserving maps. Since it is a closed subspace, it is compactly generated. It has a natural base-point (the continuous function sending all of Y to $*_X$). In the standard adjunction $f \mapsto f^\sharp$,

$$\begin{array}{ccc} \mathbf{CGHaus}\ (Z \square Y, X) & \cong & \mathbf{CGHaus}\ (Z, X^Y) \\ \cup & & \cup \\ \mathbf{CGHaus}_*(?, X) & \cong & \mathbf{CGHaus}_*(Z, X^{(*)Y}), \end{array}$$

consider on the right the indicated subset: Those $f^\sharp: Z \to X^{(*)Y}$ which preserve base-point. Thus $(f^\sharp z)*_Y = *_X$ and $(f^\sharp *_Z)y = *_X$; that is, for all

$z \in Z$ and $y \in Y$

$$f(z, *) = * = f(*, y).$$

These are exactly the continuous functions f which collapse the "wedge" $Z \vee Y = (Z \square *) \cup (* \square Y)$ to a point. The corresponding identification space is called the *smash product*

$$(Z \square Y) // [(Z \square *) \cup (* \square Y)] = Z \wedge Y$$

(or sometimes written as $Z \# Y$). This gives an adjunction

$$\mathbf{CGHaus}_*(Z \wedge Y, X) \cong \mathbf{CGHaus}_*(Z, X^{(*)Y}). \tag{1}$$

The circle S^1 may be obtained from the closed unit interval $I = \{t | 0 \leqq t \leqq 1\}$ as the identification space $S^1 = I // \{0, 1\}$; we regard it as a pointed space with base point $0 = +1$. The functors Σ (reduced *suspension*) and Ω (*loop space*) on $\mathbf{CGHaus}_*$ to $\mathbf{CGHaus}_*$ are defined as

$$\Sigma X = X \wedge S^1, \qquad \Omega X = X^{(*)S^1};$$

by the bijection above $\Sigma : \mathbf{CGHaus}_* \to \mathbf{CGHaus}_*$ has Ω as right adjoint. The points of ΩX are the loops in X at the base point; that is, the continuous maps $f : I \to X$ with $f(0) = f(1) = *_X$. On the other hand, ΣX is the cylinder $X \times I$ with top $X \times \{+1\}$, bottom $X \times \{0\}$, and generator $* \times I$ all collapsed to a single point, the (new) base point; equivalently it is the double cone over X ($X \times I$ with top and bottom collapsed) and with the generator over $*$ collapsed, as in the figure

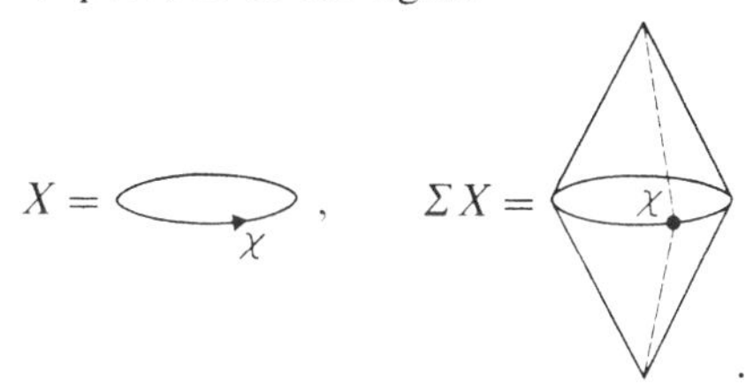

For example, $\Sigma S^1 = S^1 \wedge S^1$ is the two sphere S^2, $\Sigma^n S^1$ the $(n+1)$-sphere. The unit $X \to \Omega \Sigma X$ of the adjunction sends $x \in X$ to the function $\langle x, - \rangle : I \to \Sigma X$; it has a vivid geometric picture; it sends each point $x \in X$ to that generator of the cone which passes through x; this generator is a loop from north pole to south pole = north pole, hence a point of $\Omega \Sigma X$.

By iteration, Σ^n is the left adjoint of $\Omega^n : \mathbf{CGHaus}_* \to \mathbf{CGHaus}_*$; this adjunction has a unit $X \to \Omega^n \Sigma^n X$ which can be written as a composite

$$X \to \Omega \Sigma X \xrightarrow{\Omega \eta \Sigma X} \Omega \Omega \Sigma \Sigma X \to \cdots$$

and Ω^n, as a right adjoint, preserves products: $\Omega^n(X \square Y) \cong \Omega^n X \square \Omega^n Y$. These and similar facts can be obtained either by direct topological arguments, or by application of the properties of adjunctions.

Exercises

1. Construct a left adjoint for $\mathbf{Set}_*(S, -) : \mathbf{Set}_* \to \mathbf{Set}_*$.
2. Show that the smash product in $\mathbf{CGHaus}_*$ is commutative and associative up to natural isomorphisms which make $\mathbf{CGHaus}_*$ a symmetric monoidal category with unit the two-point space.
3. In $\mathbf{Top}_*$ show that $- \times Y$ does not have a right adjoint (because it does not preserve coproducts).
4. The Path space functor $P : \mathbf{CGHaus} \to \mathbf{CGHaus}$ has $PX = X^{(*)I}$, where 0 is taken as the base point of the interval I. For each path $f \in PX$, $f \mapsto f(1)$ defines a natural transformation $\pi : P \dot{\to} \mathrm{Id}$. Show that Ω can be obtained as the pullback of a diagram $P \dot{\to} \mathrm{Id} \dot{\leftarrow} *$ (Classically, ΩX is the "fibre" of $\pi_X : PX \to X$).
5. Describe the counit of the Σ-Ω adjunction.

Notes.

Monoidal categories were first explicitly formulated by Bénabou [1963, 1964], who called them "catégories avec multiplication" and by Mac Lane [1963b], who called them "categories with multiplication"; the renaming is due to Eilenberg. Coherence theorems were initiated by Stasheff in a 1963 treatment of higher homotopies, by Mac Lane [1963b], and by Epstein [1966], who needed them for a general definition of Steenrod operations. Coherence theorems are undergoing active development; Lambek [1968] found a fascinating connection with the cut-elimination theorems of Gentzen-style proof theory; following his lead, Kelly-Mac Lane [1970] proved a coherence theorem for closed categories. The simplicial category, long implicit in the boundary formulas of algebraic topology, became explicit in the study of Eilenberg-Mac Lane spaces and of the Eilenberg-Zilber theorem about 1950, and played a role in the development of homological algebra (see the notes to Chap. VI). Our discussion of monads and homology is only a slight introduction to the recent proliferation of conceptual schemes for the organization of homological algebra.

Compactly generated spaces first appeared in John Kelley's 1955 book on General Topology; their convenience for topology was emphasized by Steenrod [1967], Gabriel-Zisman [1967], and others. There are alternative closed categories convenient for topology, notably the quasi-topological spaces due to Spanier.

The suspension Σ of a topological space is a tool long used in homotopy theory. The Cartan-Serre attack (about 1951) on the difficult problem of computing the homotopy groups of spheres made essential use of loop spaces and suspension. These constructions originally seemed thoroughly geometric. Thus the natural map $X \to \Omega\Sigma X$ came from a topological insight, but now appears in conceptual terms, as the unit of an adjunction.

VIII. Abelian Categories

This chapter will formulate the special properties which hold in categories such as **Ab**, R-**Mod**, **Mod**-R, and R-**Mod**-S: They are all Ab-categories (the hom-sets are abelian groups and composition is bilinear), all finite limits and colimits exist, and these limits – especially kernel and cokernel – are well behaved. This leads to a set of axioms describing an "abelian" category. The axioms suffice to prove all the facts about commuting diagrams and connecting morphisms which are proved in **Ab** by methods of chasing elements. We carry the subject exactly to this point, leaving the subsequent development of homological algebra to more specialized treatments.

1. Kernels and Cokernels

Recall (§ I.5) that a null object z in a category is an object which is both initial and terminal. If C has a null object, then to any $a, b \in C$ the unique arrows $a \to z$ and $z \to b$ have a composite $0 = 0_b^a : a \to b$ called the *zero arrow* from a to b. It follows that any composite with one factor a zero is itself a zero arrow. The null object is unique up to isomorphism, and the notion of zero arrow is independent of the choice of the null.

Let C have a null object. A *kernel* of an arrow $f : a \to b$ is defined to be an equalizer of the arrows $f, 0 : a \rightrightarrows b$. Put more directly, $k : s \to a$ is a kernel of $f : a \to b$ when $fk = 0$, and every h with $fh = 0$ factors uniquely through k (as $h = kh'$)

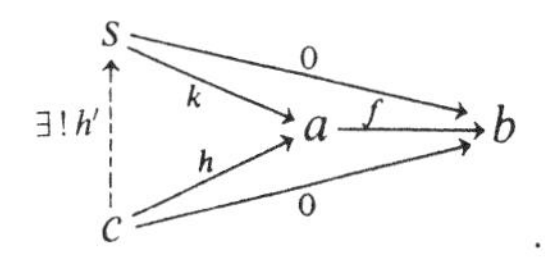

Thus any category with all equalizers (or, more generally, with all pullbacks or with all finite limits) and with a zero has kernels for all arrows, and the kernel $k : s \to a$ of f is unique, up to an isomorphism of s. Like all equalizers, a kernel k is necessarily monic ($kg' = kh'$ implies $g' = h'$, by the unique factorization requirement in the definition). Hence it is convenient to think of the kernel $k: s \to a$ as a subobject of a – that is, as an equivalence class of monics $s \to a$.

For example, in **Grp** the group 1 with just one element (the identity element) is a null object, and for any two groups the zero morphism $G \to H$ is the unique morphism which sends all of G to the identity element in H. The kernel of an arbitrary morphism $f: G \to H$ of groups is the insertion $N \to G$ of the usual kernel N, (with $N =$ all x in G with $fx = 1$). Note that N is a normal subgroup of G, so in **Grp** every kernel is monic but there are monics which are not kernels.

In the category $\mathbf{Set}_*$ of pointed sets (§ I.7), the one-point set is a null object and the zero map $P \to Q$ is the function taking all of P to the base point $*_Q$ in Q. For any morphism $f: P \to Q$ of pointed sets, the kernel $S \to P$ is the insertion of the subset S of those $x \in P$ with $fx = *_Q$, where the base point of S is identical with the base point of P. Much the same description gives kernels in $\mathbf{Top}_*$. In **Grp**, an epimorphism is determined (up to isomorphism) by its kernel, but this is by no means the case in $\mathbf{Set}_*$ or in $\mathbf{Top}_*$.

In any Ab-category A, all equalizers are kernels. Indeed, in such a category each hom-set $A(b,c)$ is an abelian group. Hence, given a parallel pair $f, g: b \to c$, a third arrow $h: a \to b$ satisfies $fh = gh$ if and only if $(f-g)h = 0$. Therefore the universal such h can be described either as the equalizer of f and g or as the kernel of $f-g$. This is the reason one usually deals with kernels and not with equalizers in R-**Mod**, **Ab**, etc.

The dual notion of cokernel has already been described, in § III.3.

Now suppose that the category C has a null object z and kernels and cokernels for all arrows. For each object $c \in C$, the set P_c of all arrows f with codomain c has a preorder $\leqq$, with $g \leqq f$ defined to mean that g factors through f (i.e., that $g = fg'$ for some arrow g'). This reflexive and transitive relation $\leqq$ defines as usual an equivalence relation $\equiv$, with $f \equiv g$ meaning that $f \leqq g$ and $g \leqq f$. The equivalence classes of arrows $f \in P_c$ under this relation form a partially ordered set, which contains the partially ordered set of subobjects of c (restrict f to be a monomorphism; then $g \leqq f$ is the inclusion relation already defined for subobjects in § V.7).

Dually, the set Q^c of all arrows u with domain c is preordered, with $u \geqq v$ when v factors through u ($v = v'u$ for some v').

Now choose a kernel for each arrow u from c and a cokernel for each arrow f to c. Then the definitions of kernel and cokernel state that

$$f \leqq \ker u \Leftrightarrow uf = 0 \Leftrightarrow \operatorname{coker} f \geqq u\,. \tag{1}$$

These logical equivalences state exactly that the functions

$$\ker: Q^c \to P_c, \quad \operatorname{coker}: P_c \to Q^c$$

define a Galois connection from the preorder Q^c to the preorder P_c, as defined in § IV.5. As for any such connection, the triangular identities read

$$\ker(\operatorname{coker}(\ker u)) = \ker u\,, \qquad \operatorname{coker}(\ker(\operatorname{coker} f)) = \operatorname{coker} f\,,$$

and g is a kernel if and only if $g = \ker(\operatorname{coker} g)$. These facts are also readily provable directly from the definitions.

If C has a null object, kernels, and cokernels, then any arrow f of C has a canonical factorization

$$f = mq\,, \qquad m = \ker(\operatorname{coker} f)\,. \tag{2}$$

Lemma 1. *If also $f = m'q'$, where m' is a kernel, then in the commutative square*

$$\begin{array}{ccc} \cdot & \xrightarrow{\;q\;} & \cdot \\ {\scriptstyle q'}\downarrow & \swarrow{\scriptstyle t} & \downarrow{\scriptstyle m} \\ \cdot & \xrightarrow[\;m'\;]{} & \cdot \end{array} \tag{3}$$

there is a (unique) diagonal arrow t with $m = m't$ and $q' = tq$. Moreover, if C has equalizers and every monic in C is a kernel, then q is epi.

Proof. By assumption, $m' = \ker\, p'$ where $p' = \operatorname{coker}\, m'$; take also

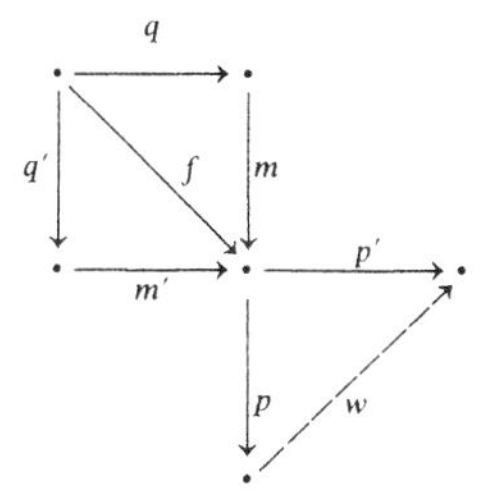

$p = \operatorname{coker}\, m = \operatorname{coker} f$. Then $p'm' = 0$, so $p'f = p'm'q' = 0$, and p' factors through p as $p' = wp$ for some w. Then $p'm = wpm = 0$, so m factors through $m' = \ker\, p'$ as $m = m't$ for a unique monic t. Moreover, $m'q' = m'tq$ and m' is monic, so $q' = tq$. This gives the desired diagram (3).

Next, to prove that q is epi, consider some parallel pair of arrows

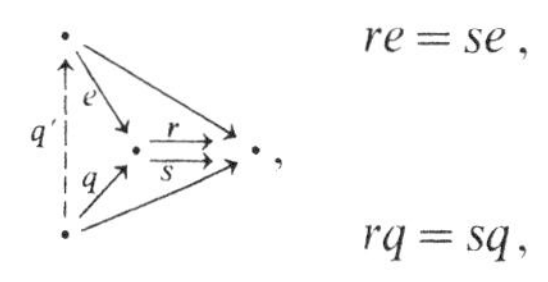

r, s with $rq = sq$. Then q factors through the equalizer e of r and s as $q = eq'$ for some q', and $f = mq = meq'$. Now $m' = me$ is monic, hence by

assumption is a kernel, so by the first conclusion of the lemma there is an arrow t with $m = m't = met$ and hence with $1 = et$. The monic e thus has a right inverse, so is an isomorphism. But e was taken to be the equalizer of r and s, so $r = s$. This proves q epi.

Thus (2) is now an epi-monic factorization of the arrow f.

2. Additive Categories

An Ab-category A, as defined in § I.8, is a category in which each hom-set $A(b,c)$ is an additive abelian group (not necessarily small) and composition of arrows is bilinear relative to this addition. Thus each abelian group $A(b,c)$ has a zero element $0: b \to c$, called the *zero arrow* (even though A may not have a null object in the previous sense). Again, a composite with a zero arrow is necessarily zero, since composition is distributive over addition.

Proposition 1. *The following properties of an object z in an Ab-category A are equivalent:* (i) *z is initial;* (ii) *z is terminal;* (iii) $1_z = 0: z \to z$; (iv) *the abelian group* $A(z,z)$ *is the zero group. In particular, any initial (or any terminal) object in A is a null object.*

Proof. If z is initial, there is a unique map $z \to z$, hence $1_z = 0$ and $A(z,z) = 0$. If $1_z = 0$, then any $f: b \to z$ has $f = 1_z f = 0f = 0: b \to z$, so there is a unique arrow, namely 0, from b to z, and z is terminal. The rest follows by duality.

When there is a null (= initial and terminal) object z in the Ab-category A, the unique maps $b \to z$ and $z \to c$ are the zero elements of $A(b,z)$ and $A(z,c)$ respectively. Hence the composite $b \to z \to c$, which is the zero morphism $0: b \to c$, as defined in § 1, is also the zero element of the abelian group $A(b,c)$.

Next we consider products and coproducts in the Ab-category A.

Definition. *A biproduct diagram for the objects* $a, b \in A$ *is a diagram*

$$a \underset{i_1}{\overset{p_1}{\leftrightarrows}} c \underset{i_2}{\overset{p_2}{\rightleftarrows}} b \tag{1}$$

with arrows p_1, p_2, i_1, i_2 *which satisfy the identities*

$$p_1 i_1 = 1_a, \quad p_2 i_2 = 1_b, \quad i_1 p_1 + i_2 p_2 = 1_c. \tag{2}$$

Theorem 2. *Two objects a and b in an Ab-category A have a product in A if and only if they have a biproduct in A. Specifically, given a biproduct diagram* (1), *the object c with the projections* p_1 *and* p_2 *is a product of a and b, while, dually, c with* i_1 *and* i_2 *is a coproduct. In particular, two objects a and b have a product in A if and only if they have a coproduct in A.*

Proof. First assume we have the biproduct diagram (1) with the condition (2). Then

$$p_1 i_2 = p_1(i_1 p_1 + i_2 p_2) i_2 = 1 \circ p_1 i_2 + p_1 i_2 \circ 1 = p_1 i_2 + p_1 i_2 ;$$

subtracting, $p_1 i_2 = 0$; symmetrically $p_2 i_1 = 0$. (These are familiar equations for the usual biproduct of modules.) Now consider any diagram $a \xleftarrow{f_1} d \xrightarrow{f_2} b$. The sum $h = i_1 f_1 + i_2 f_2 : d \to c$ then has $p_i h = f_i$; conversely, if $h' : d \to c$ has $p_i h' = f_i$ for $i = 1, 2$, then

$$h' = (i_1 p_1 + i_2 p_2) h' = i_1 p_1 h' + i_2 p_2 h' = i_1 f_1 + i_2 f_2 ,$$

so $h' = h$. This states that there is a unique $h : d \to c$ with $p_i h = f_i$ for $i = 1, 2$, so the diagram $a \xleftarrow{p_1} c \xrightarrow{p_2} b$ is indeed a product. The assignment $h \mapsto \langle f_1, f_2 \rangle$ is an isomorphism

$$A(d, c) \cong A(d, a) \oplus A(d, b)$$

of abelian groups, where $\oplus$ on the right is the direct sum of abelian groups.

Conversely, given a product diagram $a \xleftarrow{p_1} a \times b \xrightarrow{p_2} b$, the definition of this product provides a unique arrow $i_1 : a \to a \times b$ with components $p_1 i_1 = 1_a$, $p_2 i_1 = 0$ and a unique $i_2 : b \to a \times b$ with $p_1 i_2 = 0$, $p_2 i_2 = 1_b$. Then

$$p_1(i_1 p_1 + i_2 p_2) = p_1 + 0 p_2 = p_1 , \qquad p_2(i_1 p_1 + i_2 p_2) = p_2 ,$$

so $i_1 p_1 + i_2 p_2 : a \times b \to a \times b$ is the unique arrow with components p_1 and p_2, hence is the identity $1_{a \times b}$. Thus the given product diagram does indeed yield a biproduct, with (1) and (2).

In special categories, such as **Ab** and R-**Mod**, the biproduct is often called a *direct sum*. Note also that the description of the biproduct diagram is "internal", since it involves only the objects a, b, and c and the arrows between them, while the standard categorical description of the product (or the coproduct) is "external", since it refers to construction of arrows in the whole category.

Given objects a, $b \in A$, the biproduct diagram (1), if it exists, is determined uniquely up to an isomorphism of the object c. If all such biproducts exist, then a choice of $c = a \oplus b$ for each pair $\langle a, b \rangle$ determines a bifunctor $\oplus : A \times A \to A$, with $f_1 \oplus f_2$ defined for arrows $f_1 : a \to a'$ and $f_2 : b \to b'$ either by the equations

$$p'_j(f_1 \oplus f_2) = f_j p_j , \qquad j = 1, 2 , \tag{3}$$

(i.e., defined as for a product $\times = \oplus$) or by the equations

$$(f_1 \oplus f_2) i_k = i'_k f_k , \qquad k = 1, 2 , \tag{4}$$

that is, as for a coproduct $\oplus = \amalg$, with i'_1, i'_2 the injections of the second coproduct. Indeed, the first pair (3) of equations determines $f_1 \oplus f_2$ uniquely as the arrow with components f_1, f_2; then by the defining equations for the second biproduct and by $p'_1 i'_2 = 0$,

$$(f_1 \oplus f_2) i_k = (i'_1 p'_1 + i'_2 p'_2)(f_1 \oplus f_2) i_k = i'_k f_k ,$$

as in the second pair of equations, and dually.

The conclusion may also be formulated thus: The identification of the product functor $a \times b$, with mapping function defined by (3), with the coproduct functor $a \amalg b$, mapping function defined by (4), is a natural isomorphism.

Iteration, for given $a_1, \ldots, a_n \in A$, yields a biproduct $\bigoplus_j a_j$ characterized (up to isomorphism in A) by the diagram

$$a_j \xrightarrow{i_j} \bigoplus_j a_j \xrightarrow{p_k} a_k , \qquad j, k = 1, \ldots, n$$

and the equations

$$i_1 p_1 + \cdots + i_n p_n = 1 , \qquad p_k i_j = \delta_{kj} = 0 \quad k \neq j , \qquad = 1 \quad k = j . \tag{5}$$

Moreover, for given $c_1, \ldots, c_m \in A$ there is an isomorphism

$$A\Big(\bigoplus_k c_k, \bigoplus_j a_j\Big) \cong \sum_{j,k} A(c_k, a_j)$$

of abelian groups, where Σ denotes the iterated biproduct of abelian groups. This implies that each arrow $f : \oplus_k c_k \to \oplus_j a_j$ is determined by the $n \times m$ matrix of its *components* $f_{kj} = p_k f i_j : a_j \to c_k$. Composition of arrows is then given by the usual matrix product of the matrices of components. In other words, the equations (5) contain the familiar calculus of matrices (cf. § III.5).

An *additive category* is by definition an Ab-category which has a zero object 0 and a biproduct for each pair of its objects.

Proposition 3. *For parallel arrows $f, f' : a \to b$ in an additive category A,*

$$f + f' = \check{\delta}^b (f \oplus f') \delta_a : a \to b , \tag{6}$$

where $\delta_a : a \to a \times a$ is the diagonal map, $\check{\delta}^b : b \oplus b = b \amalg b \to b$ the codiagonal.

Here the diagonal is defined by $p_1 \delta_a = 1_a = p_2 \delta_a$ and the codiagonal by $\check{\delta}^b i_1 = 1_b = \check{\delta}^b i_2$. The proof is a direct calculation:

$$\begin{aligned} \check{\delta}^b (f \oplus f') \delta_a &= \check{\delta}^b (f \oplus f')(i_1 p_1 + i_2 p_2) \delta_a \\ &= \check{\delta}^b (f \oplus f') i_1 + \check{\delta}^b (f \oplus f') i_2 \\ &= \check{\delta}^b i_1 f + \check{\delta}^b i_2 f' = f + f' . \end{aligned}$$

This proposition suggests that the additive structure of A can be derived from the biproduct (cf. Exercise 4).

If A and B are Ab-categories, an *additive functor* $T: A \to B$ is a functor from A to B with

$$T(f + f') = Tf + Tf' \tag{7}$$

for any parallel pair of arrows $f, f' : b \to c$ in A. It follows that $T0 = 0$. Since the additive structure of A can be described in terms of the biproduct structure of A, this condition (7) can also be reformulated as follows:

Proposition 4. *If A and B are Ab-categories, while A has all binary biproducts, then a functor $T: A \to B$ is additive if and only if T carries each binary biproduct diagram in A to a biproduct diagram in B.*

Proof. Each of the equations $p_1 i_1 = 1$, $p_2 i_2 = 1$, and $i_1 p_1 + i_2 p_2 = 1$ describing a biproduct in terms of its insertions i_j and projections p_j is preserved by an additive functor; therefore each additive functor preserves biproducts.

Conversely, suppose that T preserves all binary biproducts. Then a parallel pair of arrows $f_1, f_2 : a \to a'$ has $T(f_1 \oplus f_2) = Tf_1 \oplus Tf_2$ and therefore $T(f_1 + f_2) = Tf_1 + Tf_2$ by the formula (6) for sum in terms of direct sum and the equations $T(\delta_a) = \delta_{Ta}$, $T(\check{\delta}^a) = \check{\delta}^{Ta}$, which follows at once from the definition of the diagonal δ and the codiagonal $\check{\delta}$ in terms of product and coproduct.

Our proposition can also be modified: T is additive if and only if T carries each binary product diagram in A to a product diagram in B, or, if and only if it carries each binary coproduct in A to a coproduct in B.

Many familiar functors for Ab-categories A are additive. For example, if A has small hom-sets each hom-functor

$$A(a, -) : A \to \mathbf{Ab}, \qquad A(-, a) : A^{op} \to \mathbf{Ab}$$

is additive. If A and B are Ab-categories, so is $A \times B$, and the projections $A \times B \to A$, $A \times B \to B$ of this product are additive functors. The tensor product of abelian groups is a functor $\mathbf{Ab} \times \mathbf{Ab} \to \mathbf{Ab}$, additive in each of its arguments, and so is the torsion product.

Exercises

1. In any additive category A, show that the canonical map

   $$\kappa : a_1 \amalg \cdots \amalg a_n \to a_1 \times \cdots \times a_n$$

   (defined in §III.5) is an isomorphism. (This is essentially a reformulation of Theorem 2.)
2. Define the corresponding canonical map κ of an infinite coproduct to the corresponding infinite product, and show by an example that it need not be an isomorphism in every additive category.

3. In an additive category, show that the biproduct is associative and commutative (up to a natural isomorphism).
4. (Alternative definition of addition of arrows, to get an additive category.)
   (a) Let A be a category with a null object, finite products, and finite coproducts in which the canonical arrow $a_1 \amalg a_2 \to a_1 \times a_2$ from the coproduct to the product (§ III.5) is always an isomorphism. For $f, f' : a \to b$ define $f + f' = \check{\delta}^b (f \times f') \delta_a$. Prove that this addition makes each set $A(a, b)$ a commutative monoid, and that composition is distributive over this addition.
   (b) If, moreover, there is for each $a \in A$ an arrow $v_a : a \to a$ with $v_a + 1_a = 0 : a \to a$, prove that each $A(a, b)$ is a group under the addition defined above, and hence that this addition gives A the structure of an additive category (Mac Lane [1950]).
5. (The free Ab-category on a given C.) Given a category C, construct an Ab-category A and a functor $C \to A$ which is universal from C to an Ab-category. (Hint: The objects of A are those of C, while $A(b, c) = \mathbf{Z}(C(b, c))$ is the free abelian group on the set $C(b, c)$.)
6. (The free additive category.)
   (a) Given an Ab-category A, construct an additive category Add (A) and an additive functor $A \to$ Add (A) which is universal from A to an additive category. (Hint: Objects of Add (A) are n-tuples of objects of A, for $n = 0, 1, \ldots,$ while arrows are matrices of arrows of A.)
   (b) If A is the commutative ring K, regarded as an additive category with one object, show that Add (K) is the category $\mathbf{Matr}_K$ described in § I.2. (Hint: Show that $\mathbf{Matr}_K$ has the desired universal property.)

3. Abelian Categories

Definition. *An abelian category A is an Ab-category satisfying the following conditions*

(i) *A has a null object,*
(ii) *A has binary biproducts,*
(iii) *Every arrow in A has a kernel and a cokernel,*
(iv) *Every monic arrow is a kernel, and every epi a cokernel.*

The first two conditions ensure that A is an additive category, as described in §2. Instead of requiring a null object in (i), we could by Proposition 2.1 require a terminal object or an initial object. Instead of requiring all biproducts $a \oplus b$, we could require all products $a \times b$ or all binary coproducts.

With (i) and (ii), the existence of kernels in condition (iii) implies that A has all finite limits. Indeed, the equalizer of $f, g : a \to b$ may be constructed as the kernel of $f - g$, (i) and (ii) give finite products, and finite products and equalizers give all finite limits. Dually, the existence of cokernels implies the existence of all finite colimits.

Condition (iv) is powerful. It implies, for example, that any arrow f which is both monic and epi is an isomorphism. For $f: a \to b$ monic means $f = \ker g$ for some g, hence $gf = 0 = 0f$. But f is epi, so cancels to give $g = 0 : b \to c$, and the kernel of $g = 0$ is equivalent to the identity of b, hence is an isomorphism.

The categories R-**Mod**, **Mod**-R, **Ab** (and many others) are all abelian, with the usual kernels and cokernels. If A is abelian, so is any functor category A^J, for arbitrary J. Specifically, if $S, T: J \to A$ are any two functors, the set $\mathrm{Nat}(S, T) = A^J(S, T)$ of all natural transformations $\alpha, \beta: S \dot{\to} T$ is an abelian group, with addition defined termwise – $(\alpha + \beta)_j = \alpha_j + \beta_j : Sj \to Tj$ for each $j \in J$. The functor $N: J \to A$ everywhere equal to the null object of A is the null functor in A^J, the biproduct $S \oplus T$ of two functors is defined termwise, as $(S \oplus T)a = Sa \oplus Ta$, and the kernel K of a natural transformation $\alpha: S \dot{\to} T$ is defined termwise, so that for each j, $Kj \to Sj$ is the kernel of α_j. All the axioms follow, to make A^J abelian.

Proposition 1. *In an abelian category A, every arrow f has a factorization $f = me$, with m monic and e epi; moreover,*

$$m = \ker (\operatorname{coker} f), \qquad e = \operatorname{coker} (\ker f). \tag{1}$$

Given any other factorization $f' = m'e'$ with m' monic and e' epi and a commutative square

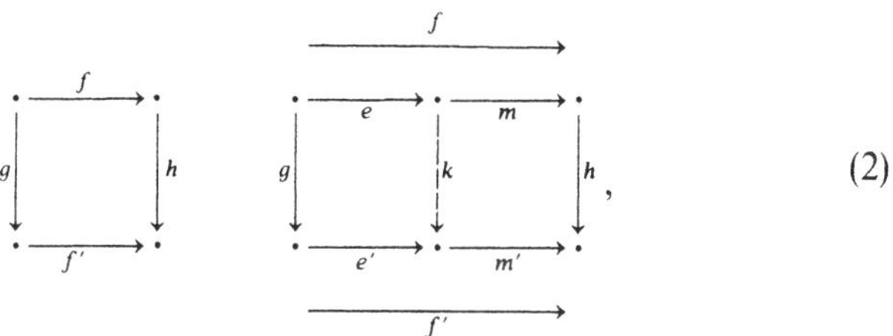

(2)

as shown at the left above, there is a unique k with $e'g = ke$, $m'k = hm$ (i.e., with the squares at the right commutative).

Proof. To construct such a factorization of f, take $m = \ker (\operatorname{coker} f)$. Since $(\operatorname{coker} f) \circ f = 0$, f factors as $f = me$ for a unique e, and by Lemma 1 of § 1, e is epi. Now m is monic, so for any composable t, $ft = 0$ if and only if $et = 0$. This implies that $\ker f = \ker e$. But e is epi, so the arrow $e = \operatorname{coker} (\ker e) = \operatorname{coker} (\ker f)$. We have proved (1).

Now regard f and f' as objects in the arrow category A^2; a morphism $\langle g, h \rangle : f \to f'$ is then just a commutative square as in (2) above. Consider the factorizations $f = me$ and $f' = m'e'$, and set $u = \ker f = \ker e$. Then $0 = hfu = m'e'gu$, so $e'gu = 0$, and $e'g$ must factor through $e = \operatorname{coker} u$ as $e'g = ke$ for a unique k. Then also $m'ke = m'e'g = hme$, so $m'k = hm$, and both squares commute in the rectangle of (2).

This completes the proof. The second part shows that any morphism $\langle g, h\rangle : f \to f'$ must carry a factorization of f to a factorization of f', so that the factorization is functorial. In particular, for the identity morphism $\langle 1, 1\rangle : f \to f$, this proves that any two monic-epi factorizations $f = me$ and $f = m'e'$ are isomorphic (k an isomorphism above).

From this factorization, we define (the usual) image and coimage of $f = me : a \to b$ as

$$m = \operatorname{im} f, \qquad e = \operatorname{coim} f, \tag{3}$$

uniquely up to isomorphism. Thus the *image* m of f is a subobject of its codomain b, its *coimage* a quotient object of its domain. More generally, if $f = m_1 t e_1$ with m_1 monic, t an isomorphism and e epi, then $m_1 \equiv \operatorname{im} f$, $e_1 \equiv \operatorname{coim} f$ and t is (the usual) isomorphism of coimage to image. This is the situation which arises in familiar concrete categories like **Ab**. If $f : B \to C$ is a morphism in **Ab** with kernel a subgroup K of B, image a subgroup S of C, then f factors as a three-fold composite

$$B \xrightarrow{e_1} B/K \xrightarrow{u} S \xrightarrow{m_1} C,$$

with e_1 the projection on the standard quotient group, m_1 the inclusion, and u the evident isomorphism of the coimage B/K to the image S. This three-fold factorization arises because each quotient object B/K has a canonical representation (by cosets).

Exact sequences work as usual in any abelian category.

Definition. *A composable pair of arrows,*

$$\cdot \xrightarrow{f} b \xrightarrow{g} \cdot\,, \tag{4}$$

is exact at b when $\operatorname{im} f \equiv \ker g$ (equivalence as subobjects of b) – or, equivalently, when $\operatorname{coker} f \equiv \operatorname{coim} g$.

Observe that $\operatorname{im} f \leqq \ker g$ if and only if $gf = 0$, while $\operatorname{im} f \geqq \ker g$ if and only if every k with $gk = 0$ factors as $k = mk'$, where m is the first factor in the monic-epi factorization $f = me$. This bipartite definition of exactness is just the usual condition (say in **Ab**): $\langle f, g\rangle$ exact means that the composite gf is zero and that every element killed by g is in the image of f.

Definition. *The diagram (with 0 the null object)*

$$0 \longrightarrow a \xrightarrow{f} b \xrightarrow{g} c \longrightarrow 0 \tag{5}$$

is a short exact sequence when it is exact at a, at b, and at c.

Since $0 \to a$ is the zero arrow, exactness at a means just that f is monic; dually, exactness at c means g epi. All told, (5) short exact thus is equivalent to

$$f = \ker g, \qquad g = \operatorname{coker} f. \tag{6}$$

Similarly, the statement that $h = \operatorname{coker} f$ becomes the statement that the sequence

$$a \xrightarrow{f} b \xrightarrow{h} c \longrightarrow 0 \qquad (7)$$

is exact at b and at c. Classically, such a sequence (7) was called a short *right exact sequence*. Similarly, $k = \ker f$ is expressed by a short left exact sequence.

The monic-epi factorization $f = me$ of any arrow f determines two short exact sequences which appear (with the bordering zeros omitted) as the top and side of the following commutative diagram:

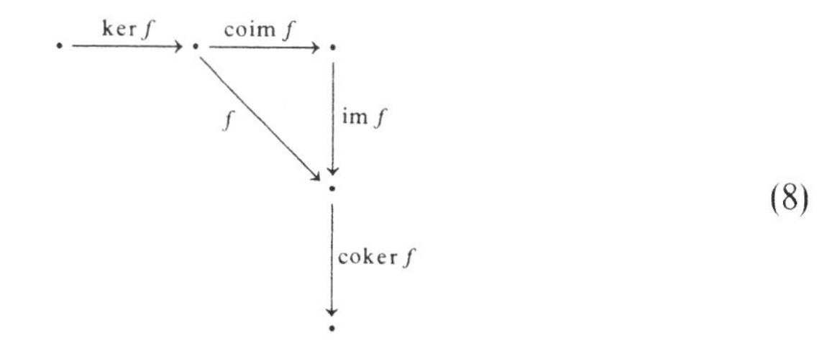

A functor $T: A \to B$ between abelian categories A and B is, by definition, *exact* when it preserves all finite limits and all finite colimits. In particular, an exact functor preserves kernels and cokernels, which means that

$$\ker(Tf) = T(\ker f), \qquad \operatorname{coker}(Tf) = T(\operatorname{coker} f)\,; \qquad (9)$$

it also preserves images, coimages, and carries exact sequences to exact sequences. By the familiar construction of limits from products and equalizers and dual constructions, $T: A \to B$ is exact if and only if it is additive and preserves kernels and cokernels.

A functor T is *left exact* when it preserves all finite limits. In other words, T is *left exact* if and only if it is additive and $\ker(Tf) = T(\ker f)$ for all f: the last condition is equivalent to the requirement that T preserves short left exact sequences.

Abelian categories have a more economical description, not involving a given abelian group structure on each hom-set. Explicitly, let A be *any* category which satisfies the axioms (i), (ii′), (iii), and (iv) just as above, except that (ii) is replaced by

(ii′) A has binary products and binary coproducts.

Then the formula (2.6) can be used to introduce an addition in each hom-set $A(a, b)$, and with this addition A is an abelian category. The somewhat fussy proof, Freyd [1964], Schubert [1970], will be omitted here because it seems of little use for the applications, where the categories usually come equipped with the needed addition in each $A(a, b)$.

Exercises

1. For A, B abelian categories, show that an additive functor $T: A \to B$ is exact if and only if it carries all short exact sequences in A to short exact sequences in B.
2. Prove: A and B abelian implies that the product category $A \times B$ is abelian.
3. Show that the category of all free abelian groups is not abelian.
4. Show that the category of all finite abelian groups (with arrows all morphisms of such) is abelian.
5. If R is a left noetherian ring, show that the category of all finitely generated left R-modules (with arrows all morphisms of such modules) is abelian.
6. For subobjects $u \leqq v$ of an object a in an abelian category, define a "quotient" object v/u (to agree with the usual notion in **Ab**). If $gf = 0$, prove that $\ker g/\operatorname{im} f$ is isomorphic to the dual object $\operatorname{coim} g/\operatorname{coker} f$.

4. Diagram Lemmas

In an abelian category A, a *chain complex* is a sequence

$$\cdots \longrightarrow c_{n+1} \xrightarrow{\partial_{n+1}} c_n \xrightarrow{\partial_n} c_{n-1} \longrightarrow \cdots \tag{1}$$

of composable arrows, with $\partial_n \partial_{n+1} = 0$ for all n. The sequence need not be exact at c_n; the deviation from exactness is measured by the n-th *homology object* (for the quotient, cf. Exercise 3.6)

$$H_n c = \operatorname{Ker}(\partial_n : c_n \to c_{n-1}) / \operatorname{Im}(\partial_{n+1} : c_{n+1} \to c_n) . \tag{2}$$

Initially in algebraic topology one used chain complexes only in **Ab** or in K-**Mod** (especially for K the integers modulo a prime), but more general considerations of sheaf theory and homological algebra use complexes in many other abelian categories. The definition (2) of homology applies in any abelian category; the development of its properties depends on certain manipulations of exact sequences, normally proved in **Ab** by chasing elements around diagrams. We will now show how the basic diagram lemmas hold in any (fixed) abelian category A.

A morphism $\langle m, e\rangle \to \langle m', e'\rangle$ of short exact sequences (in A) is by definition a triple $\langle f, g, h\rangle$ of arrows in A such that the diagram

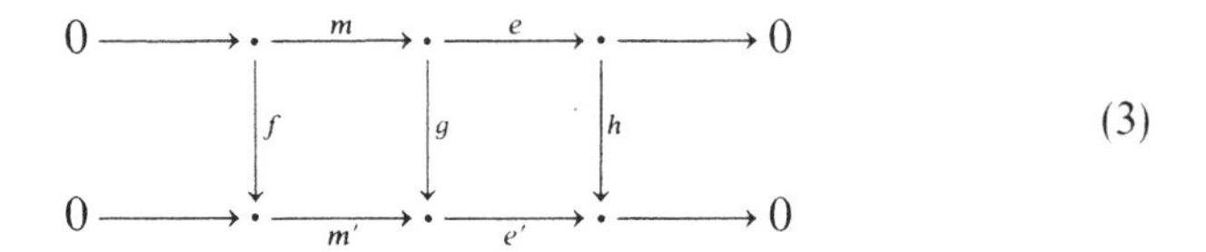

(3)

commutes. The short exact sequences with these morphisms constitute a category **Ses** A; in an evident way, it is additive. A first basic lemma is:

Lemma 1 *(The short five lemma). In any commutative diagram* (3) *with short exact rows, f and h monic imply g monic, and f and h epi imply g epi.*

In **Ab**, take any element x in ker g; then $g(x)=0$:

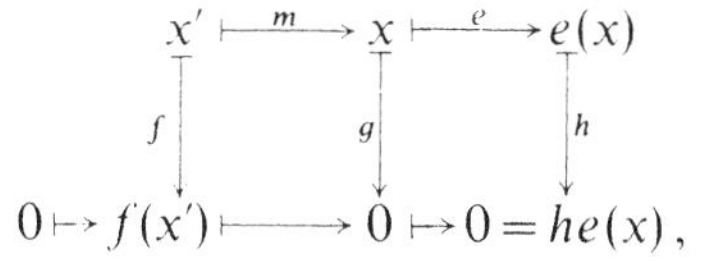

$$\begin{array}{ccccc} x' & \stackrel{m}{\longmapsto} & x & \stackrel{e}{\longmapsto} & e(x) \\ \downarrow f & & \downarrow g & & \downarrow h \\ 0 \mapsto f(x') & \longmapsto & 0 & \mapsto & 0 = he(x), \end{array}$$

so $he(x)=0$, $e(x)=0$. By exactness of the first row, there must be an element x' with $m(x')=x$. By exactness of the second row, $f(x')=0$, therefore $x'=0$, and so $x=0$. This argument is a "diagram chase" with x.

In any abelian category, the same argument can be done without elements. Take $k=\ker g$. Then $hek=e'gk=0$; since h is monic, $ek=0$. Therefore k factors through $m=\ker e$ as $k=mk'$. But $0=gk=gmk'=m'fk'$, and m' and f are monic, so $k=0$. Since $k=\ker g$, this proves g monic.

The proof that g is epi is dual.

In **Ab**, a pullback of a monic or an epi is monic or epi, respectively. This holds for pullbacks of monics in any category (Lemma V.7), and for pullbacks of epis in an abelian category, as follows.

Proposition 2. *Given a pullback square (on the right below)*

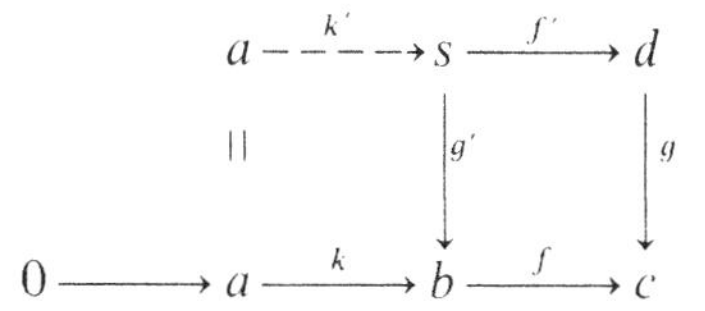

$$\begin{array}{ccccccc} & & a & \stackrel{k'}{\dashrightarrow} & s & \stackrel{f'}{\longrightarrow} & d \\ & & \| & & \downarrow g' & & \downarrow g \\ 0 & \longrightarrow & a & \stackrel{k}{\longrightarrow} & b & \stackrel{f}{\longrightarrow} & c \end{array}$$

in an abelian category, f epi implies f' epi. Also, the kernel k of f factors as $k=g'k'$ for a k' which is the kernel of f'.

In particular, given a short exact sequence $a \rightarrowtail b \twoheadrightarrow c$, each arrow $g: d \to c$ to the right-hand end object c yields by pullback a short exact sequence $a \rightarrowtail s \twoheadrightarrow d$. This operation (and its dual) is basic to the description of Ext (c, a) (the set of "all" short exact sequences from a to c) as a bifunctor for an abelian category (Mac Lane [1963] Chap. III).

Proof. The pullback s (like any pullback) is constructed from products and equalizers thus: Take $b \oplus d$ with projections p_1 and p_2, form the left exact sequence

$$0 \longrightarrow s \stackrel{m}{\longrightarrow} b \oplus d \xrightarrow{fp_1 - gp_2} c$$

(i.e., m is a kernel), and set $g'=p_1 m$, $f'=p_2 m$.

Here $fp_1 - gp_2$ is epi. For suppose $h(fp_1 - gp_2)=0$ for some h. Then, using the injection i_1 of the biproduct,

$$0 = h(fp_1 - gp_2)i_1 = hfp_1 i_1 = hf,$$

and $h=0$ because f is given to be epi.

Now suppose $uf' = 0$ for some u. Since $f' = p_2 m$, $up_2 m = 0$, so up_2 factors through $fp_1 - gp_2 = \operatorname{coker} m$ as $up_2 = u'(fp_1 - gp_2)$. But $p_2 i_1 = 0$, so

$$0 = up_2 i_1 = u'(fp_1 - gp_2)i_1 = u'fp_1 i_1 = u'f.$$

Since f is epi, $u' = 0$; therefore f' is epi, as desired.

Finally, consider $k = \ker f$. The pair of arrows $k : a \to b$ and $0 : a \to d$ have $fk = 0 = g0$, so by the definition of the pullback s there is a unique arrow $k' : a \to s$ with $g'k' = k$ and $f'k' = 0$; since k is monic, so is k'. To show it the kernel of f' consider any arrow v with $f'v = 0$. Then $fg'v = gf'v = 0$, so $g'v$ factors through $k = \ker f$ as $g'v = kv'$ for some v'. Then $g'v = g'(k'v')$ and $f'v = 0 = f'(k'v')$, so by the uniqueness involved in the definition of a pullback, $v = k'v'$. Therefore $k' = \ker f'$, as desired.

In virtue of this Proposition, diagram chases can be made in any abelian category using "members" (in A) instead of elements (in **Ab**). Call an arrow x with codomain $a \in A$ a *member* of a, written $x \in_m a$, and define $x \equiv y$ for two members of a to mean that there are epis u, v with $xu = yv$. This relation is manifestly reflexive and symmetric. To prove it transitive, suppose also that $yw = zr$ for epis w and r and form the pullback square displayed at the upper left in the diagram

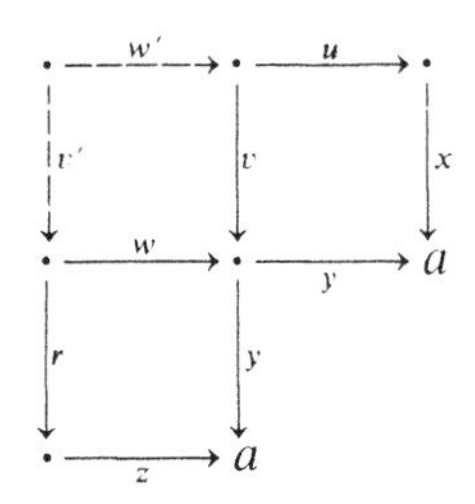

By Proposition 2, v' and w' are epi, and hence $x \equiv z$. Then a member of a is an equivalence class, for the relation $\equiv$, of arrows to a. Since every arrow x has a factorization $x = me$, every member of a is represented by a subobject (a monic m) of a, but we shall not need to use this fact. Each object a has a zero member, the (equivalence class of the) zero arrow $0 \to a$. Each member $x \in_m a$ has a "negative" $-x$.

For any arrow $f : a \to b$, each member $x \in_m a$ gives $fx \in_m b$, and $x \equiv y$ in a implies $fx \equiv fy$ in b, so any arrow from a to b carries members of a to members of b – just as if these members were elements of sets.

Theorem 3 *(Elementary rules for chasing diagrams). For the members in any abelian category*

(i) *$f : a \to b$ is monic if and only if, for all $x \in_m a$, $fx \equiv 0$ implies $x \equiv 0$;*
(ii) *$f : a \to b$ is monic if and only if, for all $x, x' \in_m a$, $fx \equiv fx'$ implies $x \equiv x'$;*
(iii) *$g : b \to c$ is epi if and only if for each $z \in_m c$ there exist a $y \in_m b$ with $gy \equiv z$;*

(iv) $h: r \to s$ *is the zero arrow if and only if, for all* $x \in_m r$, $hx \equiv 0$;

(v) *A sequence* $a \xrightarrow{f} b \xrightarrow{g} c$ *is exact at b if and only if* $gf = 0$ *and to every* $y \in_m b$ *with* $gy \equiv 0$ *there exists* $x \in_m a$ *with* $fx \equiv y$;

(vi) *(Subtraction) Given* $g: b \to c$ *and* $x, y \in_m b$ *with* $gx \equiv gy$, *there is a member* $z \in_m b$ *with* $gz \equiv 0$; *moreover, any* $f: b \to d$ *with* $fx \equiv 0$ *has* $fy \equiv fz$ *and any* $h: b \to a$ *with* $hy \equiv 0$ *has* $hx \equiv -hz$.

Proof. Rules (i) and (ii) are just the definition of a monic. In (iii), if g is epi, then one can construct $y \in_m b$ with $gy \equiv z$ by pullback (using Proposition 2); conversely, if g is not epi, the member $1_c \in_m c$ is not of the form $gy \equiv 1_c$ for any $y \in_m b$. Rule (iv) is trivial.

For rule (v), take the standard factorization $f = me$, and suppose first that the given sequence is exact at b, so that $m = \ker g$. If $gy \equiv 0$, $y \equiv my_1$ for some y_1. Form the pullback at the left of the diagram

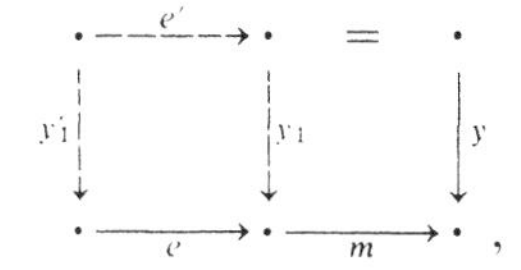

since $ye' = mey_1' = fy_1'$ and e' is epi, $y \equiv fy_1'$, as required. Conversely, given this property for all $y \in_m b$, take $k = \ker g$; then $k \in_m b$ and $gk \equiv 0$ (in c). Therefore there is a member $x \in_m a$ with $fx \equiv k$; that is, with $ku = mexv$ for suitable epis u and v. But this equation implies that the monic k factors through m, and hence that $\operatorname{im} f \geq \ker g$. Combined with $gf = 0$, this gives the desired exactness.

Rule (vi) is intended to replace the subtraction of elements in **Ab**. If $gx \equiv gy$, there are epis u, v with $gxu = gyv$, and (vi) holds with $z = yv - xu \in_m b$.

Here is an example of a diagram chase with these methods:

Lemma 4 *(The Five Lemma). In a commutative diagram*

$$\begin{array}{ccccccccc}
a_1 & \xrightarrow{g_1} & a_2 & \xrightarrow{g_2} & a_3 & \xrightarrow{g_3} & a_4 & \xrightarrow{g_4} & a_5 \\
\downarrow{\scriptstyle f_1} & & \downarrow{\scriptstyle f_2} & & \downarrow{\scriptstyle f_3} & & \downarrow{\scriptstyle f_4} & & \downarrow{\scriptstyle f_5} \\
b_1 & \xrightarrow[h_1]{} & b_2 & \xrightarrow[h_2]{} & b_3 & \xrightarrow[h_3]{} & b_4 & \xrightarrow[h_4]{} & b_5
\end{array} \tag{4}$$

with exact rows, f_1, f_2, f_4, f_5 *isomorphisms imply* f_3 *an isomorphism.*

Proof. By duality, it suffices to prove f_3 monic. In **Ab** one would "chase" an element $x \in \ker f_3$. Consider instead any member $x \in_m a_3$ with $f_3 x \equiv 0$. This gives $f_4 g_3 x \equiv 0$; since f_4 is monic, $g_3 x \equiv 0$:

$$\begin{array}{ccccccc}
z & & y & \overset{g_2}{\longmapsto} & x & \longmapsto & g_3 x \\
\downarrow{\scriptstyle f_1} & & \downarrow{\scriptstyle f_2} & & \downarrow{\scriptstyle f_3} & & \downarrow \\
y' & \xrightarrow[h_1]{} & f_2 y & \longrightarrow & 0 & \longmapsto & 0 = f_4 g_3 x .
\end{array}$$

By exactness at a_3 and Rule (v) of the theorem, there is a $y \in_m a_2$ with $g_2 y \equiv x$. Then $0 \equiv f_3 x \equiv f_3 g_2 y \equiv h_2 f_2 y$, so by exactness at b_2 there is a $y' \in_m b_1$ with $h_1 y' \equiv f_2 y$. Since f_1 is epi, there is a $z \in_m a_1$ with $h_1 f_1 z \equiv f_2 y$ or $f_2 g_1 z \equiv f_2 y$. But f_2 is monic, so, by Rule (ii), $g_1 z \equiv y$ and $x \equiv g_2 y \equiv g_2 g_1 z \equiv 0$. Since any x with $f_3 x \equiv 0$ is itself 0, f_3 is monic, as required.

As another illustration, consider any morphism $\langle f, g, h \rangle$ of short exact sequences, as in (3); add the kernels and cokernels of f, g, and h to form a diagram

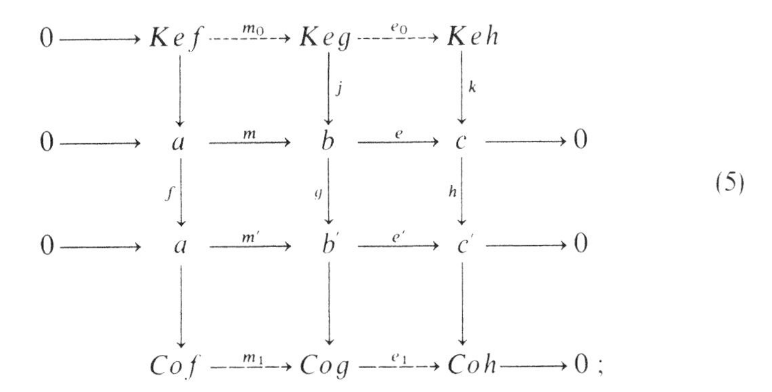

(5)

where Kef is the domain of ker f, Cof the codomain of coker f, etc. In this diagram the columns (with 0's added top and bottom) are exact sequences by construction, and both middle rows are given to be exact. By the definitions of kernel and cokernel, one may add unique arrows m_0, e_0 in the top row and m_1, e_1 in the bottom row so as to make the added squares commute. An easy diagram chase (by the method of Theorem 3) shows the first row exact at Kef and Keg; dually, the last row is exact at Cog and Coh. However, the first row is not necessarily a short exact sequence because e_0 need not be epi; moreover, this happens precisely when m_1 is not monic. An easy example of this phenomenon (in **Ab**; $g \neq 0$) is

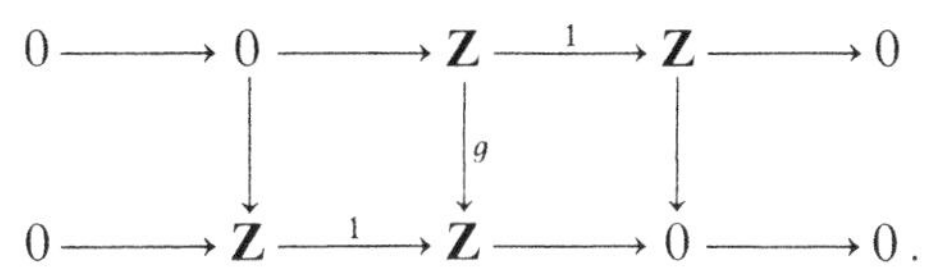

The failure of exactness can be repaired by the following striking lemma, which produces an added δ called the *connecting homomorphism* – it is essentially the connecting homomorphism used for relative homology (a complex modulo a subcomplex) and for the connecting maps between derived functors in homological algebra.

Lemma 5 (Ker-coker *sequence* = *Snake* **lemma**). *Given a morphism* $\langle f, g, h \rangle$ *of short exact sequences, as in* (3), *there is an arrow* $\delta : Keh \to Cof$

such that the following sequence is exact:

$$0 \longrightarrow Kef \xrightarrow{m_0} Keg \xrightarrow{e_0} Keh \xrightarrow{\delta} Cof \xrightarrow{m_1} Cog \xrightarrow{e_1} Coh \longrightarrow 0 \qquad (6)$$

Proof. From the map of short exact sequences we first build a different diagram; on the left in

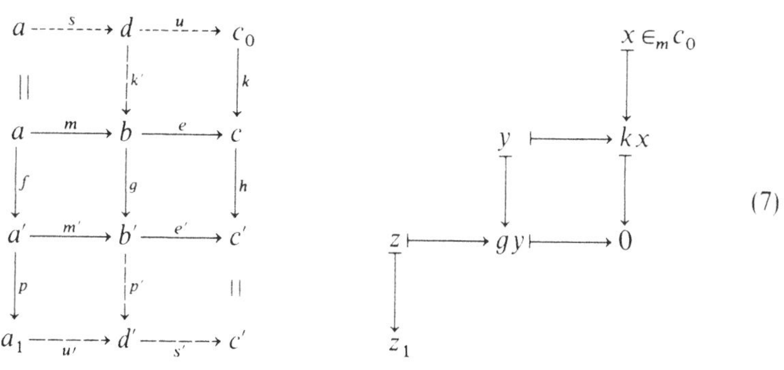

$c_0 = Keh$, d is the pullback of e and $k = \ker h$, so that u is epi with kernel s as in Proposition 2; dually, d' is the pushout of $p = \operatorname{coker} f$ and m', with cokernel s' as shown. Right down the middle runs a composite arrow $\delta_0 = p'gk' : d \to d'$, with $s'\delta_0 = hku = 0$ and $\delta_0 s = u'pf = 0$. Since $u' = \ker s'$ and $u = \operatorname{coker} s$, this means that δ_0 factors uniquely as

$$\delta_0 = u'\delta u : d \xrightarrow{u} c_0 \xrightarrow{\delta} a_1 \xrightarrow{u'} d'.$$

The middle factor is the required "connecting" arrow $\delta : c_0 \to a_1$.

The effect of this arrow δ on a member $x \in_m c_0$ can be described by the zig-zag staircase shown at the right of (7) above. Indeed, since e is epi there is a member $y \in_m b$ with $ey \equiv kx$. Then $e'gy \equiv hey \equiv hkx \equiv 0$, so by exactness there is a member $z \in_m a'$ with $m'z \equiv gy$. We claim that δx is then the member $z_1 = pz \in_m a_1$. For, d is a pullback so there is an $x_0 \in_m d$ with $ux_0 \equiv x$, $k'x_0 \equiv y$. Then

$$u'\delta x \equiv u'\delta u x_0 = \delta_0 x_0 \equiv p'gy \equiv u'z_1$$

and u' is monic, so $\delta x \equiv z_1$. This argument also proves that (the equivalence class of) the member z_1 is independent of the choices made in the construction of the zig-zag (7). This zig-zag is exactly the description usually given for the action of a connecting morphism δ on the elements of abelian groups.

Using the zig-zag description we can now prove the exactness of the ker-coker sequence (6), say the exactness at Keh. First, to show that $\delta e_0 = 0$, it suffices to show $\delta e_0 w = 0$ for any member $w \in_m b_0 = Keg$. But the member $e_0 w = x \in_m c_0$ has $kx = ke_0 w = ejw$, where $j = \ker g$ as in (5); hence in the zig-zag (7) we may choose $y = jw$. Then $gy = gjw \equiv 0$, which proves that $\delta e_0 = 0$. On the other hand, consider any $x \in_m c_0$ with $\delta x \equiv 0$. This means that the z_1 constructed in the zig-zag has $z_1 \equiv 0$; by exactness there is a member $z_2 \in_m a$ with $fz_2 \equiv z$ which means that $gmz_2 \equiv gy$. Now form the "difference" member $y_0 = y - mz_2 \in_m b$. By Rule (vi) above, this difference member has $ey_0 \equiv ey = kx$ and $gy_0 \equiv 0$. But $j : b_0 \to b$ is $\ker g$, so there is an $x_0 \in_m b_0$ with $jx_0 \equiv y_0$:

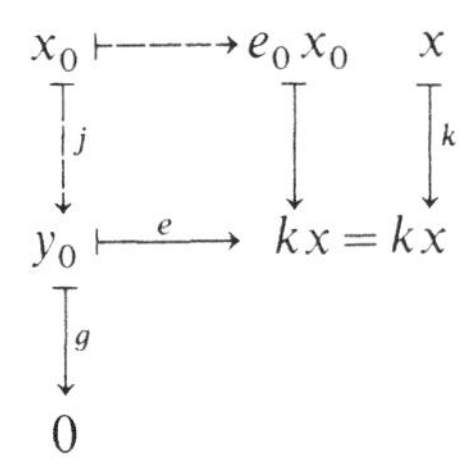

Then $ke_0x_0 = ey_0 \equiv kx$ and k is monic, so $e_0x_0 \equiv x$. We have shown that each x with $\delta x \equiv 0$ has the form $x \equiv e_0x_0$, so is in the image of e_0. This proves exactness; in fact, it is exactly like the usual exactness proof with honest elements of actual abelian groups.

Exercises

1. In the five-lemma, obtain minimal hypotheses (on f_1, f_2, and f_4 only) for f_3 to be monic.
2. In the five-lemma, prove f_3 epi using members (*not* comembers). (Hint: Rule (vi) of Theorem 3 is necessary in this proof.)
3. Complete the proof of the exactness of the ker-coker sequence.
4. Show that the connecting morphism δ is natural; i.e., that it is a natural transformation between two appropriate functors defined on a suitable category whose objects are morphisms (3) of short exact sequences.
5. A 3×3 diagram is one of the form (bordered by zeros)

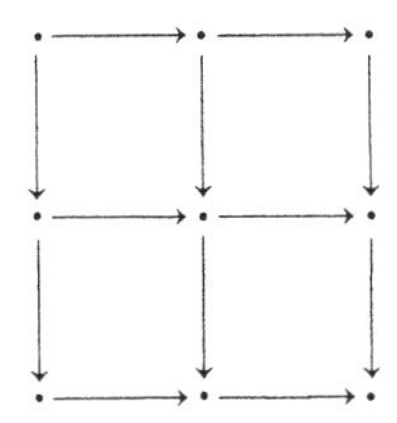

   (a) Give a direct proof of the 3×3 lemma: If a 3×3 diagram is commutative and all three columns and the last two rows are short exact sequences, then so is the first row.
   (b) Show that this lemma also follows from the ker-coker sequence.
   (c) Prove the middle 3×3 lemma: If a 3×3 diagram is commutative, and all three columns and the first and third rows are short exact sequences, then so is the middle row.
6. For two arrows $f : a \to b$ and $g : b \to c$ establish an exact sequence
$$0 \to \mathrm{Ke}\, f \to \mathrm{Ke}\, gf \to \mathrm{Ke}\, g \to \mathrm{Co}\, f \to \mathrm{Co}\, gf \to \mathrm{Co}\, g \to 0\,.$$
7. Show explicitly that the category **Ses** (A) is not in general abelian.

Notes.

Shortly after the discovery of categories, Eilenberg and Steenrod [1952] showed how the language of categories and functors could be used to give an axiomatic description of the homology and cohomology of a topological space. This, in turn, suggested the problem of describing those categories in which the values of such a homology theory could lie. After discussions with Eilenberg, this was done by Mac Lane [1948, 1950]. His notion of an "abelian bicategory" was clumsy, and the subject languished until Buchsbaum's axiomatic study [1955] and the discovery by Grothendieck [1957] that categories of sheaves (of abelian groups) over a topological space were abelian categories but not categories of modules, and that homological algebra in these categories was needed for a complete treatment of sheaf cohomology (Godement [1958]). With this impetus, abelian categories joined the establishment.

This chapter has given only an elementary theory of abelian categories – a demonstration directly from the axioms of all the usual diagram lemmas. Our method of "chasing members" is an adaptation of the method given by Mac Lane [1963, Chap. XII]; the critical point is the snake lemma, which must *construct* an arrow. Earlier proofs of this lemma in abelian categories were obscure; the present version is due to M. André (private communication). These diagram lemmas can also be proved in abelian categories from the case of R-modules by using suitable embedding theorems (Lubkin-Haron-Freyd-Mitchell). These beautiful theorems construct for any small abelian category A a faithful, exact functor $A \rightarrow \mathbf{Ab}$ and a full and faithful exact functor $A \rightarrow R\text{-}\mathbf{Mod}$ for a suitable ring R. For proofs we refer to Mitchell [1965], Freyd [1964], and Pareigis [1970].

These sources will also indicate the further elegant developments for abelian categories: A Krull-Remak-Schmidt theorem, Morita duality, the construction of "injective envelopes" in suitable abelian categories, the structure of Grothendieck categories, and the locally Noetherian categories (Gabriel [1962]).

IX. Special Limits

This chapter covers two useful types of limits (and colimits): The filtered limits, which are limits taken over preordered sets which are directed (and, more generally, over certain filtered categories), and the "ends", which are limits obtained from certain bifunctors, and which behave like integrals.

1. Filtered Limits

A preorder P is said to be *directed* when any two elements $p, q \in P$ have an upper bound in P; that is, an r with $p \leqq r$ and $q \leqq r$ (there is no requirement that r be unique). It follows that any finite set of elements of P has an upper bound in P. A directed preorder is also called a "directed set" or a "filtered set".

This notion (renamed) generalizes to categories. A category J is *filtered* when J is not empty and

(a) To any two objects $j, j' \in J$ there is $k \in J$ and arrows $j \rightarrow k$, $j' \rightarrow k$:

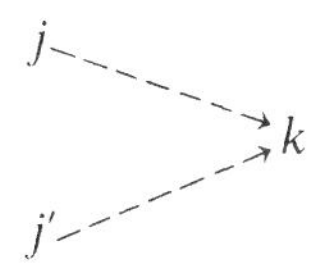

(b) To any two parallel arrows $u, v : i \rightarrow j$ in J, there is $k \in J$ and an arrow $w : j \rightarrow k$ such that $wu = wv$, as in the commutative diamond

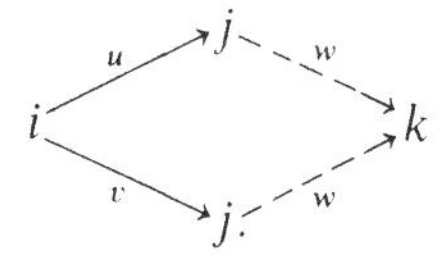

Condition (a) states that the discrete diagram $\{j, j'\}$ is the base of a cone with vertex k. Condition (b) states that $i \rightrightarrows j$ is the base of a cone. It follows that any finite diagram in a filtered category J is the base of at least one cone with a vertex $k \in J$.

Note that the terminology for "co" varies. Some authors (e.g., Mac Lane-Moerdijk [1992]) call such a filtered J "cofiltered".

A *filtered colimit* is by definition a colimit of a functor $F : J \to C$ defined on a filtered category J.

Classically, colimits were defined only over directed preorders (sometimes just over directed orders). This has proved to be a needless conceptual restriction of the notion of colimit. What does remain relevant is the interchange formulas for filtered colimits (§ 2) and the possibility of obtaining all colimits from finite coproducts, coequalizers, and colimits over directed preorders. Since we already know that (infinite) coproducts and coequalizers give all colimits (the dual of Theorem V.2.1) this needs only the following result.

Theorem 1. *A category C with finite coproducts and colimits over all (small) directed preorders has all (small) coproducts.*

Proof. We wish to construct a colimit for a functor $F : J \to C$, where J is a set ($=$ a discrete category). Let J^+ be the preorder with objects all finite subsets $S \subset J$, ordered by inclusion; clearly, J^+ is filtered. Let F^+ assign to each finite subset S the coproduct $\amalg Fs$, taken over all $s \in S$. If $S \subset T$ is an arrow $u : S \to T$ of J^+, take $F^+ u$ to be the unique (dotted) arrow which makes the diagram

$$\begin{array}{ccc} F^+S = \amalg Fs & \dashrightarrow & \amalg Ft = F^+T \\ \uparrow{\scriptstyle i_s} & & \uparrow{\scriptstyle i'_s} \\ Fs & = & Fs \end{array}$$

commute for every $s \in S$, with i and i' the injections of the coproducts. This evidently makes F^+ a functor $J^+ \to C$ which agrees on J with the given functor F, if J is included in J^+ by identifying each j with the one-point subset $\{j\}$.

Now consider any natural transformation $\theta : F^+ \dot{\to} G$ to some other functor $G : J^+ \to C$. For each $s \in S$ the diagram

$$\begin{array}{ccc} F^+S = \amalg Fs & \overset{\theta_S}{\dashrightarrow} & GS \\ \uparrow{\scriptstyle i_s} & & \uparrow{\scriptstyle G(\{s\} \subset S)} \\ Fs & \underset{\theta_s}{\dashrightarrow} & G\{s\} \end{array}$$

commutes. By the definition of coproducts, this means that θ is completely determined by the values θ_s of θ on Fs. In particular, each cone $v^+ : F^+ \dot{\to} c$ over F^+ is completely determined by its values on J, which form a cone $v : F \dot{\to} c$ over F. Moreover, v^+ is a limiting cone if and only if v is. Thus we can calculate the desired coproduct $\amalg F_j$, which is the colimit of F, as the colimit of F^+, known to exist because J^+ is a directed preorder.

As a typical application, we construct colimits in **Grp**.

Proposition 2. *The forgetful functor* **Grp** $\to$ **Set** *creates filtered colimits.*

Proof. We are given a filtered category J and a functor $G: J \to \mathbf{Grp}$; it assigns to arrows $j \to k$ group morphisms $G_j \to G_k$; we shall write G_j both for the group and its underlying set. We are also given a limiting cone μ for the composite functor $J \to \mathbf{Grp} \to \mathbf{Set}$; it has a set S as vertex and assigns to each $j \in J$ a function $\mu_j: G_j \to S$. We first show that there is a unique group structure on the set S which will make all functions μ_j morphisms of groups. First note that to each $s \in S$ there is at least one index j with a group element g_j for which $\mu_j g_j = s$; otherwise we could omit s from S to have a cone with a smaller set S' as vertex, an evident contradiction to the universality of S (there would be two functions $S \rightrightarrows S$ having the same composite with μ).

Now we define a product of any two elements $s, t \in S$. Write $s = \mu_j g_j$ and $t = \mu_k g_k$ for some $j, k \in J$; since J is filtered, there is in J a cone

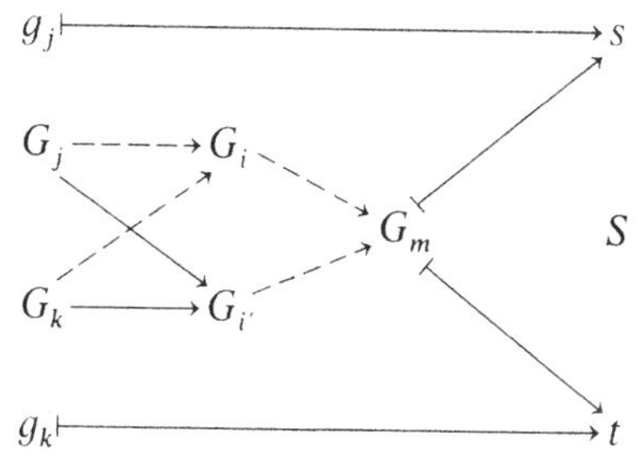

over j, k with some vertex i. The image of this cone under G is $G_j \to G_i \leftarrow G_k$, so s and $t \in S$ both come from elements of the group G_i; define their product in S to be μ_i of their product in G_i. This product is independent of the choice of i, because a different choice i' is part of a cone $G_i \to G_m \leftarrow G_{i'}$ of *group morphisms*. Also, the product of three factors r, s, t is associative, because we can choose G_i to contain pre-images of all three, and multiplication is known to be associative in G_i. Each group G_i has a unit element, and each $G_j \to G_k$ maps unit to unit; the common image of these units is a unit for the multiplication in S. Inverses are formed similarly.

We now have found a (unique) group structure on S for which $\mu_j: G_j \to S$ is a morphism of groups. This states that μ is a cone from G to S in $\mathbf{Grp}$. It is universal there: If $\nu: G \dot{\to} T$ is another cone in $\mathbf{Grp}$, it is also a cone in $\mathbf{Set}$, so there is a unique set map $f: S \to T$ with $f\mu = \nu$; one checks as above that this map f must be a morphism of groups.

This argument is clearly not restricted to $\mathbf{Grp}$; it applies to each category $\mathbf{Alg}_\tau$ of algebras of a fixed type τ (defined by operators and identities, § V.6). The same remark applies to the following corollary.

Corollary 3. $\mathbf{Grp}$ *has all (small) colimits.*

Proof. First, the one-element group is an initial object in $\mathbf{Grp}$. Next, any two groups G and H have a coproduct $G * H$. Indeed, any pair of homomorphisms $G \to L$, $H \to L$ to a third group L factors through the subgroup of L generated by the images of G and H. The cardinal number of this subgroup is bounded; this verifies the solution set condition for an application of the adjoint functor theorem to construct the coproduct $G * H$. Comparison with $G \to G \times X$ shows that $G \to G \times H$ is monic.

These two observations show that **Grp** has all finite coproducts. By Proposition 2, it has all filtered colimits. Hence, by Theorem 1, it has all small coproducts. To get all small colimits we then need only coequalizers, and the coequalizer of two homomorphisms $u, v : G \to H$ is the projection $H \to H/N$ on the quotient group by the least normal subgroup N containing all the elements $(ug)(vg)^{-1}$ for $g \in G$.

This proof gives an explicit picture of the coproduct in **Grp**. The coproduct $G * H$ of two groups is usually called the *free product*; its elements are finite words $\langle g_1, h_1, g_2, h_2, \ldots, g_n, h_n \rangle$ spelled in letters $g_i \in G$ and $h_i \in H$; these words are multiplied by juxtaposition, while equality is given by successive cancellations (if $h_i = 1$ in H, drop it and multiply $g_i g_{i+1}$ in G, etc.). A direct proof of associativity of the multiplication from this definition is fussy. By this corollary an infinite coproduct $\coprod_i G_i$ of groups G_i is obtained by pasting together all the finite coproducts

$$G_{i_1} * G_{i_2} * \cdots * G_{i_k}$$

(the inclusion maps make this a subgroup of any coproduct of more factors). Thus $\coprod_i G_i$ is the union of all these finite coproducts, identified along these inclusion maps.

Exercises

1. Use the adjoint functor theorem to prove in one step that **Grp** has all small colimits.
2. Prove that $\mathbf{Alg}_\tau$ as described in §V.6 has all small colimits; in particular, describe the initial object (when is it empty?).

2. Interchange of Limits

Consider a bifunctor $F : P \times J \to X$ to a cocomplete category X. For values $p \in P$ of the parameter p, the colimits of $F(p, -) : J \to X$ define functors $p \mapsto \mathrm{Colim}_j F(p, j)$ of p, so that the colimiting cones.

$$\mu_{p,j} : F(p, j) \to \underrightarrow{\mathrm{Colim}}_j F(p, j) \tag{1}$$

are natural in p (Theorem V.3.1). One may prove readily (§ 8 below) that

$$\underrightarrow{\mathrm{Colim}}_p \, \underrightarrow{\mathrm{Colim}}_j \, F(p, j) \cong \underrightarrow{\mathrm{Colim}}_j \, \underrightarrow{\mathrm{Colim}}_p F(p, j) \tag{2}$$

with the isomorphism given by the "canonical" map. Dually, limits commute. But limits need not commute with colimits, because the canonical map

$$\kappa : \underrightarrow{\mathrm{Colim}}_j \underleftarrow{\mathrm{Lim}}_p F(p,j) \to \underleftarrow{\mathrm{Lim}}_p \underrightarrow{\mathrm{Colim}}_j F(p,j) \tag{3}$$

need not be an isomorphism.

This *canonical map* exists as soon as all four limits and colimits in (3) exist, and is constructed as in the following diagram

$$\begin{array}{ccccc} F(p,j) & \xleftarrow{\nu_{p,j}} & \underleftarrow{\mathrm{Lim}}_p F(p,j) & \xrightarrow{\mu_j} & \underrightarrow{\mathrm{Colim}}_j \underleftarrow{\mathrm{Lim}}_p F(p,j) \\ \downarrow{\scriptstyle \mu_{p,j}} & & \downarrow{\scriptstyle \alpha_j} & & \downarrow{\scriptstyle \kappa} \\ \underrightarrow{\mathrm{Colim}}_j F(p,j) & \xleftarrow{\nu_p} & \underleftarrow{\mathrm{Lim}}_p \underrightarrow{\mathrm{Colim}}_j F(p,j) & = & \underleftarrow{\mathrm{Lim}}_p \underrightarrow{\mathrm{Colim}}_j F(p,j) \end{array} \tag{4}$$

where ν and $\nu_{-,j}$ for each j are limiting cones, and μ, $\mu_{p,-}$ for each p the colimiting cones. Since ν is a cone in p and μ is natural in p, the composite $\mu_{p,j}\nu_{p,j}$ for fixed j is a cone in p; by the universality of ν there then exist arrows α_j for each j making the left hand squares commute. Since $\mu_{p,-}$ is a cone in j, so is α; by the universality of μ, there is then a map κ making the right hand square commute. It is the desired canonical arrow.

This κ need not be an isomorphism. Consider, for example, the case when $P=\{1,2\}$ and $J=\{1,2\}$ are both discrete 2-object categories. The canonical κ when it exists (in evident notation)

$$\kappa : (A_1 \times B_1) \amalg (A_2 \times B_2) \to (A_1 \amalg A_2) \times (B_1 \amalg B_2)$$

is given by two components α_1 and α_2, where α_1 is determined by

$$\begin{array}{ccccc} A_1 & \xleftarrow{p} & A_1 \times B_1 & \xrightarrow{q} & B_1 \\ \downarrow{\scriptstyle i} & & \downarrow{\scriptstyle \alpha_1} & & \downarrow{\scriptstyle i'} \\ A_1 \amalg A_2 & \xleftarrow{p} & (A_1 \amalg A_2) \times (B_1 \amalg B_2) & \xrightarrow{q} & B_1 \amalg B_2 . \end{array}$$

In **Ab**, κ is evidently an isomorphism, but in **Set** it is not – the domain of κ is a disjoint union of two sets, while the codomain of κ is the four-fold disjoint union

$$(A_1 \times B_1) \amalg (A_1 \times B_2) \amalg (A_2 \times B_1) \amalg (A_2 \times B_2) .$$

We now turn to conditions which suffice to make κ an isomorphism.

Theorem 1. *If the category P is finite while J is small and filtered, then for any bifunctor $F : P \times J \to$ **Set** the canonical arrow*

$$\kappa : \underrightarrow{\mathrm{Colim}}_j \underleftarrow{\mathrm{Lim}}_p F(p,j) \to \underleftarrow{\mathrm{Lim}}_p \underrightarrow{\mathrm{Colim}}_j F(p,j)$$

as in (iv) is an isomorphism.

This states that finite limits commute with filtered colimits in **Set**.

Proof. By the construction of colimits in terms of coproducts and coequalizers (dual of Theorem V.2.2 with J filtered),

$$\underrightarrow{\mathrm{Colim}}_j F(p,j) = \amalg_j F(p,j)/E , \tag{5}$$

where $\amalg_j$ is the disjoint union and E is the equivalence relation defined for elements $x \in F(p,j)$ and $x' \in F(p,j')$ in that union by xEx' if and only if there are arrows $u: j \to k$, $u': j' \to k$ with $F(p,u)x = F(p,u')x'$. Write (x,j) for the E-equivalence class of an element $x \in F(p,j)$. Now J is filtered; condition (a) in the definition of "filtered" implies that any finite list $(x_1, j_1), \ldots, (x_m, j_m)$ of such elements can be written as a list $(y_1, k), \ldots, (y_m, k)$ with one second index k. Condition (b) in the definition implies that every equality between elements of this list takes place after application of a suitable one arrow $w: k \to k'$.

For any functor $G: P \to \mathbf{Set}$, $\underleftarrow{\mathrm{Lim}}_p Gp = \mathrm{Cone}(*, G)$, the set of cones τ over G with vertex a point $*$. If $Gp = \underrightarrow{\mathrm{Colim}}_j F(p,j)$ and P is finite, each such cone consists of a finite number of elements of $\underrightarrow{\mathrm{Colim}}_j F(p,j)$ and the conditions that τ be a cone involve a finite number of equations between these elements. Since J is directed, the observations above now mean that each cone τ can consist of elements $\tau_p = (y_p, k')$ for some one index k', where the $y_p \in F(p,k')$ already constitute a cone $y: * \dot{\to} F(-, k')$. This cone y is an element of $\mathrm{Lim}_p F(p,k')$; its equivalence class (y,k') is an element of $\mathrm{Colim}_j \mathrm{Lim}_p$. The map

$$\tau \mapsto (y, k') \in \underrightarrow{\mathrm{Colim}}_j \underleftarrow{\mathrm{Lim}}_p F(p,j),$$

which is independent of the choices made, is the desired (two-sided) inverse of the canonical arrow κ.

Exercises

1. Show that κ of (3) is natural for arrows $\sigma: F \dot{\to} F'$ in $X^{P \times J}$.
2. (Verdier.) A category J is *pseudo-filtered* when it satisfies condition (b) for filtered categories and the following condition (a'): Any two arrows $i \to j$, $i \to j'$ with the same domain can be embedded in a commutative diamond

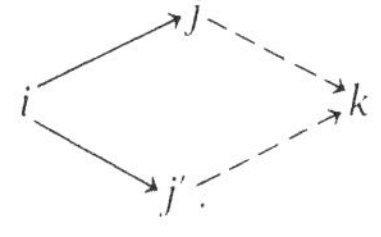

Prove that a category J is filtered if and only if it is connected and pseudo-filtered. Prove that a category is pseudo-filtered if and only if its connected components are filtered.
3. In **Set**, show that coproducts commute with pullback.
4. Using Exercises 2 and 3, show that pseudo-filtered colimits commute with pullbacks in **Set**.

3. Final Functors

Colimits may often be computed over subcategories. For example, the colimit of a functor $F : \mathbf{N} \to \mathbf{Cat}$, where $\mathbf{N}$ is the linearly ordered set of natural numbers, is clearly the same as the colimit of the restriction of F to any infinite subset S of $\mathbf{N}$ (i.e., to any subcategory which contains at least one object "beyond" each object of $\mathbf{N}$). In classical terminology, such a subset S was called "cofinal" in $\mathbf{N}$; it now seems preferable to drop the "co", as not related to dualizations. Also, we will replace the subset S first by the inclusion functor $S \to \mathbf{N}$ and then by an arbitrary functor.

A functor $L : J' \to J$ is called *final* if for each $k \in J$ the comma category $(k \downarrow L)$ is non-empty and connected. This means that to each k there is an object $j' \in J'$ and an arrow $k \to Lj'$, and that any two such arrows can be joined to give finite commutative diagram of the form

$$\begin{array}{ccccccccc} k & = & k & = & k & & k & = & k \\ \downarrow & & \downarrow & & \downarrow & \cdots & \downarrow & & \downarrow \\ Lj' & \xrightarrow[Lf_1]{} & \cdot & \xleftarrow[Lf_2]{} & \cdot & & \cdot & \xleftarrow[Lf_t]{} & Li'. \end{array}$$

A subcategory is called *final* when the corresponding inclusion functor is final. For example, if J is a linear order, $J' \subset J$ and L the inclusion, then L final means simply that to each $k \in J$ there is $j' \in J'$ with $k \leqq j'$.

For $L : J' \to J$ and $F : J \to X$ there is a canonical map

$$h : \underrightarrow{\mathrm{Colim}}\, FL \to \underrightarrow{\mathrm{Colim}}\, F \tag{1}$$

defined when both colimits exist; if $\mu' : FL \dot{\to} \underrightarrow{\mathrm{Colim}}\, FL$ and μ are the colimiting cones, h is the unique arrow of X with $h\mu'_{j'} = \mu_{Lj'}$ for all $j' \in J'$.

The main theorem now is:

Theorem 1. *If $L : J' \to J$ is final and $F : J \to X$ is a functor such that $x = \underrightarrow{\mathrm{Colim}}\, FL$ exists, then $\underrightarrow{\mathrm{Colim}}\, F$ exists and the canonical map (1) is an isomorphism.*

Proof. Given a colimiting cone $\mu : FL \dot{\to} \mathrm{Colim}\, FL = x$, we construct arrows $\tau_k : Fk \to x$ for each $k \in J$ by choosing an arrow $u : k \to Lj'$ and taking τ_k to be the composite

$$Fk \xrightarrow{Fu} FLj' \xrightarrow{\mu_{j'}} x .$$

Since μ is a cone and $(k \downarrow L)$ is connected, the connectivity diagram above readily shows τ_k independent of the choice of u and j'. It follows at once that $\tau : F \dot{\to} x$ is a cone with vertex x and base F. On the other hand, if $\lambda : F \dot{\to} y$ is another cone with this base F, then $\lambda L : FL \dot{\to} y$ is a cone with base FL, so by the universal property of μ there exists a unique $f : x \to y$

with $f\mu = \lambda L$, and hence (because $\lambda_k = \lambda_{Lj'} \cdot Fu$) with $f\tau = \lambda$. This shows that τ is a limiting cone and hence that $x = \operatorname{Colim} F$; clearly this also makes the canonical map h an isomorphism.

The condition that L be final is necessary for the validity of this theorem (cf. Exercise 5). The dual of this result (the dual of final is "initial") is useful for limits.

Exercises

1. If $j \in J$ and $\{j\}$ is the discrete subcategory of J with just the one object j, show that the inclusion $\{j\} \to J$ is final if and only if j is a terminal object in J. What does this say about colimits and terminal objects?
2. Prove that a composite of final functors is final.
3. If J is filtered, $L: J' \to J$ is full, and each $(k \downarrow L)$ is non-empty $(k \in J)$ prove that L is final.
4. For the covariant hom-functor $J(k, -): J \to \mathbf{Set}$, use the Yoneda Lemma to show that $\underrightarrow{\operatorname{Colim}}\, J(k, -)$ is the one-point set.
5. (Converse of the Theorem of the text). Let $L: J' \to J$ be a functor, where J' and J have small hom-sets, such that for every $F: J \to X$ with X cocomplete the canonical map $\underrightarrow{\operatorname{Colim}}\, FL \to \underrightarrow{\operatorname{Colim}}\, F$ is an isomorphism. Prove that L must be final. (Hint: Use $F = J(k, -)$, $X = \mathbf{Set}$, and Exercise 4.)

4. Diagonal Naturality

We next consider an extension of the concept of naturality. Given categories C, B and functors $S, T: C^{\mathrm{op}} \times C \to B$, a *dinatural transformation* $\alpha: S \ddot{\to} T$ is a function α which assigns to each object $c \in C$ an arrow $\alpha_c: S(c, c) \to T(c, c)$ of B, called the *component* of α at c, in such a way that for every arrow $f: c \to c'$ of C the following hexagonal diagram

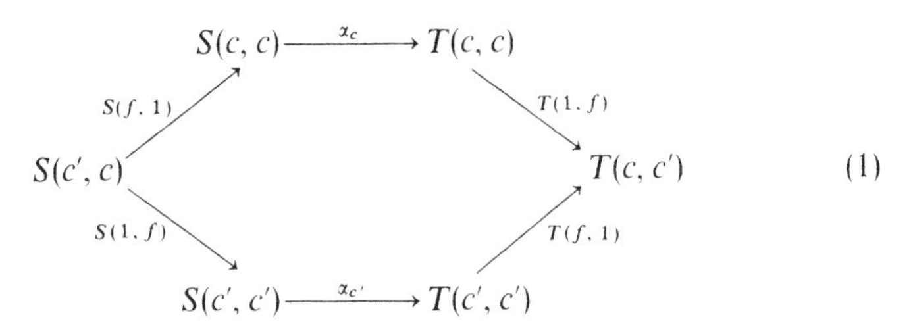

is commutative. Observe that the contravariance of S and T in the first argument is used in forming the arrows $S(f, 1)$ and $T(f, 1)$ in this diagram.

Every ordinary natural transformation $\tau: S \dot{\to} T$ between the bifunctors S and T, with components $\tau_{c,c'}: S(c, c') \to T(c, c')$, will yield a dinatural transformation $\alpha: S \ddot{\to} T$ between the same bifunctors, with components just the diagonal components of τ; thus $\alpha_c = \tau_{c,c}$. More interesting examples arise from functors which are "dummy" in one or more variables. For example, $T: C^{\mathrm{op}} \times C \to B$ is dummy in its first variable if it is a composite

$$C^{\mathrm{op}} \times C \xrightarrow{Q} C \xrightarrow{T_0} B,$$

where Q is the projection on the second factor and T_0 is some functor (of one variable). Put differently, each functor $T_0 : C \to B$ of one variable may be treated as a bifunctor $C^{\mathrm{op}} \times C \to B$, dummy in the first variable. Again, a functor dummy in *both* variables is in effect a constant object $b \in B$ with $T(c, c') = b$ for all objects $c, c' \in C$ and $T(f, f') = 1_b$ for all arrows f and f' in C.

The following types of dinatural transformations $S \dashrightarrow T$ arise. If S is dummy in its second variable and T dummy in its first variable, a dinatural transformation $\alpha : S \dashrightarrow T$ sends a functor $S_0 : C^{\mathrm{op}} \to B$ to a covariant $T_0 : C \to B$ by components $\alpha_c : S_0 c \to T_0 c$ which make the diagrams

$$\begin{array}{ccc} S_0 c & \xrightarrow{\alpha_c} & T_0 c \\ {\scriptstyle S_0 f}\uparrow & & \downarrow{\scriptstyle T_0 f} \\ S_0 c' & \xrightarrow{\alpha_{c'}} & T_0 c' \end{array} \tag{2}$$

commute for each arrow $f : c \to c'$ of C. Such a dinatural transformation might be called a natural transformation of the contravariant functor S_0 to the covariant functor T_0. (Dually, of a covariant to a contravariant one.)

If $T = b : C^{\mathrm{op}} \times C \to B$ is dummy in both variables, a dinatural transformation $\alpha : S \dashrightarrow b$ consists of components $\alpha_c : S(c, c) \to b$ which make the diagram

$$\begin{array}{ccc} S(c', c) & \xrightarrow{S(1, f)} & S(c', c') \\ {\scriptstyle S(f, 1)}\downarrow & & \downarrow{\scriptstyle \alpha_{c'}} \\ S(c, c) & \xrightarrow[\alpha_c]{} & b \end{array} \tag{3}$$

commute for every $f : c \to c'$. (The right hand side of the hexagon (1) has collapsed to one object b.) Such a transformation $\alpha : S \dashrightarrow b$ is called an *extranatural* transformation, a "supernatural" transformation or a *wedge* from S to b. The same terms are applied to the dual concept $\beta : b \dashrightarrow T$, given by components $\beta_c : b \to T(c, c)$ such that every square

$$\begin{array}{ccc} b & \xrightarrow{\beta_c} & T(c, c) \\ {\scriptstyle \beta_{c'}}\downarrow & & \downarrow{\scriptstyle T(1, f)} \\ T(c', c') & \xrightarrow[T(f, 1)]{} & T(c, c') \end{array} \tag{4}$$

is commutative. (The left hand side of the hexagon (1) has collapsed.)

We give an example of each type. A Euclidean vector space E is a vector space over the field $\mathbf{R}$ of real numbers equipped with an inner product function $(\ ,\ ): E \times E \to \mathbf{R}$ which is bilinear, symmetric, and positive definite. These spaces are the objects of a category **Euclid**, with arrows those linear maps which preserve inner product. There are two functors

$$U : \mathbf{Euclid} \to \mathbf{Vct_R}, \qquad * : (\mathbf{Euclid})^{\mathrm{op}} \to \mathbf{Vct_R}$$

to the category of real vector spaces: The (covariant) forgetful functor U "forget the inner product" and the contravariant functor "take the dual space". Now for each Euclidean vector space E the assignment $e \mapsto (e, -)$, for $e \in E$, is a linear function $\kappa_E : E \to E^*$; these functions κ_E are the components for a transformation κ which is dinatural from U to $*$ (dual of type (2)): This is the fact, familiar in Riemannian geometry, that each Euclidean vector space is *naturally* isomorphic to its dual – and we need the notion of dinaturality to express this fact categorically.

Evaluation. V_X, for X a (small) set, takes the value of each function $h: X \to A$ at each argument $x \in X$. If the (small) set A is fixed, we may regard V_X as a function

$$V_X : \hom(X, A) \times X \to A, \qquad \langle h, x\rangle \mapsto hx,$$

defined for each object $X \in \mathbf{Set}$. For two small sets X and Y, $\hom(X, A) \times Y$ is the (object function of a) functor $\mathbf{Set}^{\mathrm{op}} \times \mathbf{Set} \to \mathbf{Set}$, while for every arrow $f: Y \to X$ the obvious property $h(fx) = (hf)x$ of evaluation states that the square like (3) always commutes. Hence the functions V_X are the components of an extranatural transformation

$$V : \hom(-, A) \times (-) \xrightarrow{\cdot\cdot} A.$$

Observe that V is also natural (in the usual sense) in the argument A; we say that $\hom(X, A) \times X \to A$ by evaluation is dinatural (= extranatural) in X and natural in A.

Counits. For functors $F : X \times P \to A$ and $G : P^{\mathrm{op}} \times A \to X$ a bijection

$$A(F(x, p), a) \cong X(x, G(p, a)) \tag{5}$$

natural in x, p, and a is an adjunction with parameter p (Theorem IV.7.3); its counit, obtained by setting $x = G(p, a)$ in (5), is a collection of components

$$\varepsilon_{\langle p, a\rangle} : F(G(p, a), p) \to a \tag{6}$$

natural in a and dinatural (= extranatural) in p. This includes the case of evaluation above.

Here is an example of the dual type of dinaturality. In any category C the identity function assigns to each object c the identity arrow $1_c : c \to c$, which may be regarded as an element $1_c \in \hom(c, c)$ or as an arrow

$1_c : * \to \hom(c, c)$, where $*$ is the one-point set. Now $\hom(c, c')$ is (the object function of) a functor $C^{\mathrm{op}} \times C \to \mathbf{Set}$, and for each arrow $f : c \to c'$ the identity function 1 has the evident property $f \circ 1_c = 1_{c'} \circ f$, which states in the present language that 1 is a dinatural transformation

$$1 : * \ddot{\to} \hom(-, -) .$$

All three types of dinatural transformations occur in combination with natural transformations in the previous sense (and indeed we will usually simply call all three types "natural transformations", dropping the "di" except where it is needed for emphasis). Thus given categories and functors

$$S : C^{\mathrm{op}} \times C \times A \to B , \qquad T : A \times D^{\mathrm{op}} \times D \to B$$

a natural transformation $\gamma : S \ddot{\to} T$ is a function which assigns to each triple of objects $c \in C$, $a \in A$, and $d \in D$ an arrow

$$\gamma(c, a, d) : S(c, c, a) \to T(a, d, d)$$

of B such that (i) for c and d fixed, $\gamma(c, -, d)$ is natural in a, in the usual sense; (ii) for a and d fixed, $\gamma(-, a, d)$ is dinatural in c; (iii) for c and a fixed, $\gamma(c, a, -)$ is dinatural in d. In the description of these natural transformations any one of the categories A, B, or C may be replaced by a product of several categories, and in each case naturality in a product argument $c \in C = C' \times C''$ may be replaced by naturality in each argument of the pair $c = \langle c', c'' \rangle$ when the other is fixed (see Exercise 3 below). For example, in any category the operation of composition

$$\hom(b, c) \times \hom(a, b) \to \hom(a, c)$$

is natural; i.e., natural in a, dinatural in b, and natural in c.

The composite of two dinatural transformations need not be dinatural at all, but any dinatural transformation $\alpha : S \ddot{\to} T$ may be composed on either side with transformations which are natural in both arguments. If $\sigma : S' \dot{\to} S$ and $\tau : T \dot{\to} T'$ are natural transformations, the composite arrows

$$S'(c, c) \xrightarrow{\sigma_{c,c}} S(c, c) \xrightarrow{\alpha_c} T(c, c) \xrightarrow{\tau_{c,c}} T'(c, c)$$

are the components of a dinatural transformation $S' \ddot{\to} T'$. Here is a more interesting case (easily proved).

Proposition 1. *Given functors*

$$R : C \to B , \qquad S : C \times C^{\mathrm{op}} \times C \to B , \qquad T : C \to B$$

and functions (for all $c, d \in C$)

$$\varrho(c, d) : R(c) \to S(c, d, d) , \qquad \sigma(d, c) : S(d, d, c) \to T(c)$$

which are natural in c and dinatural in d, the function which assigns to each $c \in C$ the composite arrow

$$R(c) \xrightarrow{\varrho(c,c)} S(c,c,c) \xrightarrow{\sigma(c,c)} T(c)$$

is a natural transformation $R \stackrel{\cdot}{\to} T$.

Exercises

1. Prove that the unit $\eta_x : x \to G(p, F(x, p))$ of an adjunction with parameter is dinatural in p, and that this property is equivalent to the naturality of the adjunction itself in p (cf. IV.7., Exercise 2). Dualize.
2. Formulate the triangular identities for an adjunction with parameter.
3. (Naturality by separation of arguments.) Given $b \in B$, a functor

$$S : (C \times D)^{\mathrm{op}} \times C \times D \to B,$$

   and a function β assigning to $c \in C$, $d \in D$ an arrow

$$\beta_{c,d} : S(c, d, c, d) \to b$$

   of B, show that $\beta : S \stackrel{\cdot\cdot}{\to} b$ is dinatural if and only if it is dinatural in c (for each fixed d) and dinatural in d (for each fixed c). State the dual result.
4. Extend the composition rule of Proposition 1 to the case when S is a functor $C \times C^{\mathrm{op}} \times C \times C^{\mathrm{op}} \times C \to B$. Do the same for any odd number of factors C.
5. For $S : C^{\mathrm{op}} \times C \to B$ and $b, b' \in B$, show that dinatural transformations $b \stackrel{\cdot\cdot}{\to} S$ and $S \stackrel{\cdot\cdot}{\to} b'$ do not in general have a well defined composite $b \to b'$.
6. Extending Exercises 3 and 4, find a general rule for the composition of natural transformations in many variables.

5. Ends

An "end" is a special (and especially useful) type of limit, defined by universal wedges in place of universal cones.

Definition. *An end of a functor $S : C^{\mathrm{op}} \times C \to X$ is a universal dinatural transformation from a constant e to S; that is, an end of S is a pair $\langle e, \omega \rangle$, where e is an object of X and $\omega : e \stackrel{\cdot\cdot}{\to} S$ is a wedge (a dinatural transformation) with the property that to every wedge $\beta : x \stackrel{\cdot\cdot}{\to} S$ there is a unique arrow $h : x \to e$ of B with $\beta_a = \omega_a h$ for all $a \in C$.*

Thus for each arrow $f : b \to c$ of C there is a diagram

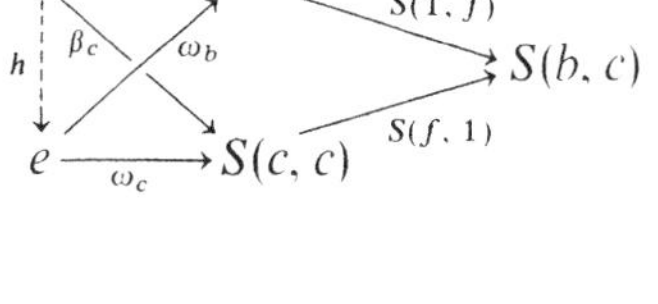

$$\begin{array}{ccccc} x & \xrightarrow{\beta_b} & S(b,b) & & \\ {\scriptstyle h}\downarrow & \beta_c \searrow \nearrow \omega_b & & \searrow^{S(1,f)} & \\ & & & & S(b,c) \\ & \nearrow \searrow & & \nearrow_{S(f,1)} & \\ e & \xrightarrow[\omega_c]{} & S(c,c) & & \end{array} \qquad (1)$$

such that both quadrilaterals commute (these are the dinaturality conditions); the universal property of ω states that there is a unique h such that both triangles (at the left) commute.

The uniqueness property which applies to any universal states in this case that if $\langle e, \omega \rangle$ and $\langle e', \omega' \rangle$ are two ends for S, there is a unique isomorphism $u: e \to e'$ with $\omega' \circ u = \omega$ (i.e., with $\omega'_c \circ u = \omega_c$ for each $c \in C$). We call ω the *ending wedge* or the *universal wedge*, with *components* ω_c, while the object e itself, by abuse of language, is called the "end" of S and is written with integral notation as

$$e = \int_c S(c, c) = \text{End of } S .$$

Note that the "variable of integration" c appears twice under the integral sign (once contravariant, once covariant) and is "bound" by the integral sign, in that the result no longer depends on c and so is unchanged if "c" is replaced by any other letter standing for an object of the category C. These properties are like those of the letter x in the usual integral $\int f(x)\,dx$ of the calculus.

Natural transformations provide an example of ends. Two functors $U, V: C \to X$ define a functor $\hom_X(U -, V -)$: $C^{\text{op}} \times C \to \mathbf{Set}$, and if Y is any set, a wedge $\tau: Y \ddot{\to} \hom_X(U -, V -)$, with components

$$\tau_c : Y \to \hom_X(Uc, Vc), \qquad c \in C ,$$

assigns to each $y \in Y$ and to each $c \in C$ an arrow $\tau_{c,y}: Uc \to Vc$ of X such that for every arrow $f: b \to c$ one has the "wedge condition" $Vf \circ \tau_{b,y} = \tau_{c,y} \circ Uf$. But this condition is just the commutativity of the square

$$\begin{array}{ccc} Ub & \xrightarrow{\tau} & Vb \\ {\scriptstyle Uf}\downarrow & & \downarrow{\scriptstyle Vf} \\ Uc & \xrightarrow{\tau} & Vc \end{array}$$

which asserts that $\tau_{-,y}$ for fixed y is a natural transformation $\tau_{-,y}: U \dot{\to} V$. Thus, if we write $\text{Nat}(U, V)$ for the set of all such natural transformations, the assignment $y \mapsto \tau_{-,y}$ is the unique function $Y \to \text{Nat}(U, V)$ which makes the following diagram commute.

$$\begin{array}{ccc} Y & \xrightarrow{\tau_c} & \hom(Uc, Vc) \\ \downarrow & & \| \\ \text{Nat}(U, V) & \xrightarrow{\omega_c} & \hom(Uc, Vc) , \end{array}$$

where ω_c assigns to each natural $\lambda: U \dot{\to} V$ its component $\lambda_c: Uc \to Vc$. This states exactly that ω is a universal wedge. Hence

$$\text{Nat}(U, V) = \int_c \hom(Uc, Vc); \qquad U, V: C \to X . \tag{2}$$

Every end is manifestly a limit – specifically, a limit of a suitable diagram in X made up of pieces like those pieces $S(b,b) \to S(b,c) \leftarrow S(c,c)$, one for each f in C, which come up in the diagram (1) defining an end. This can be stated formally in terms of the following construction (to be used only in this section) of a category $C^{\S}$ depending on C. The objects of $C^{\S}$ are all symbols $c^{\S}$ and $f^{\S}$ for $c \in C$ and f an arrow in C (note especially that $c^{\S}$ and $(1_c)^{\S}$ are different objects). The arrows of $C^{\S}$ are the identity arrows for these objects, plus for each arrow $f: b \to c$ in C two arrows

$$b^{\S} \to f^{\S} \leftarrow c^{\S}$$

in $C^{\S}$. The only meaningful compositions for these arrows in $C^{\S}$ are compositions with one factor an identity arrow. Thus we have defined a category $C^{\S}$, called the *subdivision category* of C.

Each functor $S: C^{\mathrm{op}} \times C \to X$ defines a functor $S^{\S}: C^{\S} \to X$ by the assignments indicated (from top to bottom) in the following figure for a typical $f: b \to c$ in C:

$$\begin{array}{ccccc} C^{\S} & & b^{\S} \longrightarrow f^{\S} \longleftarrow c^{\S} \\ S^{\S}\downarrow & & \downarrow \quad \downarrow \quad \downarrow \quad \downarrow \quad \downarrow \\ X & & S(b,b) \xrightarrow[S(b,f)]{} S(b,c) \xleftarrow[S(f,c)]{} S(c,c) \end{array}$$

Inspection of this figure shows that a cone $\tau: x \dot{\to} S^{\S}$ is exactly the same thing as a wedge $\omega: x \ddot{\to} S$. This proves that a limit of $S^{\S}$ is an end of S, in the following sense.

Proposition 1. *For any functor $S: C^{\mathrm{op}} \times C \to X$ and the associated functor $S^{\S}: C^{\S} \to X$, as defined above, there is an isomorphism*

$$\theta: \int_c S(c,c) \cong \underleftarrow{\mathrm{Lim}}\, [S^{\S}: C^{\S} \to X]\,. \tag{3}$$

In more detail, if either the indicated end or the indicated limit exists, then both exist, and there is a unique arrow θ in X such that the diagram

$$\begin{array}{ccc} \int_c S(c,c) & \xrightarrow{\omega_c} & S(c,c) \\ \theta \downarrow & & \| \\ \underleftarrow{\mathrm{Lim}}\, S^{\S} & \xrightarrow{\lambda_c} & S^{\S}(c) \end{array}$$

commutes for every $c \in C$, where ω is the ending wedge and λ the limiting cone; moreover, this arrow θ is an isomorphism.

Corollary 2. *If X is small-complete and C is small, every functor $S: C^{\mathrm{op}} \times C \to X$ has an end in X.*

The Proposition above has reduced ends to limits. The converse is easier: Every limit may be regarded as an end!

Proposition 3. *For each functor $T: C \to X$ let S be the composite functor*

$$C^{\mathrm{op}} \times C \xrightarrow{Q} C \xrightarrow{T} X$$

where Q is the second projection $\langle c, c' \rangle \mapsto c'$ of the product. Then $\langle e, \tau : e \xrightarrow{\cdot} T \rangle$ is a limit for T in X if and only if $\langle e, \tau : e \xrightarrow{\cdot\cdot} S \rangle$ is an end for S in X.

Proof. The components τ_c of a cone $e \xrightarrow{\cdot} T$ make the triangle $\tau_c = Tf \circ \tau_b$ commute (naturality condition!) for each $f: b \to c$ in C. This amounts to saying that every square

$$\begin{array}{ccc} e & \xrightarrow{\tau_b} & Tb = S(b, b) \\ \downarrow{\scriptstyle \tau_c} & & \downarrow{\scriptstyle S(1, f) = T(f)} \\ Tc = S(c, c) & \xrightarrow[S(f, 1) = 1]{} & Tc = S(b, c) \end{array}$$

commutes ($S(-, -)$ is "dummy" in the first variable), and this in turn states exactly that $\tau : e \xrightarrow{\cdot\cdot} S$ is a wedge. It follows that τ is universal as a cone if and only if it is universal as a wedge.

This conclusion reads: There is an isomorphism

$$\int_c S(c, c) = \int_c Tc \cong \operatorname{Lim} T$$

valid when either the end *or* the limit exists, carrying the ending wedge to the limiting cone; the indicated notation thus allows us to write any limit as an integral (an end) without explicitly mentioning the dummy variable (the first variable c of S).

A functor $H: X \to Y$ is said to *preserve the end* of a functor $S: C^{\mathrm{op}} \times C \to X$ when $\omega : e \xrightarrow{\cdot\cdot} S$ an end of S in X implies that $H\omega : He \xrightarrow{\cdot\cdot} HS$ is an end for HS; in symbols

$$H \int_c S(c, c) = \int_c HS(c, c).$$

Similarly, H *creates* the end of S when to each end $\nu : y \xrightarrow{\cdot\cdot} HS$ in Y there is a unique wedge $\omega : e \xrightarrow{\cdot\cdot} S$ with $H\omega = \nu$, and this wedge ω is an end of S. Since an end (of the functor S) is the same thing as a limit (of the corresponding $S^{\S}$) the properties we have established for the preservation of limits carry over to the preservation of ends. For example, the hom-functors preserve (and reverse, see § 6) ends:

$$X\left(x, \int_c S(c, c)\right) = \int_c X(x, S(c, c)), \tag{4}$$

$$X\left(\int^c S(c, c), x\right) = \int_c X(S(c, c), x). \tag{5}$$

6. Coends

The definition of the coend of a functor $S: C^{op} \times C \to X$ is dual to that of an end. A *coend* of S is a pair, $\langle d, \zeta: S \xrightarrow{..} d \rangle$, consisting of an object $d \in X$ and a dinatural transformation ζ (a wedge), universal among dinatural transformations from S to a constant. The object d (when it exists, unique up to isomorphism) will usually be written with an integral sign *and* with the bound variable c as superscript; thus

$$S(c, c) \xrightarrow{\zeta_c} \int^{c} S(c, c) = d .$$

The formal properties of coends are dual to those of ends.

Coends are familiar under other names. For example, the tensor product of modules over a ring R is a coend. Specifically, a ring R is an Ab-category with one object (which we call R again) and with arrows the elements $r \in R$, composition of arrows being their product in R. A left R-module B is an additive functor $R \to \mathbf{Ab}$ which sends the (one) object R to the abelian group B and each arrow r in R to the scalar multiplication $r_*: b \mapsto rb$ in B. Similarly, a right R-module A is an additive functor $R^{op} \to \mathbf{Ab}$ (contravariant on R to $\mathbf{Ab}$). If $\otimes$ is the usual tensor product in $\mathbf{Ab}$, then $R \mapsto A \otimes B$ is a bifunctor $R^{op} \times R \to \mathbf{Ab}$. Moreover, the coend

$$\int^{R} A \otimes B = A \otimes_R B$$

is exactly the usual tensor product over R. Indeed, a wedge ζ from the bifunctor $A \otimes B$ to an abelian group M is precisely a (single) morphism $\varrho: A \otimes B \to M$ of abelian groups such that the diagram

$$\begin{array}{ccc} A \otimes B & \xrightarrow{1_A \otimes r_*} & A \otimes B \\ {\scriptstyle r_* \otimes 1_B} \downarrow & & \downarrow {\scriptstyle \varrho} \\ A \otimes B & \xrightarrow{\varrho} & M \end{array}$$

commutes for every arrow r in R. With the above interpretation of modules as functors, this means for elements $a \in A$ and $b \in B$ that

$$\varrho(ar \otimes b) = \varrho(a \otimes rb) .$$

Therefore M is an end precisely when M is $A \otimes B$ modulo all $ar \otimes b - a \otimes rb$, and this is precisely the usual description of the tensor product $M = A \otimes_R B$.

The point of these observations is not the reduction of the familiar to the unfamiliar (tensor products to coends) but the extension of the familiar to cover many more cases. If B is any monoidal category with multiplication $\square$, as in Chapter VII, then any two functors $T: P^{op} \to B$ and $S: P \to B$ have a "tensor product"

$$T \square_P S = \int^{P} (Tp) \square (Sp) ,$$

an object of B. The simplicial category Δ of § VII.5 has a functor $\Delta : \Delta \to \mathbf{Top}$ (each ordinal $n+1$ realized by the n-dimensional affine simplex), while any $S : \Delta^{\mathrm{op}} \to \mathbf{Set}$ is called a simplicial set. Now the copower $S \cdot X$ (set S times space X) is just the disjoint union $\coprod_S X$ of S copies of X. Hence $(n, m) \mapsto Sn \cdot \Delta m$ is a functor $\Delta^{\mathrm{op}} \times \Delta \to \mathbf{Top}$ and the coend

$$\int^n (Sn) \cdot \Delta n \tag{1}$$

is the usual *geometric realization* (Lamotke [1968], p. 34; May [1967], p. 55) of the simplicial set S. The coend formula describes the geometric realization in one gulp: Take the disjoint union of affine n-simplices, one for each $t \in Sn$, and paste them together according to the given face and degeneracy operations (arrows of Δ). There is a similar efficient description of the (Stasheff-Milgram) classifying space of a topological monoid (best situated in the category **CGHaus** of § VII.8); see Mac Lane [1970].

Exercises

1. For $S : C^{\mathrm{op}} \times C \to \mathbf{Set}$, prove that the set Wedge $(*, S)$ of all wedges $\omega : * \ddot{\to} S$ from the one-point set $*$ to S is an end of S, with ending wedge given by $\omega \mapsto \omega_c(*) \in S(c, c)$. Compare with the explicit description of a limit in **Set** as a set of cones.
2. Show directly (without using limits) that a category X with all small products and with equalizers has all small ends (cf. the corresponding proof for limits in § V.2).
3. To each category C there is a "twisted arrow category" $C_\#$ with objects the arrows $f : a \to b$ of C and arrows $\langle h, k \rangle : f \to f'$ the arrows $h : a' \to a$ (note the twist!) and $k : b \to b'$ such that $f' = kfh$. Then $\langle f : a \to b \rangle \mapsto \langle a, b \rangle$ is a functor $K : C_\# \to C^{\mathrm{op}} \times C$. For any $S : C^{\mathrm{op}} \times C \to B$, prove that cones $e \dot{\to} SK$ correspond to wedges $e \ddot{\to} S$, and use this fact to give another proof of the reduction of ends to limits (Proposition 5.1).
4. Let **Fin** (the skeletal category of finite non-empty sets) be the category with objects finite nonzero ordinals n and arrows all functions $f : n \to m$. For each set X, $n \mapsto X^n$ defines (the object function of) a functor $(\mathbf{Fin})^{\mathrm{op}} \to \mathbf{Set}$. For each ring R, the assignment $n \mapsto R^n$ becomes a *covariant* functor $R_{(\ )} : \mathbf{Fin} \to R\text{-}\mathbf{Mod}$ if each function $f : n \to m$ (arrow in **Fin**) takes a list $a_0, \ldots, a_{n-1} \in R^n$ to $b_0, \ldots, b_{m-1} \in R^m$, where $b_i = \Sigma a_j$, the sum over all those $j \in n$ with $f_j = i$. Show that the free R-module generated by the set X is the coend

   $$\int^n X^n \cdot R_{(n)},$$

   and show that this formula is essentially the usual description of the elements of the free module as finite formal sums $\Sigma x_i a_i$, $i = 1, \ldots, n$.
5. If D is cocomplete, functors $S : C^{\mathrm{op}} \to \mathbf{Set}$ and $T : C \to D$ have a tensor product defined as the coend $\int^c (Sc) \cdot (Tc)$, where $\cdot$ denotes the copower. Show that the tensor product is a functor $D^C \times \mathbf{Set}^{C^{\mathrm{op}}} \to D$.

7. Ends with Parameters

The basic formal properties for ends are much like those for integrals in calculus. All these properties will apply equally well to limits (regarded as ends with a dummy variable).

Proposition 1 *(End or limit of a natural transformation). Given a natural transformation* $\gamma : S \dashrightarrow S'$ *between functors* $S, S' : C^{op} \times C \to X$ *which both have ends* $\langle e, \omega \rangle$ *and* $\langle e', \omega' \rangle$, *respectively, there is a unique arrow* $g = \int_c \gamma_{c,c} : e \to e'$ *in* X *such that the following diagram commutes for every* $c \in C$:

$$\begin{array}{ccc} \int_c S(c,c) & \xrightarrow{\omega_c} & S(c,c) \\ {\scriptstyle g = \int_c \gamma_{c,c}} \downarrow & & \downarrow {\scriptstyle \gamma_{c,c}} \\ \int_c S'(c,c) & \xrightarrow{\omega'_c} & S'(c,c)\,. \end{array} \tag{1}$$

Proof. The composites $\gamma_{c,c} \circ \omega_c$ define a wedge, so g exists and is unique by the universality of the wedge ω'.

We call the arrow g the *end* of the natural transformation γ.

Composing γ with another $\gamma' : S' \dashrightarrow S''$ yields the rule

$$\int_c (\gamma' \cdot \gamma)_{c,c} = \left[\int_c \gamma'_{c,c}\right] \circ \left[\int_c \gamma_{c,c}\right] . \tag{2}$$

By this composition rule, a limit (or an end) involving a parameter p (in some category P) can be shown to be a functor of that parameter in the following sense.

Theorem 2 *(Parameter Theorem for ends and limits). Let* $T : P \times C^{op} \times C \to X$ *be a functor such that* $T(p, -, -)$ *for each object* $p \in P$ *has an end*

$$\omega_p : \int_c T(p,c,c) \stackrel{\cdot\cdot}{\to} T(p, -, -) \tag{3}$$

in X. *Then there is a unique functor* $U : P \to X$ *with object function* $Up = \int_c T(p,c,c)$ *such that the components of the wedge* (3) *for each* $c \in C$ *define a transformation* $(\omega_p)_c : Up \to T(p,c,c)$, *natural in* p.

Proof. Each arrow $s : p \to q$ of P defines a natural transformation $\gamma = T(s, -, -) : T(p, -, -) \dashrightarrow T(q, -, -)$. Hence the arrow function of the desired functor U must have $Us = \int_c T(s,c,c)$, defined as in (1), and the composition rule (2) shows that this definition of Us does determine a functor $U : P \to X$.

The functor U will be written $U = \int_c T(-,c,c)$; thus

$$\left[\int_c T(-,c,c)\right] p = \int_c T(p,c,c)\,, \qquad \left[\int_c T(-,c,c)\right] s = \int_c T(s,c,c)\,. \tag{4}$$

The notation suggests that this functor U is itself an end. Indeed, regard $\int_c T(-,c,c)$ as an object of the functor category X^P and rewrite $T: P \times C^{\text{op}} \times C \to X$ as the functor $T^\sharp: C^{\text{op}} \times C \to X^P$ given on arrows (or objects) f, f' of C by

$$T^\sharp(f,f') = T(-,f,f'): P \to X .$$

Put differently, $T^\sharp$ is the image of T under the standard adjunction

$$\mathbf{Cat}(P \times C^{\text{op}} \times C, X) \cong \mathbf{Cat}(C^{\text{op}} \times C, X^P) .$$

Theorem 3 *(Parameter Theorem, continued). Under the same hypotheses on T, the functor $T^\sharp$ has the end*

$$\omega^\sharp : \int_c T(-,c,c) \dot{\to} T^\sharp$$

where $(\omega_c^\sharp)_p = (\omega_p)_c$ for all $p \in P$ and $c \in C$.

Proof. The end $\int_c T(-,c,c)$ is an object of X^P, while $T^\sharp$ is a functor with codomain X^P. By the previous theorem, the arrows $(\omega_p)_c$ of X provide for each c an arrow of X^P (a natural transformation)

$$\omega_c^\sharp : \int_c T(-,c,c) \dot{\to} T(-,c,c) ;$$

its component at p is $(\omega_c^\sharp)_p = (\omega_p)_c$. Moreover, varying c, $\omega^\sharp$ is a wedge $\int_c T(-,c,c) \dot{\to} T^\sharp$. It is a universal wedge, for, given any object $F \in X^P$ and any wedge $\beta: F \dot{\to} T^\sharp$, each component β_p factors uniquely through the corresponding component ω_p, so β itself factors uniquely through $\omega^\sharp$. This gives the end for $T^\sharp$, as required.

This theorem can also be formulated wholly in terms of the functor category X^P, as was done in the case of limits in Theorem V.3.1.

Exercises

1. (Dubuc.) Construct a functor category X^P and a functor $T: C \to X^P$ which has a limit not a pointwise limit. (Suggestion: Take $C = \mathbf{2}$.)
2. State and prove the parameter theorem for coends.
3. If X is small complete and C is small, use Proposition 1 to prove that $\underleftarrow{\mathrm{Lim}} : X^C \to X$ is a functor (cf. Ex. V.2.3).
4. For any categories X and P, show that the functor $X^P \to X^{|P|}$ induced by inclusion (of the discrete subcategory $|P|$) creates ends and coends (cf. Theorem V.3.2).

8. Iterated Ends and Limits

We now describe when the "double integral" can be obtained as an "iterated" integral (Fubini!).

Proposition. *Let $S : P^{\mathrm{op}} \times P \times C^{\mathrm{op}} \times C \to X$ be a functor such that the end $\int_c S(p, q, c, c)$ exists for all pairs $\langle p, q\rangle$ of objects of P; by the parameter theorems, regard these ends as a bifunctor $P^{\mathrm{op}} \times P \to X$, and regard S as a bifunctor $(P \times C)^{\mathrm{op}} \times (P \times C) \to X$. Then there is an isomorphism*

$$\theta : \int_{\langle p,c\rangle} S(p, c, p, c) \cong \int_p \left[\int_c S(p, p, c, c)\right].$$

Indeed, the "double end" on the left exists if and only if the end $\int_p$ on the right exists, and then there is a unique arrow θ in X such that the diagram

$$\begin{array}{ccccc}
\int_{\langle p,c\rangle} S(p, p, c, c) & & \xrightarrow{\xi_{\langle p,c\rangle}} & & S(p, p, c, c) \\
\downarrow \theta & & & & \| \\
\int_p \left[\int_c S(p, p, c, c)\right] & \xrightarrow{\varrho_p} & \int_c S(p, p, c, c) & \xrightarrow{\omega_{p,p,c}} & S(p, p, c, c)
\end{array}$$

commutes, where the horizontal arrows ξ, ϱ, and ω are the universal wedges belonging to the corresponding ends; moreover, the arrow θ is an isomorphism.

Proof. For each $\langle p, q\rangle \in P \times P$ we are given the end

$$\omega_{p,q} : \int_c S(p, q, c, c) \xrightarrow{\cdot\cdot} S(p, q, -, -).$$

For any $x \in X$ each P-indexed family $\rho_p : x \to \int_c S(p,p,c,c)$ of arrows of X determines a $(P \times C)$-indexed family $\xi_{p,c}$ as the composites

$$\xi_{p,c} : x \xrightarrow{\varrho_p} \int_c S(p,p,c,c) \xrightarrow{\omega_{p,p,c}} S(p, p, c, c);$$

for p fixed, $\xi_{\langle p,-\rangle}$ is trivially a wedge in c. Conversely, since $\omega_{p,p}$ is universal, every $(P \times C)$-indexed family which is natural in c for each p is such a composite, for a unique family ϱ. Now ϱ or ξ is extranatural in p (the latter for some c) if and only if the corresponding square below

$$\begin{array}{ccc}
x & \xrightarrow{\varrho_p} & \int_c S(p, p, c, c) \\
\downarrow \varrho_q & & \downarrow \int_c S(p,s,c,c) \\
\int_c S(q, q, c, c) & \xrightarrow{\int_c -} & \int_c S(p, q, c, c),
\end{array}
\qquad
\begin{array}{ccc}
x & \xrightarrow{\xi_{p,c}} & S(p, p, c, c) \\
\downarrow \xi_{q,c} & & \downarrow S(p,s,c,c) \\
S(q, q, c, c) & \xrightarrow{S(s,q,c,c)} & S(p, q, c, c)
\end{array}$$

commutes for each arrow $s : p \to q$ in P. Also, the first square commutes precisely when it commutes after composition with the arrows $\omega_{p,q,c}$ for all objects c. Form the cubical diagram with these two squares as front and back faces and with edges 1_x, $\omega_{p,p,c}$, $\omega_{p,q,c}$, and $\omega_{q,q,c}$ (front to back). By our definitions the four side faces involving these edges commute; hence the front square commutes if and only if the back square commutes for all c. Therefore ϱ is a wedge (in p) if and only if ξ is a wedge (in $\langle p, c\rangle$), so that wedges from x to $\int_c S(-, -, c, c)$ correspond one-one to wedges from x to S. Since the end is a universal wedge, and since a universal is determined up to isomorphism, this gives the isomorphism θ of the proposition.

Note one essential point: This proposition reduces double to iterated integrals provided the inner integral $\int_c S(p, q, c, c)$ exists for *all* pairs $\langle p, q\rangle$ (not just for $p = q$). The case of limits involves no such refinement.

The familiar result on change of order of integrals follows from this one, expanding a double integral in two ways.

Corollary. *Let $S : P^{\mathrm{op}} \times P \times C^{\mathrm{op}} \times C \to X$ be a functor such that the ends $\int_c S(p, q, c, c)$ and $\int_p S(p, p, b, c)$ exist, for all $p, q \in P$ and $b, c \in C$. By the parameter theorems regard these ends as bifunctors (of p, q or b, c) respectively; then there is an isomorphism*

$$\theta : \int_p \left[\int_c S(p, p, c, c)\right] \cong \int_c \left[\int_p S(p, p, c, c)\right].$$

Indeed, the (outside) iterated end on the left exists if and only if the (outside) iterated end on the right exists, and the isomorphism θ is the unique arrow in X such that the diagrams

$$\begin{array}{ccccc}
\int_p \int_c S(p, p, c, c) & \longrightarrow & \int_c S(p, p, c, c) & \longrightarrow & S(p, p, c, c) \\
\downarrow \theta & & & & \| \\
\int_c \int_p S(p, p, c, c) & \longrightarrow & \int_p S(p, p, c, c) & \longrightarrow & S(p, p, c, c)
\end{array}$$

commute for all $p \in P$ and $c \in C$, where the horizontal arrows are the appropriate components of the universal wedges for the integrals involved.

These results include the corresponding facts for limits and colimits. Thus, for a functor $F : P \times C \to X$ with P and C small, X complete

$$\mathrm{Lim}_p \mathrm{Lim}_c F(p, c) \cong \mathrm{Lim}_{\langle p, c\rangle} F(p, c) \cong \mathrm{Lim}_c \mathrm{Lim}_p F(p, c),$$

by Proposition 5.3, with the corresponding formula for colimits.

Notes.

A systematic treatment of all possible properties of limits was contained in a manuscript by Chevalley on category theory; the manuscript was unfortunately lost by some shipping company.

Eilenberg and Kelly discovered the extranatural transformations (and all the rules for their composition) in [1966b], while diagonally natural transformations are due to Dubuc and Street [1970]. The tensor product of functors was first defined by Kan [1958, § 14]; these products have been further developed in unpublished work of F. Ulmer and Allen Clark.

The idea of an end was discovered by Yoneda [1960], and its efficient utilization is due to Day and Kelly [1969], who observed that this notion is essential in categories based not on **Set** but on other closed categories. See also Kelly [1982].

X. Kan Extensions

If M is a subset of C, any function $t: M \to A$ to a non-empty set A can be extended to all of C in many ways, but there is no canonical or unique way of defining such an extension. However, if M is a subcategory of C, each functor $T: M \to A$ has in principle *two* canonical (or extreme) "extensions" from M to functors $L, R: C \to A$. These extensions are characterized by the universality of appropriate natural transformations; they need not always exist, but when M is small and A is complete and cocomplete they do exist, and can be given as certain limits or as certain ends. These "Kan extensions" are fundamental concepts in category theory. With them we find again that each fundamental concept can be expressed in terms of the others. This chapter begins by expressing adjoints as limits and ends by expressing "everything" as Kan extensions.

1. Adjoints and Limits

Limits and colimits, if they exist for all functors $J \to C$, provide respectively right and left adjoints for the diagonal functor Δ:

$$\underset{\longrightarrow}{\mathrm{Lim}} \uparrow \; \downarrow \Delta \; \uparrow \underset{\longleftarrow}{\mathrm{Lim}} \quad \text{between } C \text{ and } C^J \qquad (=\text{right adjoint of } \Delta). \tag{1}$$

Conversely, left adjoints can be interpreted as limits. First note that an initial object in any category C is a limit:

$$\text{Initial object } C = \mathrm{Colim}(\mathbf{0} \to C) = \underset{\longleftarrow}{\mathrm{Lim}}(\mathrm{Id} : C \to C)\,, \tag{2}$$

where $\mathbf{0}$ denotes the empty category (the ordinal 0) and $\mathbf{0} \to C$ is the empty functor. The definition of the initial object e states exactly that it is the colimit of the empty functor. Moreover, the unique arrows $\mu_c: e \to c$, one for each c, define a cone $e \dot{\to} \mathrm{Id}_C$. If $\lambda: d \dot{\to} \mathrm{Id}_C$ is a cone from some other vertex, then there is a unique $f: d \to e$ with all $\mu_c f = \lambda_c$; indeed, this equation for $c = e$ shows that f must be λ_e, and for $f = \lambda_e$ this equation $\mu_c \lambda_e = \lambda_c$ does hold because λ is a cone over Id_C. This proves $e = \mathrm{Lim}\,\mathrm{Id}_C$. The converse property, that any limit of Id is initial, is a special case of

Lemma 1. *If $\lambda : d \dot{\to} \mathrm{Id}_C$ is a cone over the identity functor and $F : J \to C$ is a functor such that $\lambda F : d \dot{\to} F$ is a limiting cone for F, then d is initial in C.*

Proof. Since λ is a cone, the triangles

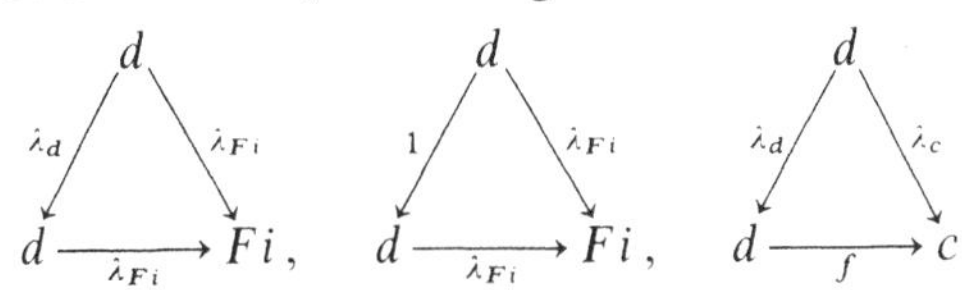

commute for each $i \in J$ and each arrow f in C. But λF is a limiting cone, so the first two triangles prove $\lambda_d = 1$. Then by the third triangle, $f = \lambda_c$: There is a unique arrow f from d to each c, and d is indeed initial.

This result reduces initial objects to limits. Now a functor $G : A \to X$ has a left adjoint precisely when for each x the comma category $(x \downarrow G)$ of all pairs $\langle g : x \to Ga, a \rangle$ has an initial object. In this way we can express the left adjoint by limits. Recall that $\langle g, a \rangle \mapsto a$ defines the (second) projection $Q : (x \downarrow G) \to A$ of the comma category.

Theorem 2 *(Formal criterion for the existence of an adjoint). A functor $G : A \to X$ has a left adjoint if and only if both*

(i) *G preserves all limits which exist in A;*

(ii) *For each $x \in X$, $\mathrm{Lim}(Q : (x \downarrow G) \to A)$ exists in A.*

When this is the case, a left adjoint F is given on each $x \in X$ as

$$Fx = \underleftarrow{\mathrm{Lim}}(Q : (x \downarrow G) \to A), \tag{3}$$

and the left adjunct of each arrow $g : x \to Ga$ is the component $\lambda_g : Fx \to Qg = a$ of the limiting cone λ for the limit (3).

Proof. Since right adjoints preserve all limits, (i) is necessary. Since a left adjoint F to G has each $\langle \eta_x : x \to GFx, Fx \rangle$ an initial object in $(x \downarrow G)$, any functor on this comma category has a limit (namely, its value on that initial object). Hence (ii) is necessary.

This motivates the converse. By hypothesis (ii) the composite functor

$$(y \downarrow G) \xrightarrow{\mathrm{Id}} (y \downarrow G) \xrightarrow{Q} A \tag{4}$$

has a limit in A for each $y \in X$. By hypothesis (i), G preserves all limits; hence, using the Lemma of § V.6, Q creates all limits. Therefore Id has a limit on $(y \downarrow G)$. This limit is, by (2), an initial object there, say $y \to Ga$. But then a is a value $a = Fy$ for a left adjoint F, and

$$Fy = Q[\underleftarrow{\mathrm{Lim}}(y \downarrow G) \to (y \downarrow G)] = \underleftarrow{\mathrm{Lim}}(Q : (y \downarrow G) \to A)$$

(since Q preserves this limit which it has created!). This is the desired formula; the rule for finding the left adjuncts follows at once.

Exercises

1. State the dual of Theorem 2.
2. (Bénabou, formal criterion for representability). Let C have small hom-sets, while $*$ is the one-point set. Prove: A functor $K : C \to \mathbf{Set}$ is representable if and only if (i) K preserves all limits which exist in C, and (ii) the projection $Q : (* \downarrow K) \to C$ of the comma category has a limit in C. When this is the case, the limiting cone $\lambda : r \dot{\to} Q$ for this projection assigns to each $h \in K_c$ an arrow $\lambda_h : r \to c$ and $h \mapsto \lambda_h$ is a representation $K \cong C(r, -)$.
3. (Formal criterion for a universal arrow.) Let X have small hom-sets. Prove that there is a universal arrow from $x \in X$ to $G : A \to X$ if and only if (i) $X(x, G-) : A \to \mathbf{Set}$ preserves all limits and (ii) $\varprojlim Q : ((x \downarrow G) \to A)$ exists in A.
4. (Refinements of formal existence criteria.)
   (a) In the Theorem, show that condition (i) may be replaced by "G preserves the limits required to exist in (ii)".
   (b) In Ex. 2, show that condition (i) may be replaced by "K preserves the limit of Q".
5. (Representables and adjoints; Bénabou.) Let C have small hom-sets, and construct from each $K : C \to \mathbf{Set}$ the category C_K obtained by adjoining to C one new object ∞ with new hom-sets $C_K(\infty, c) = Kc$, $C_K(\infty, \infty) = *$, the one-point set and $C_K(c, \infty) = \emptyset$, the empty set, with appropriate composition. Let $J_K : C \to C_K$ be the inclusion. Prove that K is representable if and only if J_K has a left adjoint.

2. Weak Universality

Given a functor $G : A \to X$ and an object $x \in X$, a *weak universal* arrow from x to G is a pair $\langle r, w : x \to Gr \rangle$ consisting of an object $r \in A$ and an arrow w of X, as indicated, such that for every arrow $f : x \to Ga$ there exists an arrow $f' : r \to a$ with $f = Gf' \circ w$. This is just the definition of universal arrow, except that f' is *not* required to be unique. By the same device (Freyd) we can modify all the various types of universals, defining weak products, weak limits, weak coproducts (existence but not uniqueness in each case).

As an application, we give a second proof of the Freyd existence theorem for an initial object (Theorem V.6.1).

Theorem 1. *If D is a small complete category with small hom-sets, then D has an initial object if and only if it has a small set S of objects which is weakly initial: For every $d \in D$ there exists $s \in S$ and an arrow $s \to d$.*

Proof. Let S also denote the full subcategory of D with the objects s; since D has small hom-sets, S is still small, so by completeness the inclusion functor $F : S \to D$ has a limiting cone $\mu : v \dot{\to} F$. We shall prove $v = \operatorname{Lim} F$ initial in D.

First, for every $d \in D$ we choose $s \to d$ and define γ_d as the composite $\gamma_d : v \xrightarrow{\mu_s} s \to d$. We claim that $\gamma : v \to \mathrm{Id}_D$ is a cone. For, take any arrow $f : d \to d'$ and form the diagram

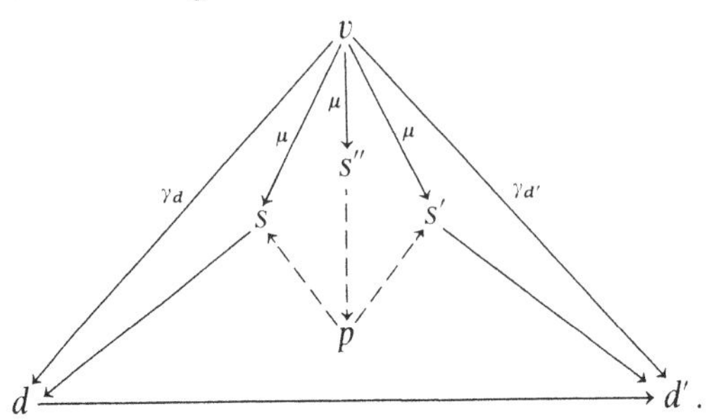

Since S is small complete, there is a pullback of $s \to d' \geq s'$ with vertex p; since S is weakly initial, there is an arrow $v \xrightarrow{\mu_{s''}} s'' \to p$. The two composite arrows $s'' \to p \to s$ and $s'' \to p \to s'$ are in S because S is full, so the two upper quadrilaterals commute (μ is a cone), while the pentagon commutes because p is a pullback. This proves γ a cone.

If, in defining γ, we choose $v \to s \to s$ to be μ_s, then γ is a cone such that the composite $\gamma F : v \dot{\to} F$ is the limiting cone μ. By Lemma 1.1, v is initial in D, q.e.d.

Carefully examined, this proof is just a refinement of the previous one (§ V.6), where we took first a product Πs (to get a single weakly initial object) and then a suitable equalizer. In this proof, these operations are combined to one: Lim F for $F : S \subset D$.

3. The Kan Extension

Given a functor $K : M \to C$ and a category A we consider the functor category A^C, with objects the functors $S : C \to A$ and arrows the natural transformations $\sigma : S \dot{\to} S'$, and we define the functor $A^K : A^C \to A^M$ by the assignments

$$\langle \sigma : S \dot{\to} S' \rangle \mapsto \langle \sigma K : SK \dot{\to} S'K \rangle .$$

The problem of Kan extension is to find left and right adjoints to A^K. We consider this problem first for right adjoints.

Definition. *Given functors $K : M \to C$ and $T : M \to A$, a right Kan extension of T along K is a pair $R, \varepsilon : RK \dot{\to} T$ consisting of a functor $R \in A^C$ and a natural transformation ε which is universal as an arrow from $A^K : A^C \to A^M$ to $T \in A^M$.*

As always, this universality determines the functor $R = \mathrm{Ran}_K T$ uniquely, up to natural isomorphism. In detail, this universality means that for each pair $S, \alpha : SK \dot\to T$ there is a unique natural transformation $\sigma : S \dot\to R$ such that $\alpha = \varepsilon \cdot \sigma K : SK \dot\to T$. The diagram is

$$\begin{array}{ccc} C & & \\ {\scriptstyle K}\uparrow & \overset{\sigma\cdot}{\searrow} \; R \;\; S & \\ M & \xrightarrow[T]{\;\varepsilon\cdot\;} & A \end{array} \qquad \varepsilon : RK \dot\to T \tag{1}$$

The assignment $\sigma \mapsto \varepsilon \cdot \sigma K$ is a bijection

$$\mathrm{Nat}(S, \mathrm{Ran}_K T) \cong \mathrm{Nat}(SK, T), \tag{2}$$

natural in S; again, this natural bijection determines $\mathrm{Ran}_K T$ from K and T. It is a *right* Kan extension because it appears at the right in the hom-set "Nat" (But note that some authors call this R a "left" Kan extension).

By the general result that universal arrows from the functor A^K to all the objects T together constitute a left adjoint to the functor A^K, it follows that if *every* functor $T \in A^M$ has a right Kan extension $\langle R, \varepsilon_T : RK \dot\to T \rangle$, then $T \mapsto R$ is (the object function of) a right adjoint to A^K and ε is the unit of this adjunction. In the sequel, we shall construct right Kan extensions for individual functors T, which may exist when (the whole of) the right adjoint of A^K does not exist.

A useful case is that in which M is a subcategory of C and $K : M \to C$ the inclusion $M \subset C$; in this case, A^K is the operation which restricts the domain of a functor $S : C \to A$ to the subcategory M. Conversely, for given $T : M \to A$ we consider extensions $E : C \to A$ of T to C. Then $Ec \in A$ must have for each arrow $f : c \to m$ in C an arrow $Ef : Ec \to Tm$ in A, and these arrows must constitute a cone from the vertex Ec to the base T, where T is regarded as a functor on the category of arrows $f : c \to m$ (fixed c to variable m). These arrows f are the objects of the comma category $(c \downarrow K)$, so a natural choice of Ec is the limit (with Ef the limiting cone) of the functor $T : (c \downarrow K) \to A$:

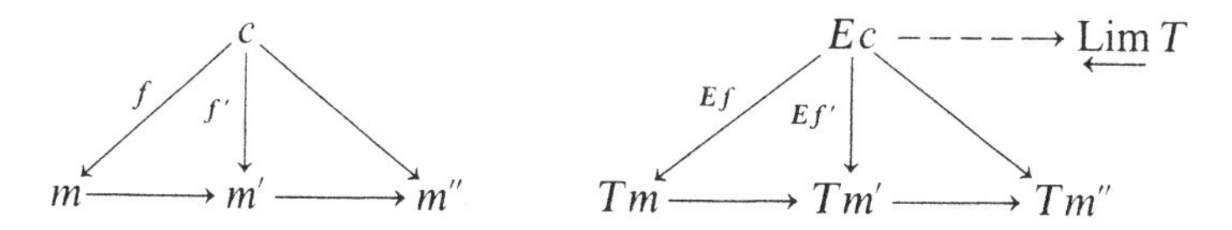

This procedure (compare (1.3)) works in general. For each $c \in C$, the comma category $(c \downarrow K)$ has the objects $\langle f, m \rangle$, written f for short, where $f : c \to Km$ in C, while $\langle f, m \rangle \mapsto m$ is (the object function of) the projection functor $Q : (c \downarrow K) \to M$.

Theorem 1 *(Right Kan extension as a point-wise limit). Given* $K : M \to C$, *let* $T : M \to A$ *be a functor such that the composite*

$(c\downarrow K)\to M\to A$ has for each $c\in C$ a limit in A, with limiting cone λ, written

$$Rc=\varprojlim((c\downarrow K)\xrightarrow{Q}M\xrightarrow{T}A)=\mathrm{Lim}_f\,Tm,\qquad f\in(c\downarrow K). \tag{3}$$

Each $g:c\to c'$ induces a unique arrow

$$Rg:\varprojlim TQ\to\varprojlim TQ' \tag{4}$$

commuting with the limiting cones. These formulas define a functor $R:C\to A$, and for each $n\in M$, the components $\lambda_{1_{Kn}}=\varepsilon_n$ of the limiting cones define a natural transformation $\varepsilon:RK\dot{\to}T$, and R,ε is a right Kan extension of T along K.

Proof. First, Rg is defined in (4) by the fact that the limit is a functor of $(c\downarrow K)$ and hence of c. Specifically, given $g:c\to c'$ and the projection $Q':(c'\downarrow K)\to A$, each $f':c'\to Km$ determines $f'g:c\to Km\in(c\downarrow K)$, the components $\lambda_{f'g}:Rc\to Tm$ form a cone from Rc, and since the cone λ' is universal, there is a unique arrow Rg which makes

$$\begin{array}{ccc} Rc=\varprojlim TQ & \xrightarrow{\lambda_{f'g}} & Tm \\ \downarrow{\scriptstyle Rg} & & \| \\ Rc'=\varprojlim TQ' & \xrightarrow{\lambda'_{f'}} & Tm \end{array} \tag{5}$$

commute for all f'. (Actually, $f'\mapsto f'g$ defines the functor $(g\downarrow K):(c'\downarrow K)\to(c\downarrow K)$, so that $TQ'=TQ\circ(g\downarrow K)$, and Rg is the canonical comparison (cf. "final functors")). This choice of Rg clearly makes R a functor.

For each $n\in M$, 1_{Kn} is an object of $(Kn\downarrow K)$, so the limiting cone λ has a component $\lambda_{1_{Kn}}:RKn\to Tn$, called ε_n. For each $h:n\to n'$ form the diagram

$$\begin{array}{ccc} RKn & \xrightarrow{\lambda_{1_{Kn}}} & Tn \\ {\scriptstyle RKh}\downarrow & \searrow{\scriptstyle \lambda_{Kh}} & \downarrow{\scriptstyle Th} \\ RKn' & \xrightarrow[\lambda'_{1_{Kn'}}]{} & Tn'\,; \end{array} \tag{6}$$

the lower triangle commutes by the definition of Rg for $g=Kh$, and the upper triangle commutes because λ is a cone. Therefore the square commutes; this states that $\varepsilon:RK\dot{\to}T$ is natural.

Now let $S:C\to A$ be another functor, with $\alpha:SK\dot{\to}T$ natural. We construct $\sigma_c:Sc\to Rc$ from the diagram for $f:c\to Km$

$$\begin{array}{ccccc} Rc=\mathrm{Lim}_f\,Tm & \xrightarrow{\lambda_f} & Tm & \xrightarrow{Th} & Tm' \\ \uparrow{\scriptstyle \sigma_c} & \nearrow & \uparrow{\scriptstyle \alpha_m} & & \uparrow{\scriptstyle \alpha_{m'}} \\ Sc & \xrightarrow[Sf]{} & SKm & \xrightarrow[SKh]{} & SKm' \end{array} \tag{7}$$

For each arrow $h:\langle f,m\rangle\to\langle f',m'\rangle$ of $(c\downarrow K)$, where $f'=Kh\circ f$, the right-hand square commutes because α is natural. This shows that the diagonal arrows $\alpha_m\circ Sf:Sc\to Tm$ form a cone from Sc. Hence there is a unique arrow σ_c, as shown in (7). To prove σ natural for $g:c\to c'$, form the diagram

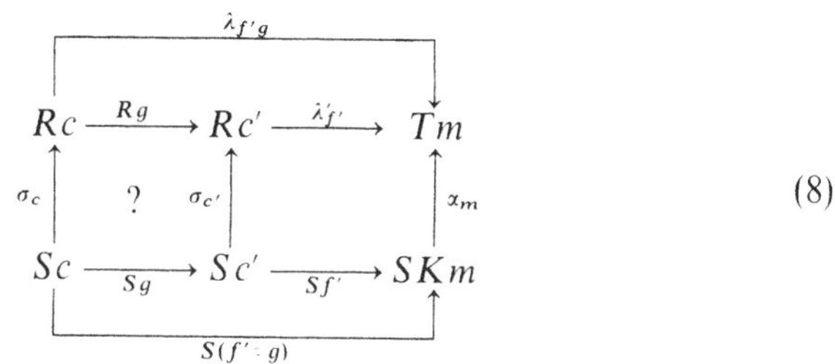

(8)

for each $f':c'\to Km$ in $(c'\downarrow K)$. The right-hand square and the outer square commute by the definition of σ, and the top box by the definition (5) of Rg. Therefore the left-hand (inner) square commutes after both legs are composed with $\lambda'_{f'}$ – and this for all f'. But λ' is a limiting cone, so the left-hand square commutes. Therefore σ is natural.

The definition (7) of σ for $c=Kn$, $f=1_{Kn}$, and $m=n$ shows that $\alpha_n=\lambda_{1_{Kn}}\sigma_{Kn}$, hence that $\alpha=\varepsilon\cdot\sigma K$. This proves that $\varepsilon:RK\dot{\to}T$ gives every α as $\alpha=\varepsilon\cdot\sigma K$ for some σ. The diagram (8) shows that σ is unique with this property. Indeed, this property determines the components σ_{Kn} of σ; to determine other components, set $c'=Kn$, $f'=1_{Kn}$, and $m=n$ in (8). The lefthand square commutes if σ is natural, and then $\lambda_g\circ\sigma_c$ is determined for all $g:c\to Kn$. But λ is a limiting cone, so σ_c is determined. This shows that ε is universal, q.e.d.

Corollary 2. *If M is small and A complete, any functor $T:M\to A$ has a right Kan extension along any $K:M\to C$, and A^K has a right adjoint.*

This applies in particular when $A=\mathbf{Set}$; this is the case originally studied by Kan [1958].

Corollary 3. *If the functor K in the theorem is full and faithful, then the universal arrow $\varepsilon:RK\dot{\to}T$ for the Kan extension R of T along K is a natural isomorphism $\varepsilon:RK\cong T$.*

Proof. For $n\in M$, RKn is obtained from a limit over the comma category $(c\downarrow K)$ with $c=Kn$. Because K is full and faithful, every object $f:Kn\to Km$ in this comma category can be written as $f=Kh$ for a unique $h:n\to m$. This states that $1:Kn\to Kn$ is an initial object in this comma category and hence that $RKn=\mathrm{Lim}_f TQ$ can be found by evaluating TQ just at this initial object: Thus $RKn=Tn$, $\varepsilon_n=1$, q.e.d.

This also gives a case in which a Kan extension is an actual extension:

Corollary 4. *If M is a full subcategory of a category C and $T:M\to A$ is a functor such that each composite $(c\downarrow K)\to M\to A$ has a limit in A,* *then there is a functor $R: C \to A$ with $RK = T$ (i.e., R extends T) such that the identity natural transformation $1: RK \dot{\to} T$ makes R the right Kan extension of T along the insertion $K: M \to C$.*

Proof. Apply Corollary 3 to the insertion $M \to C$.

The left Kan extension $L = \operatorname{Lan}_K T$ is described similarly, as a pair $L, \eta: T \dot{\to} LK$ with η universal from T to A^K; this gives a bijection

$$\operatorname{Nat}(\operatorname{Lan}_K T, S) \cong \operatorname{Nat}(T, SK) \tag{9}$$

natural in $S \in A^C$. When the requisite colimits exist, L is given by

$$Lc = \underrightarrow{\operatorname{Colim}}((K \downarrow c) \xrightarrow{P} M \xrightarrow{T} A), \tag{10}$$

where P is the projection $\langle m, Km \to c \rangle \mapsto m$.

Exercises

Exercises 1–4 refer to the data for a Kan extension:

$$K: M \to C, \qquad T: M \to A.$$

1. If A is the arrow category $\mathbf{2}$, and M and C are sets, then a functor $T: M \to \mathbf{2}$ can be regarded as a subset of M. Show that $\mathbf{2}^M$ is the contravariant power set $\mathscr{P}M$, that $\operatorname{Lan}_K T$ is the direct image of $T \subset M$ under the function K, and describe $\operatorname{Ran}_K T$.
2. (Kan extensions of representable functors.) If $A = \mathbf{Set}$, and M, C have small hom-sets, show that the left Kan extension of $M(m, -)$ is $C(Km, -)$ with unit $\eta: M(m, -) \dot{\to} C(Km, K-)$ given by $\eta m = 1_{Km}$.
3. If M, C, and A are all sets, while A has at least two elements and K is not surjective, prove that neither $\operatorname{Lan}_K T$ nor $\operatorname{Ran}_K T$ exists.
4. (Ulmer.) Show that Corollary 3 still holds if the hypothesis "K is full and faithful" is replaced by "K is full, and as faithful as T". Here K "is as faithful as T" when, for arrows $h, h': m \to n$ in M, $Kh = Kh'$ implies $Th = Th'$.
5. For any category M, let M_∞ be the category formed by adding to M one new object ∞, terminal in M_∞. For $T: M \to A$, prove (from first principles) that a colimiting cone for T is a left Kan extension of T along the inclusion functor $M \subset M_\infty$, and conversely.

4. Kan Extensions as Coends

The calculus of coends gives an elegant formula for Kan extensions; for variety we treat the left Kan.

Theorem 1. *Given functors $K: M \to C$ and $T: M \to A$ such that for all $m, m' \in M$ and all $c \in C$ the copowers $C(Km', c) \cdot Tm$ exist in A, then T has a left* Kan *extension $L = \operatorname{Lan}_K T$ along K if for every $c \in C$* the *following coend exists, and when this is the case, the object functor of L is this coend*

$$Lc = (\operatorname{Lan}_K T)c = \int^m C(Km, c) \cdot Tm, \qquad c \in C. \tag{1}$$

Proof. By the parameter theorem, we may regard this coend as a functor of c. Compare it with any other functor $S: C \to A$. Then

$$A(Tm, SKm) \cong \mathrm{Nat}(C(Km, -), A(Tm, S-)) \cong \int_c \mathbf{Ens}(C(Km, c), A(Tm, Sc)) \tag{2}$$

by the Yoneda lemma and the representation of the set of natural transformations as an end (in a sufficiently large full category **Ens** of sets). Now we can write down in succession the following isomorphisms

$$\begin{aligned}
\mathrm{Nat}(L, S) &\cong \int A(Lc, Sc) && \text{(end formula for Nat)} \\
&\cong \int_c A\left(\int^m C(Km, c)\cdot Tm, Sc\right) && \text{(Definition (1) of } L) \\
&\cong \int_c \int_m A(C(Km, c)\cdot Tm, Sc) && \text{(Continuity of } A(-, Sc)) \\
&\cong \int_c \int_m \mathbf{Ens}(C(Km, c), A(Tm, Sc)) && \text{(Definition of copowers)} \\
&\cong \int_m \int_c \mathbf{Ens}(C(Km, c), A(Tm, Sc)) && \text{(Fubini)} \\
&\cong \int_m A(Tm, SKm) && \text{(by (2) above)} \\
&\cong \mathrm{Nat}(T, SK) && \text{(end formula for Nat).}
\end{aligned}$$

Here the Fubini theorem (interchange of ends) applies because both indicated ends $\int_m$ and $\int_c$ exist, while **Ens** must be a sufficiently large category of sets (to contain all hom-sets for A and C and all sets $\mathrm{Nat}(L, S)$, $\mathrm{Nat}(T, SK)$ for all $S: C \to A$). Since each step is natural in S, the composite isomorphism is natural in S and proves that $L = \mathrm{Lan}_K T$.

Note that we do not assert the converse: That if $\mathrm{Lan}_K T$ exists, it must be given for each c as the coend (1).

The unit η of this Kan extension is obtained by setting $S = L$ and following the chain of isomorphisms. We record the result:

Theorem 2 *(Kan extensions as coends, continued.) For the Kan extension* (1) *above the universal arrow $\eta: T \dot{\to} LK$ is given for each $n \in M$ as the composite of an injection $i_{1_{Kn}}$ of the copower (for $f = 1_{Kn}: Kn \to Kn$) with a component of the ending wedge ω:*

$$\begin{array}{ccc}
Tn & \xrightarrow{\;i_{1_{Kn}}\;} & C(Kn, Kn)\cdot Tn \\
& \searrow^{\eta} & \downarrow \omega_{n, Kn} \\
& & \int^m C(Km, Kn)\cdot Tm = (\mathrm{Lan}_K T)(Kn).
\end{array}$$

For the left Kan extension we thus have two formulas – (1) above by coends, and (3.10) by colimits. They are closely related, and simply constitute two ways of organizing the same colimit information (see Exercise 1 below). The corollaries of § 3 can be deduced from either formula. Also right Kan extensions are given by a formula

$$(\mathrm{Ran}_K T)c = \int_m Tm^{C(c,Km)}, \tag{3}$$

valid when the indicated power (powers X, $a \mapsto a^X$ in A) and its end exist.

Consider additive Kan extensions: M, C, and A are Ab-categories and the given functors K and T are both additive. Then we can describe a right Ab-Kan extension of T along K as an additive functor $R'\colon C \to A$ with a bijection (3.2) given and natural for *additive* functors S. This functor R' need not agree with the ordinary right Kan extensions $\mathrm{Ran}_K T$ obtained by forgetting that K, T (and S) are additive. However, R' can still be given by a formula (3) with an end, provided the power a^C involved (for $a \in A$, $C \in \mathbf{Ab}$) is replaced by a "cotensor" a^C defined by the adjunction

$$A(b, c^C) \cong \mathbf{Ab}(C, A(b, c)) \tag{4}$$

for all $b \in A$ (see Day and Kelly [1969], Dubuc [1970]). For example, if $A = R\text{-}\mathbf{Mod}$, this makes $a^C = \mathbf{Ab}(C, a)$ with the evident R-module structure (induced from that of $a \in R\text{-}\mathbf{Mod}$).

Derived functors are an example. If $T\colon R\text{-}\mathbf{Mod} \to \mathbf{Ab}$ is right exact, its left derived functors $T_n\colon R\text{-}\mathbf{Mod} \to \mathbf{Ab}$ come equipped with certain connecting morphisms, which make them what is called a connected sequence of functors (Mac Lane [1963a], Cartan-Eilenberg [1956]); basic example: If A is a right R-module, the left-derived functors of the tensor product $A \otimes_R -\colon R\text{-}\mathbf{Mod} \to \mathbf{Ab}$ are the torsion products $\mathrm{Tor}_n(A, -)\colon R\text{-}\mathbf{Mod} \to \mathbf{Ab}$.

The left-derived functors T_n of T can be described by the following "universal" property: $T_0 = T$, and if S_n is any connected sequence of (additive) functors, each natural transformation $S_0 \dashrightarrow T_0$ extends to a unique morphism $\{S_n \mid n \geqq 0\} \to \{T_n \mid n \geqq 0\}$ of connected sequences of functors.

This property may be rewritten thus. Embed R-**Mod** in a larger Ab-category E with objects $\langle C, n\rangle$, C an R-module and n a nonnegative integer, while the hom-groups are $E(\langle C, n\rangle, \langle B, m\rangle) = \mathrm{Ext}_R^{n-m}(C, B)$, with composites given by the Yoneda product. Then $C \mapsto \langle C, 0\rangle$ is a functor $K\colon R\text{-}\mathbf{Mod} \to E$. A connected sequence of additive functors $\{T_n \mid n \geqq 0\}$ is then the same thing as a single additive functor $T_*\colon E \to \mathbf{Ab}$ with $T_*(C, n) = T_n(C)$, while T_* on the morphisms of E gives the connecting morphisms. The universal property stated above for the sequence T_* of left derived functors of T now reads:

$$\mathrm{Nat}(S_* K, T) \cong \mathrm{Nat}(S_*, T_*).$$

This states exactly that T_* is the right Ab-Kan extension of $T = T_0$ along $K: R\text{-}\mathbf{Mod} \to E$ (and that the unit $\varepsilon: T_* K \dot{\to} T$ of this Kan extension is the identity).

For details we refer to Cartan-Eilenberg or to Mac Lane [1963a] (where the category E is treated in a different but equivalent way, as a "graded Ab-category").

Exercises

1. If the coends in Theorem 1 exist, prove that these coends do give the colimits required in the formula (3.10) for Lan_K.
2. For fixed K, describe $\mathrm{Lan}_K T$ and $\mathrm{Ran}_K T$, when they exist, as functors of T.
3. (Dubuc.) If $\mathrm{Ran}_K T$ exists, while $L: C \to D$ is any functor, prove that $\mathrm{Ran}_{LK} T$ exists if and only if $\mathrm{Ran}_L \mathrm{Ran}_K T$ exists and that then these two functors (and their universal arrows) are equal.
4. (Ulmer; Day-Kelly; Kan extensions as a coend in a functor category A^C.) If $C(Km', c) \cdot Tm$ exists for all $m', m \in M$ and all $c \in C$, show that $\langle m', m \rangle \mapsto C(Km', -) \bullet Tm$ is (the object function of) a functor $M^{\mathrm{op}} \times M \to A^C$. Prove that T has a left Kan extension along K if and only if this bifunctor has a coend, and that then this coend is the Kan extension
$$\mathrm{Lan}_K T = \int^m C(Km, -) \cdot Tm\,.$$
Describe the universal arrow for $\mathrm{Lan}_K T$ in terms of the coend.
5. (Ulmer.) As in Ex. 4, obtain a necessary and sufficient condition for the existence of $\mathrm{Ran}_K T$ in terms of the limit formula, interpreted in the functor category A^C.

5. Pointwise Kan Extensions

Given functors
$$C \xleftarrow{K} M \xrightarrow{T} A \xrightarrow{G} X \tag{1}$$
and a right Kan extension $\mathrm{Ran}_K T$ with counit $\varepsilon: (\mathrm{Ran}_K T) K \dot{\to} T$, we say that G *preserves* this right Kan extension when $G \circ \mathrm{Ran}_K T$ is a right Kan extension of GT along K with counit $G\varepsilon: G(\mathrm{Ran}_K T)K \dot{\to} GT$. This implies (but is stronger than)
$$G \circ \mathrm{Ran}_K T \cong \mathrm{Ran}_K(GT)\,.$$

We already know that right adjoints G preserve limits. We now show that they also preserve Kan extensions.

Theorem 1. *If $G: A \to X$ has a left adjoint F, it preserves all right Kan extensions which exist in A.*

Proof. First a preliminary, for an adjunction
$$A(Fx, a) \cong X(x, Ga)\,, \qquad x \in X\,,\ a \in A\,.$$

If in place of x we have a functor $H : C \to X$ and in place of a a functor $L : C \to A$, then applying this adjunction at every Lc and Hc gives a bijection,

$$\mathrm{Nat}(FH, L) \cong \mathrm{Nat}(H, GL). \tag{2}$$

(As usual the adjunction switches F on the left to G on the right.)

Now assume the adjunction and a right Kan extension $\mathrm{Ran}_K T$ for some K and $T: M \to A$. Then for any functor $H : C \to X$ we have the following bijections

$$\mathrm{Nat}(H, G \circ \mathrm{Ran}_K T) \cong \mathrm{Nat}(FH, \mathrm{Ran}_K T)$$
$$\cong \mathrm{Nat}(FHK, T) \cong \mathrm{Nat}(HK, GT),$$

natural in H; the first and third are instances of (2), and the second is the definition of the right Kan extension. The composite bijection (for all H) shows that $G \circ \mathrm{Ran}_K T$ is the right Kan extension $\mathrm{Ran}_K GT$. To get its counit, we set $H = G \circ \mathrm{Ran}_K T$ and take the image of the identity; we get $G\varepsilon$, where $\varepsilon : (\mathrm{Ran}_K T)K \dot{\to} T$ is the counit of the given Kan extension.

Corollary 2. *If R, $\varepsilon : RK \dot{\to} T$ is a right Kan extension and A has small hom-sets and all small copowers, then for each $a \in A$, $A(a, R-) : C \to \mathbf{Set}$, is the right Kan extension of $A(a, T-) : M \to \mathbf{Set}$, with counit $A(a, \varepsilon -)$.*

Proof. The functor $A(a, -): A \to \mathbf{Set}$ has the left adjoint $X \mapsto X \cdot a$, the copower.

Definition. *Given $C \xleftarrow{K} M \xrightarrow{T} A$, where A has small hom-sets, a right Kan extension R is point-wise when it is preserved by all representable functors $A(a, -): A \to \mathbf{Set}$, for $a \in A$.*

Theorem 3. *A functor $T: M \to A$ has a pointwise right Kan extension along $K: M \to C$ if and only if the limit of $(c \downarrow K) \to M \to A$ exists for all c. When this is the case, $\mathrm{Ran}_K T$ is given by the formulas of Theorem 3.1.*

Proof. Since $A(a, -)$ preserves limits, any Kan extension given by the limit formula is pointwise.

Conversely, suppose for each $a \in A$ that $A(a, T-): M \to \mathbf{Set}$ has a right Kan extension $R^a = A(a, R-)$, as in the figure

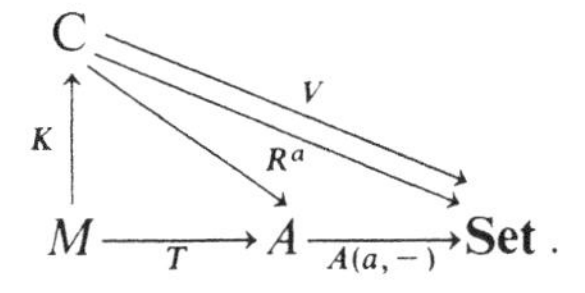

Then for each functor V, as shown, there is a bijection

$$\mathrm{Nat}(V, R^a) \cong \mathrm{Nat}(VK, A(a, T-)),$$

natural in V. This holds in particular when $V = C(c, -)$ for some $c \in C$, so

$$\mathrm{Nat}(C(c, -), A(a, R-)) \cong \mathrm{Nat}(C(c, K-), A(a, T-)).$$

We reduce the left hand side by the Yoneda Lemma and the right hand side by the lemma below to get, with Q the projection $(c \downarrow K) \to M$

$$A(a, Rc) \cong \mathrm{Cone}(a, TQ : (c \downarrow K) \to A) .$$

This states that the set of cones is representable, hence that the limit of TQ exists, q.e.d.

The missing lemma is

Lemma. *Given $K : M \to C$, there is a bijection*

$$\mathrm{Cone}(a, (c \downarrow K) \to M \to C) \cong \mathrm{Nat}(C(c, K-), A(a, T-)) .$$

Proof. A cone $\tau : a \dot{\to} TQ$ assigns to each $f : c \to Km$ an arrow $\tau(f, m) : a \to Tm$ subject to the cone conditions; for each $h : m \to m'$,

$$\tau(Kh \circ f, m') = Th \circ \tau(f, m) .$$

A natural transformation $\beta : C(c, K-) \dot{\to} A(a, T-)$ assigns to each $m \in M$ and to each $f : c \to Km$ an arrow $\beta_m f : a \to Tm$, subject to the naturality condition, for each $h : m' \to m$, that

$$\beta_{m'}(Kh \circ f) = Th \circ \beta_m f .$$

The bijection $\tau \leftrightarrow \beta$ is now evident.

This proof of the theorem also shows

Corollary 4. *$R, \varepsilon : RK \dot{\to} T$ is a pointwise Kan extension of T along K if and only if, for all $a \in A$ and $c \in C$,*

$$A(a, Rc) \to \mathrm{Nat}(C(c, K-), A(a, T-))$$

sending $g : a \to Rc$ to the transformation with the component

$$C(c, Km) \xrightarrow{R} A(Rc, RKm) \xrightarrow{A(g, \varepsilon_m)} A(a, Tm)$$

at $m \in M$ is a bijection.

Exercise

1. In the situation (1), if $\mathrm{Ran}_K T$ and $\mathrm{Ran}_K GT$ both exist, with counits ε and ε', prove that there is a unique natural transformation (the canonical map) $w : G \circ \mathrm{Ran}_K T \dot{\to} \mathrm{Ran}_K GT$ with $\varepsilon' \cdot wK = G\varepsilon$, and prove that G preserves $\mathrm{Ran}_K T$ if and only if w is an isomorphism.

6. Density

A subcategory M of C is said to be *dense* in C if every object of C is a colimit of objects of M; more exactly, a colimit in a canonical way, for which the colimiting cone consists of all arrows $m \to c$ to c from an $m \in M$. More generally, density can be defined not only for an inclusion

$M \subset C$, but for any functor $K: M \to C$. The arrows $m \to c$ are then replaced by the objects $\langle m, f: Km \to c\rangle$ of the comma category $(K \downarrow c)$. Recall that the projections P^c, Q^c of this comma category are given by $P^c\langle m, f\rangle = m$, $Q^c\langle m, f\rangle = f$, and observe that (the object function of) Q^c may also be regarded as a cone $Q^c: KP^c \to c$.

Definition. *A functor $K: M \to C$ is dense when for each $c \in C$*

$$\operatorname{Colim}\left((K \downarrow c) \xrightarrow{P^c} M \xrightarrow{K} C\right) = c, \tag{1}$$

with colimiting cone the "canonical cone" Q^c. In particular, a subcategory M of C is dense in C when the inclusion functor $M \to C$ is dense in the sense just defined.

The definition (1) is sometimes phrased, "The canonical map $\operatorname{Colim} KP^c \to c$ is an isomorphism"; here the canonical map is the unique arrow $k: \operatorname{Colim} KP^c \to c$ which carries the colimiting cone to Q^c.

For example, the one-point set $*$ is dense in **Set**: For each set X, the comma category $(* \downarrow X)$ is just the set (discrete category) of elements $x \in X$, each regarded as a function $x: * \to X$, while (1) becomes statement that each X is the coproduct $\amalg x$ of its elements (i.e., that a function f with domain X can be uniquely determined by specifying the value fx at each $x \in X$).

Dually, a functor $K: M \to C$ is *codense* when for each $c \in C$

$$\operatorname{Lim}\left((c \downarrow K) \xrightarrow{Q} M \xrightarrow{K} C\right) = c, \tag{2}$$

with limiting cone the canonical cone sending $\langle f: c \to Km, m\rangle$ to f. But this limit is precisely the one involved in the definition of $\operatorname{Ran}_K K$. Hence

Proposition 1. *The functor $K: M \to C$ is codense if and only if* Id_C, *together with the identity natural transformation* $\mathrm{Id}_K: K \dot{\to} K$, *is the pointwise right Kan extension of K along K.*

In this case Corollary 5.4 simplifies (ε is the identity) to the correspondence sending each $f: a \to c$ to the natural transformation

$$f^* = C(f, K-): C(c, K-) \dot{\to} C(a, K-) \tag{3}$$

(the transformation f^* is "composition with f on the right"). Hence

Proposition 2. *The functor $K: M \to C$ is codense if and only if the correspondence $f \mapsto C(f, K-)$ above is for all a and $c \in C$ a bijection*

$$C(a, c) \cong \operatorname{Nat}(C(c, K-), C(a, K-)); \tag{4}$$

that is, if and only if the functor $C^{\mathrm{op}} \to \mathbf{Ens}^M$ defined by

$$c \mapsto C(c, K-): M \to \mathbf{Ens}^M \tag{5}$$

is full and faithful, where the hom-sets of M lie in **Ens**.

Corollary 3. *If the hom-sets of M lie in a full category* **Ens** *of sets, then Yoneda embedding $Y : M \to (\mathbf{Ens}^M)^{\mathrm{op}}$, given by $Ym = M(m, -)$ is codense.*

Proof. By the Yoneda Lemma itself, for each $F : M \to \mathbf{Ens}$,

$$(\mathbf{Ens}^M)^{\mathrm{op}}(F, Ym) = \mathbf{Ens}^M(Ym, F) \cong Fm .$$

Thus the right side of (4) above, with $C = (\mathbf{Ens}^M)^{\mathrm{op}}, a = F$ and $c = G$ becomes

$$\mathrm{Nat}(G, F) = (\mathbf{Ens}^M)^{\mathrm{op}}(F, G) = C(F, G) ,$$

and (4) becomes an identity.

This result is often stated thus: Any functor $M \to \mathbf{Ens}$ is a canonical limit of representable functors.

The dual of Proposition 2 states that $K : M \to C$ is dense if and only if $c \mapsto C(K-, c)$ is a full and faithful functor $C \to \mathbf{Ens}^{M^{\mathrm{op}}}$. As an application, we show that the full subcategory of finitely generated abelian groups is dense in **Ab**. We need only show that for K and any two abelian groups A and B the map

$$\mathbf{Ab}(A, B) \to \mathrm{Nat}(\mathbf{Ab}(K-, A), \mathbf{Ab}(K-, B))$$

is a bijection. First, it is injective: Two homomorphisms $f, g : A \to B$ which agree on cyclic subgroups of A must agree everywhere. Also, it is surjective: Given $\tau : \mathbf{Ab}(K-, A) \dot\to \mathbf{Ab}(K-, B)$, we define a function $f : A \to B$ by taking fa for each $a \in A$ to be the value of τ on the map $\mathbf{Z} \to A$ taking 1 to a. Because $\mathbf{Z} \oplus \mathbf{Z}$ is a finitely generated group, this function must be a homomorphism. Its image under the map in question agrees with τ; the proof is complete. Note that the argument proves more: The full subcategory with one object $\mathbf{Z} \oplus \mathbf{Z}$ is dense in **Ab**. (There are two summands $\mathbf{Z}$ required because abelian groups are algebraic systems defined by binary operations.)

Exercises

1. In R-**Mod**, show that the full subcategory with one object $R \oplus R$ is dense.
2. Show that the full subcategory with one object $\mathbf{Z}$ is not dense in **Ab.**
3. Let the *image category* KM for $K : M \to C$ be the subcategory of C with objects all Km for $m \in M$ and arrows all Kh, h in M. Prove that K dense implies KM a dense subcategory of C.
4. Prove that the objects of a subcategory M generate C if and only if the functor $C \to \mathbf{Ens}^{M^{\mathrm{op}}}$ given by $c \mapsto C(K-, c)$ is faithful.
5. If all copowers $C(Km', c) \cdot Km$ exist in C, prove that $K : M \to C$ is dense if and only if each object $c \in C$ is the coend

   $$c = \int^m C(Km, c) \cdot Km$$

   with coending wedge $\omega^c_m : C(Km, c) \cdot Km \to c$ given on the injections i_f of the copower as $\omega^c_m i_f = f : Km \to c$.

7. All Concepts Are Kan Extensions

The notion of Kan extensions subsumes all the other fundamental concepts of category theory.

Theorem 1. *A functor $T: M \to A$ has a colimit if and only if it has a left Kan extension along the (unique) functor $K_1 : M \to \mathbf{1}$, and then* Colim T *is the value of* $\mathrm{Lan}_{K_1} T$ *on the unique object of* $\mathbf{1}$.

Proof. A functor $S: \mathbf{1} \to A$ is just an object $a \in A$, and a natural transformation $\alpha\colon T \dot{\to} SK_1$, for $K_1 : M \to \mathbf{1}$, is just a cone with base T and vertex a. Since the left Kan extension $L = \mathrm{Lan}_{K_1} T$ is constructed to provide the universal natural $\eta : T \dot{\to} LK_1$, it also provides the universal cone with base T, and hence the colimit of T.

Dually, right Kan extensions along the same functor K_1 give limits.

Theorem 2 *(Formal criteria for the existence of an adjoint). A functor $G: A \to X$ has a left adjoint if and only if the right Kan extension $\mathrm{Ran}_G 1_A : X \to A$ exists and is preserved by G; when this is the case, this right Kan extension is a left adjoint $F = \mathrm{Ran}_G 1_A$ for G, and the counit transformation $\varepsilon : (\mathrm{Ran}_G 1_A)G \dot{\to} 1_A$ for the Kan extension is the counit $\varepsilon : FG \dot{\to} 1$ of the adjunction.*

Proof. If G has a left adjoint F, with unit $\eta : 1_X \dot{\to} GF$ and counit $\varepsilon : FG \dot{\to} 1_A$, then we can construct for all functors $H : A \to C$ (in particular, for the identity functor 1_A) a bijection

$$\mathrm{Nat}(S, HF) \cong \mathrm{Nat}(SG, H), \tag{1}$$

natural in $S: X \to C$, by the assignments

$$\{\sigma : S \dot{\to} HF\} \mapsto \{SG \xrightarrow{\sigma G} HFG \xrightarrow{H\varepsilon} H\},$$
$$\{\tau : SG \dot{\to} H\} \mapsto \{S \xrightarrow{S\eta} SGF \xrightarrow{\tau F} HF\}.$$

The first followed by the second is the identity $\sigma \mapsto \sigma$, because the diagram

$$\begin{array}{ccccc} S & \xrightarrow{\sigma} & HF & = & HF \\ {\scriptstyle S\eta}\downarrow & & \downarrow{\scriptstyle HF\eta} & & \| \\ SGF & \xrightarrow{\sigma GF} & HFGF & \xrightarrow[H\varepsilon F]{} & HF \end{array}$$

is commutative (the first square represents the horizontal composite $\sigma\eta$ in two ways, and the second square is H applied to one of the two triangular identities for η and ε). The composite in the other order is also an identity, by a similar diagram. Hence we have the asserted bijection, clearly natural in S. If we take $H = 1_A$, this bijection shows that $F = \mathrm{Ran}_G 1_A$, its unit is the image of $\sigma = 1_F$, so is ε. If we take $H = G$, this bijection shows that $GF = \mathrm{Ran}_G G$, with unit $G\varepsilon$. Hence G preserves the right Kan extension $\mathrm{Ran}_G 1_A$. We have proved the first half of the theorem.

We have proved more: For any H, $HF = \mathrm{Ran}_G H$, with unit $H\varepsilon$. Thus $\mathrm{Ran}_G 1_A$ is preserved by any functor whatever (it is an *absolute* Kan extension). This is formulated as follows:

Proposition 3. *If $G : A \to X$ has a left adjoint F with counit $\varepsilon : FG \dot{\to} 1$, then $\mathrm{Ran}_G 1_A$ exists, is equal to F with counit ε, and is preserved by any functor whatever.*

Now suppose conversely that 1_A has a right Kan extension R along G, and that this extension is preserved by G. We then have bijections

$$\varphi = \varphi_S : \mathrm{Nat}(S, R) \cong \mathrm{Nat}(SG, 1_A), \quad \varphi(S \xrightarrow{\ \varrho\ } R) = \varepsilon \cdot \varrho G,$$
$$\psi = \psi_H : \mathrm{Nat}(H, GR) \cong \mathrm{Nat}(HG, G), \quad \psi(H \xrightarrow{\ \sigma\ } GR) = G\varepsilon \cdot \sigma G,$$

natural in $S : X \to A$ and $H : X \to X$, with counit $\varphi_R 1 = \varepsilon : RG \dot{\to} 1_A$ and $\psi_{GR} 1 = G\varepsilon : GRG \dot{\to} G$. Define $\eta : 1 \dot{\to} GR$ to be $\psi_{\mathrm{id}}^{-1}(1 : G \dot{\to} G)$. Then $\psi\eta = 1$, so

$$G\varepsilon \cdot \eta G = 1_G .$$

This is one of the two triangular identities for the proposed adjunction $\varepsilon : RG \dot{\to} 1_A, \eta : 1_X \dot{\to} GR$. The other would be $\varepsilon R \cdot R\eta = 1_R$. Applying the bijection φ_R, it will suffice to prove instead $\varphi(\varepsilon R \cdot R\eta) = \varepsilon$. Putting in the definition of φ in terms of ε, we are to prove the following square commutative:

$$\begin{array}{ccc} RG & \underset{RG\varepsilon}{\overset{R\eta G}{\rightleftarrows}} & RGRG \\ \downarrow{\scriptstyle \varepsilon} & & \downarrow{\scriptstyle \varepsilon RG} \\ 1_A & \xleftarrow{\ \varepsilon\ } & RG . \end{array}$$

Insert the dotted arrow at the top and use R of the (known) triangular identity $G\varepsilon \cdot \eta G = 1$. The square then reduces to the equivalence of two expressions for $\varepsilon\varepsilon : RGRG \to 1$, q.e.d.

The arguments so far in this section have not used either formula for Kan extensions. We now examine the meaning of these formulas in the simple case of Kan extensions along the identity functor $I : C \to C$. The universal property defining Kan extensions shows at once for each $T : C \to A$ that

$$\mathrm{Lan}_I T = T, \qquad \mathrm{Ran}_I T = T .$$

Consider in particular $T : C \to \mathbf{Set}$, and assume that C has small hom-sets. Then, in the formula for Ran_I as an end, all the powers involved exist, so for every $c \in C$

$$Tc = (\mathrm{Ran}_I T)c = \int_m Tm^{C(c,m)} .$$

But in **Set**, $X^Y = \mathbf{Set}(Y, X)$, and by (IX.5.2) the end reduces to a set of natural transformations

$$Tc = \int_m \mathbf{Set}(C(c,m), Tm) \cong \mathrm{Nat}(C(c,-), T).$$

The result is just the Yoneda Lemma.

Exercises

1. Show that the bijection (1) (and (5.2) as well) is a special case of a bijection defined for an adjoint square (Exercise IV.7.4)

$$\mathrm{Nat}(HG, G'K) \cong \mathrm{Nat}(F'H, KG).$$

2. Obtain the Yoneda Lemma from the limit formula for $\mathrm{Ran}_K T$. (This gives an independent proof of the Yoneda Lemma, which was not used in the proof of § 3).
3. (a) If $K: M \to C$ has a right Kan extension R, along itself, $\varphi: \mathrm{Nat}(S, R) \cong \mathrm{Nat}(SK, K)$, prove that $\langle R, \eta, \mu \rangle$ is a monad in C, where $\eta = \varphi^{-1}(\mathrm{Id}_K)$, $\mu = \varphi^{-1}(\varepsilon \cdot R\varepsilon)$. (This is called the *codensity monad* of K.)
   (b) Show that K is codense if and only if η is an isomorphism.
   (c) If $G: A \to X$ has a left adjoint $F: X \to A$ with unit $\eta: Id \dot{\to} GF$ and counit $\varepsilon: FG \dot{\to} Id$, then its codensity monad exists and is $\langle GF, \eta, G\varepsilon F \rangle$. (The monad defined by the adjunction.)

Notes.

The formal criteria for adjoints are due to Bénabou [1965]. The construction of Kan extensions by limits and colimits, in the critical case when the receiving category A is **Set**, was achieved by Kan in [1960]. The impact of this construction was understood only gradually. In 1963 Lawvere used these extensions in functorial semantics. Ulmer emphasized their importance, and in an unpublished paper gave the coend formula (without the name coend) for $\mathrm{Lan}_K T$. Bénabou (unpublished) and Day-Kelly [1969] describe Kan extensions in relative categories (including *Ab*-categories). This idea is further developed by Dubuc [1970]; here the coend formula for Kan extensions plays a central role.

The Cartan-Eilenberg notion of derived functors is, as noted in § 4, the original and decisive example of a Kan extension. Verdier, by embedding each abelian category in a suitable derived category, has achieved an elegant form of this interpretation of derived functors by Kan extensions. For an exposition, see Quillen [1967].

Isbell, in a pioneering paper [1960], defined a functor $K: M \to C$ to be "left adequate" when $c \mapsto C(K-, c)$ is full and faithful. This assignment is the functor of the dual of Proposition 6.2; hence by that theorem "left adequate" and "dense" agree. Isbell has developed the ideas further in characterizing categories of algebras [1964].

The ubiquity of Kan extensions has developed gradually; I have learned much in this chapter from my student Eduardo Dubuc; and Max Kelly has suggested major improvements, notably the use of pointwise Kan extensions.

XI. Symmetry and Braidings in Monoidal Categories

A monoidal category, as introduced in Chapter VII, is a category equipped with binary "tensor" products, associative up to a natural isomorphism α. A principal result for these categories was a "coherence" theorem: If a certain pentagonal diagram (§ VII.1.5) in α commutes, then all diagrams involving this α must commute. We now consider various extensions of this result.

First, we observe that this coherence theorem really amounts to an assertion that the monoidal category is equivalent to a "strict" one; that is, to one in which the associativity map as well as the maps λ and ρ for the unit object are always identities. Next, a symmetric monoidal category (§ VII.7) is one in which the tensor product is not only associative but also commutative up to a suitable natural isomorphism $\gamma : a \,\Box\, b \cong b \,\Box\, a$. Again, a coherence theorem holds, in that all diagrams involving α and γ commute; however, it is not always possible to make γ the identity (i.e., to strictify). These symmetric monoidal categories have $\gamma^2 = 1$ (that is, γ is its own inverse), but there are other cogent examples of monoidal categories where γ is a "twist" with $\gamma^2 \neq 1$. These are the "braided" monoidal categories, they (§ 4) arise in applications to quantum mechanics and to knot theory.

1. Symmetric Monoidal Categories

A monoidal category M is a category with a bifunctor, $\otimes$ or $\Box$,

$$\Box : M \times M \to M$$

written for objects a, b of M variously as a "product"

$$(a, b) \to a \,\Box\, b, a \otimes b, \text{or } a\, b$$

which is associative up to a natural isomorphism

$$\alpha : a(b\, c) \cong (a\, b)c \tag{1}$$

and is equipped with an element e, which is unit up to natural isomorphisms

$$\lambda : ea \cong a\,, \qquad \rho : ae \cong e\,. \tag{2}$$

These maps must satisfy certain commutativity requirements; for α, a pentagonal diagram

$$\begin{array}{ccccc} a(b(cd)) & \xrightarrow{\alpha} & (ab)(cd) & \xrightarrow{\alpha} & ((ab)c)d \\ {\scriptstyle 1\alpha}\downarrow & & & & \downarrow{\scriptstyle \alpha 1} \\ a((bc)d) & & \xrightarrow{\qquad\alpha\qquad} & & (a(bc))d\,, \end{array} \tag{3}$$

as in § VII.1.(5), and for λ and ρ the two commutativities

$$\begin{array}{ccc} a(ec) & \xrightarrow{\alpha} & (ae)c \\ {\scriptstyle 1\lambda}\downarrow & & \downarrow{\scriptstyle \rho 1} \\ ac & = & ac\,, \end{array} \qquad \lambda = \rho : ee \to e\,. \tag{4}$$

The category of all vector spaces over a given field F, with the usual tensor product $\otimes$ of vector spaces as the product $\Box$ and with the one-dimensional vector space F as unit, is a standard example of a monoidal category M; with this in mind, monoidal categories are often called *tensor categories*.

The assumed commutativities (3) and (4) suffice to show, as in the Corollary of Theorem VII.2.1, that "every" diagram of α's, λ's, and ρ's commutes; that is, given any word w in letters $a, b, \ldots, e$, there is a unique composite of α, λ, and ρ mapping w to a word with all parentheses starting in front and all e's removed. (For example, by (4), any e can be removed before or after the application of an associatvity α, with equal results.)

Examples to be presented later suggest the idea of a "braiding".

A *braiding* for a monoidal category M consists of a family of isomorphisms

$$\gamma_{a,b} : a \Box b \cong b \Box a \tag{5}$$

natural in a and $b \in M$, which satisfy for e the commutativity

$$\begin{array}{ccc} a \Box e & \xrightarrow{\gamma} & e \Box a \\ {\scriptstyle \rho}\downarrow & & \downarrow{\scriptstyle \lambda} \\ a & = & a \end{array} \tag{6}$$

and which, with the associativity α, make both the following hexagonal diagrams commute (with the symbol $\Box$ omitted):

$$\begin{array}{ccc} (ab)c & \xrightarrow{\gamma} & c(ab) \\ \downarrow{\scriptstyle \alpha^{-1}} & & \downarrow{\scriptstyle \alpha} \\ a(bc) & & (ca)b \\ {\scriptstyle 1\cdot\gamma}\downarrow & & \downarrow{\scriptstyle \gamma\cdot 1} \\ a(cb) & \xrightarrow{\alpha} & (ac)b , \end{array} \qquad \begin{array}{ccc} a(bc) & \xrightarrow{\gamma} & (bc)a \\ \downarrow{\scriptstyle \alpha} & & \downarrow{\scriptstyle \alpha^{-1}} \\ (ab)c & & b(ca) \\ \downarrow{\scriptstyle \gamma\cdot 1} & & \downarrow{\scriptstyle 1\cdot\gamma} \\ (ba)c & \xrightarrow{\alpha^{-1}} & b(ac) . \end{array} \tag{7}$$

Note that the first diagram replaces each $\gamma_{ab,c}$ which has a product ab as first index by two γ's with single indices, while the second hexagonal diagram does the same for $\gamma_{a,bc}$ with a product as second index. Note also that the first hexagon of (7) for γ implies the second diagram for γ^{-1}, and conversely. Thus, when γ is a braiding for M, then γ^{-1} is also a braiding for M.

A *symmetric monoidal category*, as already defined in § VII. 7, is a category with a braiding γ such that every diagram

$$\begin{array}{ccc} ab & \xrightarrow{\gamma_{a,b}} & ba \\ & \searrow & \downarrow{\scriptstyle \gamma_{b,a}} \\ & & ab \end{array} \tag{8}$$

commutes. For this case, either one of the hexagons (7) implies the other. The coherence theorem for monoidal categories, as proved in Chapter VII, will now be extended to the symmetric case, using the symmetric group S_n on n letters. As in § VII.2, we will consider $\Box$-words w in n letters and also permutations τ of S_n. For each symmetric monoidal category M, a "permuted" word $w\tau$ determines a functor $(w\tau)_M : M^n \to M$, defined by permuting the arguments of w by τ, as in

$$(w\tau)_M(a_1, \ldots, a_n) = w(a_{\tau 1}, \ldots, a_{\tau n}) , \quad a_i \in M .$$

Theorem 1. *In each symmetric monoidal category M there is a function which assigns to each pair $(v\sigma, w\tau)$ of permuted words of the same length n a (unique) natural isomorphism*

$$\mathrm{can}_M(v\sigma, w\tau) : (v\sigma)_M \to (w\tau)_M : M^n \to M , \tag{9}$$

called the canonical map from $v\sigma$ to $w\tau$, in such a way that the identity of M and all instances of α and γ are canonical, and the composite as well as the $\Box$-product of two canonical maps is canonical.

Proof. There is always at least one such map between different permuted words, since we can use instances of α to rearrange the parentheses and instances of γ to transpose adjacent arguments. This will provide for any desired permutations of the arguments, since all the permutations of the symmetric group can be achieved by successive transpositions. The identities (7) show that interchanging a single argument a with a product of arguments can always be replaced by successive interchanges of individual arguments.

It remains to show that any two such composites ("paths") from $(v\,\sigma)_M$ to a $(w\,\tau)_M$ are equal. From the monoidal coherence theorem (§ VII. 2) for associativity alone, we already know that any two sequences of applications of α to get from a $(v\,\sigma)_M$ to a $(w\,\tau)_M$ will be equal. Hence, we might as well assume that the product $\square$ is strictly associative and that α is the identity. In this case, the two hexagons (7) can be replaced by two triangles

$$\begin{array}{ccc} abc & \xrightarrow{\gamma} & cab \\ & {\scriptstyle 1\cdot\gamma}\searrow \quad \swarrow{\scriptstyle \gamma\cdot 1} & \\ & acb\,, & \end{array} \qquad \begin{array}{ccc} abc & \xrightarrow{\gamma} & bca \\ & {\scriptstyle \gamma\cdot 1}\searrow \quad \swarrow{\scriptstyle 1\cdot\gamma} & \\ & bac\,. & \end{array} \tag{7a}$$

These identities show that we need only consider successive steps which interchange two adjacent letters a, b. Now the symmetric group S_n is generated by the transpositions $\tau_i = (i, 1+i)$ of successive letters for $i = 1, \dots, n-1$. And any closed path consisting of such transpositions will correspond to a relation between these generators τ_i. It is known that all such relations are products of conjugates of a number of the known "defining relations", which (for S_n) can be taken to be just the relations

$$\begin{aligned} \tau_i^2 &= 1\,, & i &= 1, \dots, n-1\,, \\ (\tau_i \tau_{i+1})^3 &= 1\,, & i &= 1, \dots, n-2\,, \\ \tau_i \tau_j &= \tau_j \tau_i\,, & 1 &\leqq i < j-1 \leqq n-2. \end{aligned} \tag{10}$$

Hence, to prove coherence, we need only show that for each such relation the corresponding diagram of paths is commutative.

The first relation $\tau_i^2 = 1$ matches the assumed property $\gamma^2 = 1$ of (8). For the third relation, the naturality of γ suffices. For the second relation $(\tau_1 \tau_2)^3 = 1$, the naturality of $\gamma_{a,b}$ and the two triangles (7a), relabelled, give a commutative diagram

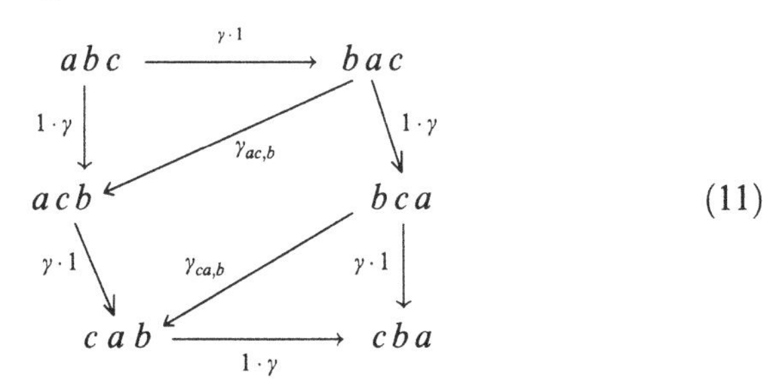

(11)

The perimeter here reads $\tau_1\,\tau_2\,\tau_1 = \tau_2\,\tau_1\,\tau_2$, as desired for (10).

This coherence theorem also extends to include the use of the maps λ and ρ which remove units. The assumption (6) provides that we can remove any unit before or after an application of γ, and the corresponding result for α is already known. The precise formulation of the resulting theorem is left to the reader; it requires consideration of words with more than n arguments, some of which are taken to be the unit. We note also that the statement of the corollary of § VII.2 requires similar adjustment in the use of "words" involving e. The result still expresses the fact that "all formal diagrams involving just α, γ, ρ, and γ will commute".

2. Monoidal Functors

For any category with added categorical structure it is in order to define the corresponding structure for functors and for natural transformations. Here we consider the monoidal case again (§ VII. 1).

A *monoidal functor* $(F, F_2, F_0) : M \to M'$ between monoidal categories M and M' consists of the following three items:

(i) An (ordinary) functor $F : M \to M'$ between categories;

(ii) For objects a, b in M morphisms

$$F_2(a, b) : F(a) \,\Box\, F(b) \to F(a \,\Box\, b) \tag{1}$$

in M' which are natural in a and b;

(iii) For the units e and e', a morphism in M'

$$F_0 : e' \to F e\,. \tag{2}$$

Together, these must make all the following three diagrams, involving the structural maps α, λ, and ρ, commute in M':

$$\begin{array}{ccc}
F(a) \,\Box\, (F(b) \,\Box\, F(c)) & \xrightarrow{\alpha'} & (F(a) \,\Box\, F(b)) \,\Box\, F(c) \\
\downarrow{\scriptstyle 1 \Box F_2} & & \downarrow{\scriptstyle F_2 \Box 1} \\
F(a) \,\Box\, (F(b \,\Box\, c)) & & (F(a \,\Box\, b) \,\Box\, F(c)) \\
\downarrow{\scriptstyle F_2} & & \downarrow{\scriptstyle F_2} \\
F(a \,\Box\, (b \,\Box\, c)) & \xrightarrow{F(\alpha)} & F((a \,\Box\, b) \,\Box\, c)\,,
\end{array} \tag{3}$$

$$\begin{array}{ccc} F(b)\,\square\, e' & \xrightarrow{\rho} & F(b) \\ \downarrow{\scriptstyle 1\square F_0} & & \uparrow{\scriptstyle F(\rho)} \\ F(b)\,\square\, F(e) & \xrightarrow[F_2]{} & F(b\,\square\, e)\,, \end{array} \qquad \begin{array}{ccc} e'\,\square\, F(b) & \xrightarrow{\lambda} & F(b) \\ \uparrow{\scriptstyle F_0\square 1} & & \uparrow{\scriptstyle F(\lambda)} \\ F(e)\,\square\, F(b) & \xrightarrow[F_2]{} & F(e\,\square\, b)\,. \end{array} \tag{4}$$

The evident composite of two monoidal functions is monoidal.

A monoidal functor is said to be *strong* when F_0 and all the $F_2(a, b)$ are isomorphisms, and *strict* when F_0 and all $F_2(a, b)$ are identities (recall that a monoidal category is strict when α, λ, and ρ are identities).

A monoidal natural transformation $\theta : (F, F_2, F_0) \to (G, G_2, G_0) : M \to M'$ between two monoidal functors is a natural transformation between the underlying ordinary functors $\theta : F \to G$ such that all the diagrams

$$\begin{array}{ccc} F(a)\,\square\, F(b) & \xrightarrow{F_2} & F(a\,\square\, b) \\ \downarrow{\scriptstyle \theta_a\square\theta_b} & & \downarrow{\scriptstyle \theta_{a\square b}} \\ G(a)\,\square\, G(b) & \xrightarrow{G_2} & G(a\,\square\, b) \end{array} \tag{5}$$

and

$$\begin{array}{ccc} e' & \xrightarrow{F_0} & F e \\ \| & & \downarrow{\scriptstyle \theta_e} \\ e' & \xrightarrow{G_0} & G e \end{array} \tag{6}$$

commute in M'. The evident composite of two monoidal natural transformations is natural.

For a monoidal function F, the maps F_2 for the product and F_0 for the unit can be extended to the functors defined by arbitrary tensor words v in n letters (as these words are defined in § VII.2). This will give for each such functor v a transformation

$$F_v : v(F a_1, \dots, F a_n) \to F v(a_1, \dots, a_n) \tag{7}$$

natural in $a_1, \dots, a_n$, such that $F_\square$ is F_2 and F_e is F_0. Indeed, words are defined inductively as tensor products $v \,\square\, v'$ of shorter words and we take $F_{v\square v'}$ as $F_2(F_v \,\square\, F_{v'})$. With this definition it is evident that all diagrams in these natural transformations commute. Specifically, if v and w are two such $\square$-words in n letters, the coherence theorem gives a unique natural transformation $\eta : v \to w$ constructed out of the maps α, ρ, and λ. Thus by induction, the diagram

$$\begin{array}{ccc} v(F\,a_1,\ldots,F\,a_n) & \xrightarrow{F_v} & F_v(a_1,\ldots,a_n) \\ \downarrow{\scriptstyle \eta} & & \downarrow{\scriptstyle F\eta} \\ w(F\,a_1,\ldots,F\,a_n) & \xrightarrow{F_w} & F\,w(a_1,\ldots,a_n) \end{array} \tag{8}$$

commutes. (This is just the extension of conditions (3) and (4) to arbitrary words.) Moreover, for any monoidal natural transformation $\theta : F \to G$ between two monoidal functors and for any word v, the diagram

$$\begin{array}{ccc} v(F\,a_1,\ldots,F\,a_n) & \xrightarrow{F_v} & F\,v(a_1,\ldots,a_n) \\ \downarrow{\scriptstyle v(\theta_{a_1},\ldots,\theta_{a_n})} & & \downarrow{\scriptstyle \theta_v(a_1,\ldots,a_n)} \\ v(G\,a_1,\ldots,G\,a_n) & \xrightarrow[G_v]{} & G\,v(a_1,\ldots,a_n) \end{array} \tag{9}$$

commutes; this condition generalizes the conditions (5) and (6).

If B, B' are braided (or even symmetric) monoidal categories, a braided monoidal functor is a monoidal functor $(F, F_2, F_0) : B \to B'$ which commutes with the braidings γ and γ' in the following sense:

$$\begin{array}{ccc} F\,a \,\Box\, F\,b & \xrightarrow{\gamma'} & F\,b \,\Box\, F\,a \\ \downarrow{\scriptstyle F_2} & & \downarrow{\scriptstyle F_2} \\ F(a \,\Box\, b) & \xrightarrow{F(\gamma)} & F(b \,\Box\, a)\,; \end{array} \tag{10}$$

here γ and γ' are the braidings of B and B', respectively. The category of braided monoidal categories has morphisms these braided monoidal functors.

3. Strict Monoidal Categories

Theorem 1. *Any monoidal category M is categorically equivalent, via a strong monoidal functor $G : M \to S$ and a strong monoidal functor $F : S \to M$, to a strict monoidal category S.*

(Recall that the monoidal category is said to be "strict" when the structure maps α, γ, and ρ are all identities.)

Proof. The coherence theorem yields unique "canonical" maps between words; hence, the plan of the proof is to embed the given category M in a larger strict monoidal category S consisting of iterated formal products (where all pairs of parenthesis start in front) of elements of M.

S will be the free monoid generated by the elements of M. Specifically, take the objects s of S to be all finite strings $s = [b_1, \ldots, b_k]$ of objects of M, including the empty string $\varnothing$. A product $s \,\square\, t$ of strings s and t is then defined by concatenation of strings, as $s \cdot t$. This product is associative, so the associativity map α for S can be the identity. Also, the empty string $\varnothing$ acts as a unit for this product, so the maps ρ and λ for S can both be the identity. With these agreements, S is an associative monoid with a unit, but not yet a monoidal category.

Now define a map $F : S \to M$ on strings in S by setting

$$\begin{aligned} &F(\varnothing) = e\,, \\ &F(s) = F[b_1, \ldots, b_k] = (\ldots(b_1 \,\square\, b_2) \,\square\, b_3) \ldots) \,\square\, b_k)\,, \end{aligned} \tag{1}$$

where on the right all pairs of parenthesis begin in front. Now define the arrows $s \to t$ between strings s and t in S to be exactly the arrows between the corresponding objects in M,

$$F(s) \to F(t), \tag{1'}$$

with composition just as in M. This convention clearly makes S into a category. Then the concatenation product $s \cdot u$ can be extended to a corresponding product $f \cdot g$ of arrows $f : s \to t$ and $g : u \to v$, where u and v, like s and t, are finite strings of objects of M. Specifically, this means that we define $f \cdot g$ as the following composite in M:

$$F(s \cdot u) \to F(s) \,\square\, F(u) \xrightarrow{f \,\square\, g} F(t) \,\square\, F(v) \to F(t \cdot v)\,;$$

here the two outer arrows are the canonical maps in M. For a triple product with a map $h : w \to y$ of strings, iteration of this definition gives $(f \cdot g) \cdot h$ as the composite of canonical maps, therefore also canonical. The coherence theorem for monoidal categories then shows that this product $f \cdot g$ of arrows is strictly associative. Hence, S is a strict monoidal category. Moreover, F is a strong monoidal functor if we take F_0 to be the identity $e \to e$ and $F_2(s,\ t)$ to be the unique canonical map (move all parentheses to the front)

$$F_2(s,\ t) : F(s) \,\square\, F(t) \to F(s \cdot t)\,. \tag{2}$$

With this definition, the requirements (3) and (4) of §2 for a monoidal functor follow from the coherence theorem for the monoidal category M.

A strong monoidal functor $G : M \to S$ in the opposite direction with $G(b) = [b]$ is defined for b, c in M by noting that $G(b) \cdot G(c) = [b][c] = [b,\ c]$ and by setting

$$\begin{aligned} G(b) &= [b], \qquad G(f) = f\,, \\ G_0 &= 1 : \varnothing \to [e]\,, \\ G_2(b, c) &= 1 : [b, c] \to [b \,\square\, c]\,. \end{aligned} \tag{3}$$

Here the last map is 1 because by the definition (1) above a map of strings $[b, c] \to [b \,\square\, c]$ is just a map $b \,\square\, c \to b \,\square\, c$ in M. The conditions (3), (4), and (4′) of §2 on G then follow. In the case of (3), observe that the map $G_2 \,\square\, 1$ is

$$\begin{array}{ccc} (G(a) \cdot G(b)) \cdot G(c) & \xrightarrow{G_2 \square 1} & G(a \,\square\, b) \cdot G(c)\,, \\ [a, b, c] & \longmapsto & [a \,\square\, b,\, c]\,, \end{array}$$

therefore, by the definition (1) of maps in S, must be the map $\alpha : a \,\square\, (b \,\square\, c) \to (a \,\square\, b) \,\square\, c$ in B as a map in S; this matches the map $G(\alpha)$ at the base of (3) of 2, while $\alpha' = 1$ is at the top. The composite functor $F\,G : M \to M$ is the identity, while the composite $G\,F$ is naturally isomorphic to the identity. Hence, the monoidal category M is indeed categorically equivalent (by monoidal functors) to the strict monoidal category S, as claimed.

Conversely, the equivalence given in the conclusion of this theorem will yield the coherence theorem as an easy consequence:

Theorem 2. *If the monoidal category M is equivalent by a strong monoidal functor $G : M \to S$ to a strict monoidal category S, then coherence holds for the associativity of the tensor product $\square$ in M.*

Proof. Suppose that v and w are two tensor words in k letters, while θ and $\theta' : v \to w$ are two natural transformations between the corresponding functors, both constructed as combinations of the associativity transformation α in M. Now use the natural transformations G_v and G_w constructed as in (2.7) from G_2 and G_0. As in (2.8), the diagram

$$\begin{array}{ccc} v(G a_1, \ldots, G a_n) & \xrightarrow{\theta_s} & w(G a_1, \ldots, G a_n) \\ \big\downarrow{\scriptstyle G_v} & & \big\downarrow{\scriptstyle G_w} \\ G v(a_1, \ldots, a_n) & \xrightarrow{G\theta_m} & G w(a_1, \ldots, a_n) \end{array}$$

commutes, as does the corresponding diagram for θ'. Here, θ_s is short for $\theta(G a_1, \ldots, G a_n)$ and θ_m short for $\theta(a_1, \ldots, a_n)$. But since the monoidal category S, with $G : M \to S$, is strict and θ and θ' are both constructed from α, we have $\theta = \theta'$ in S. Then comparing the diagrams above for θ and for θ', with G_v and G_w known to be isomorphisms, we find that

$G\theta_M = G\theta'_M$. But G is an equivalence of categories, so there is also a functor $F : S \to M$ in the opposite direction with $FG \cong 1$. But we have $FG\theta_m = FG\theta'_m$. With $FG \cong 1$, this implies $\theta = \theta'$ in M. In other words, coherence for associativity (and likewise for ρ and λ) holds in M, as claimed.

Exercises

1. For any category C, show that the functor category C^C with composition as tensor product and 1_C as the unit is a strict monoidal category.
2. If, in Exercise 1, $C = M$ is a monoidal category, show that there is a strong monoidal functor $(T, T_2, T_0)M \to M^M$ in which, for a, b, c, in M,

$$T(a) = a \,\square\, - \,,$$

$$T_2(a,b)_c = \alpha_{a,b,c} : a \,\square\, (b \,\square\, c) \to (a \,\square\, b) \,\square\, c) \,,$$

$$(T_0)a = \lambda(a)^{-1} : a \to e \,\square\, a \,.$$

   In particular, note that the conditions (3) and (4) above for this monoidal functor T became the conditions (5), (7), and (9) of § VII.1 in the definition of a monoidal category.
3. Use the above results to give another proof, independent of this coherence theorem, of Theorem 1 above. Note that this gives an independent proof of coherence, as in Theorem 2 above.

4. The Braid Groups B_n and the Braid Category

Now we introduce the promised actual braids and the resulting category of braids.

A braid on three strings, such as the following one

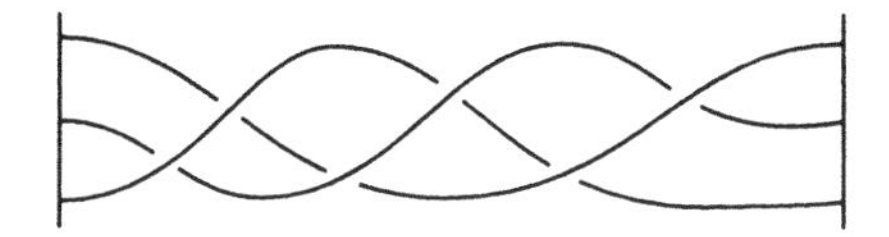

is formed by twisting the strings around each other in space without cutting or tying them. One such braid can be multiplied by a second one by attaching the right-hand ends of the first strings, in order, to the left-hand ends of the second set of strings. Two braids are said to be equal when the first one can be continuously deformed into the second without crossing or cutting strings. Here are some labelled examples of braids on three strings, including an inverse and two products:

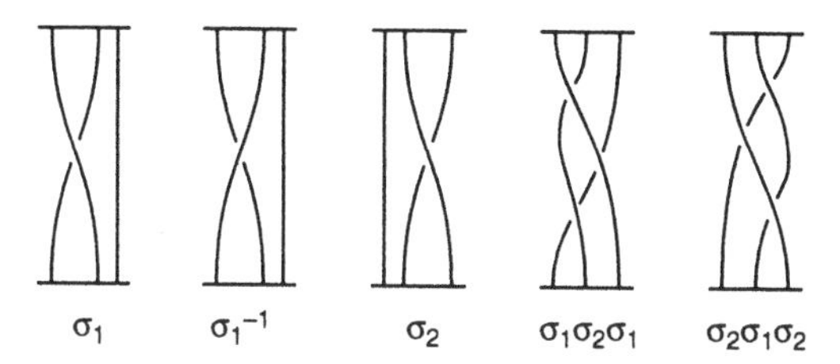

An evident deformation of the last diagram into the preceding one suggests the equality

$$\sigma_1\,\sigma_2\,\sigma_1 = \sigma_2\,\sigma_1\,\sigma_2\;. \tag{1}$$

The multiplication of these braids is clearly associative and has an identity (three untwisted strings) and an inverse. This may serve to indicate the definition of the Artin braid group B_3. It is generated by σ_1 and σ_2, subject only to the relation (1). A corresponding description yields the braid group B_n on n strings.

More formally, this braid group B_n can be defined as the fundamental group of a suitable space T_n, that of n-tuples of distinct points. Indeed, let P be the Euclidean plane and take the space T_n to be the set of all n-tuples of n distinct points of P, with the evident topology. Thus, the Artin braid group B_n can be defined formally as the fundamental group of this space T_n.

The braid group B_n clearly can be generated by the $n-1$ braids σ_i, where σ_i twists the i-th string once under the $(i+1)$-st string. Its inverse σ_i^{-1} is indicated above, with a suggested deformation of $\sigma_1\,\sigma_2\,\sigma_1$ into $\sigma_2\,\sigma_1\,\sigma_2$. Indeed, the defining relations for these generators σ_i of B_n are as follows:

$$\sigma_i\,\sigma_{i+1}\,\sigma_i = \sigma_{i+1}\,\sigma_i\,\sigma_{i+1}\;, \quad \text{all } i = 1, \ldots, n-1\;, \tag{2}$$

$$\sigma_i\,\sigma_j = \sigma_j\,\sigma_i\;, \quad |i-j| \neq 1\;. \tag{3}$$

The braid group B_2 is simply the infinite cyclic group on the (single) generator σ_1. The braid group B_1 consists of just the identity.

Each braid on n strings determines a permutation of the n end-points and hence a homomorphism $B_n \to S_n$ onto the symmetric group S_n on n letters. We recall that S_n is generated by the $n-1$ transpositions $\tau_i = (i, i+1)$ which interchange the letters i and $i+1$, and that S_n is defined by these generators and the following relations:

$$\tau_i^2 = 1 \ , \qquad \tau_i \tau_j = \tau_j \tau_i \ , \quad |i-j| \neq 1 \ , \tag{4}$$

$$\tau_i \tau_{i+1} \tau_i = \tau_{i+1} \tau_i \tau_{i+1} \ , \quad i = 1, \ldots, n-1 \ . \tag{5}$$

This again shows the homomorphism $\sigma_i \mapsto \tau_i$ of B_n onto S_n.

All the braid groups may be combined to form the braid category B. The objects are all the natural numbers $n = 0, 1, 2, \ldots$ (including zero) and the arrows are the braids $n \to n$; there are no arrows $n \to m$ for $n \neq m$ and only the identity arrow $\varnothing \to \varnothing$). This defines a monoidal category, with the box product $\square \cdot B \times B \to B$ given by "addition" $\square = +$; here the sum of two objects (natural numbers) m and n is the usual sum of numbers, while addition of braids is the operation: lay the braids side by side:

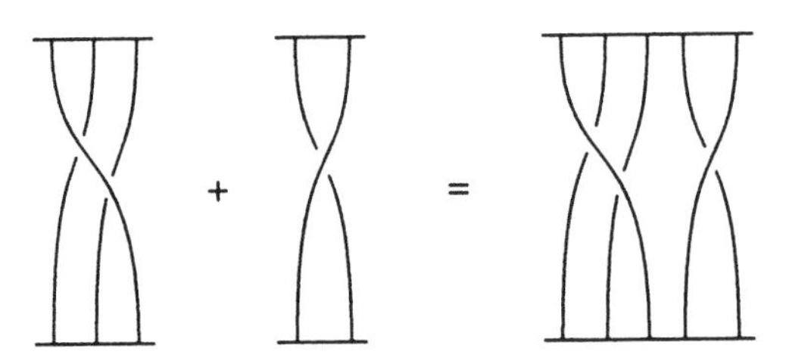

This operation is clearly (strictly) associative and has the empty braid on $\varnothing$ as unit. Hence, the braid category B is, under $+$, a strict monoidal category. It is almost a symmetric monoidal category; the addition of objects $m+n$ is commutative and one can define a transformation $\gamma_{m,n} : m+n \to n+m$ by crossing m strings over n strings, as in the following figure (for $m = 3$, $n = 2$):

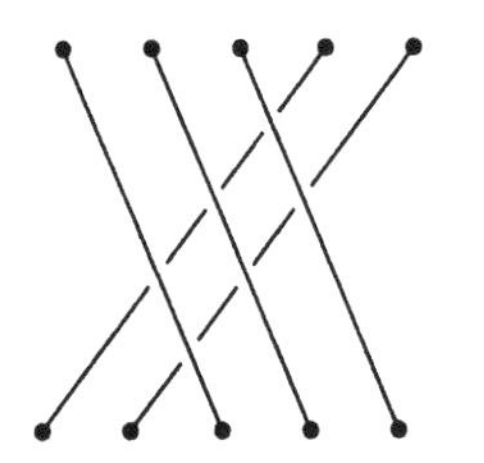

This γ natural in m and n, as one can see pictorally (Joyal-Street, [1993]) for braids $\xi : m \to m$ and $\eta : n \to n$ in the following schematic diagram:

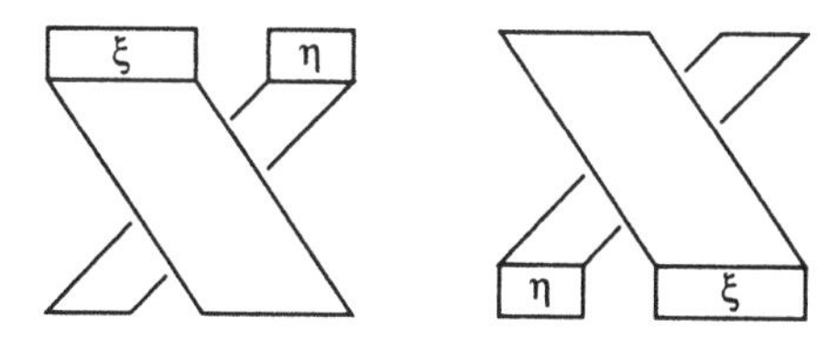

The symmetry requirement $\gamma^2 = 1$ fails, but both hexagons apply, as suggested in the following diagram (where the associativity is evidently not visible):

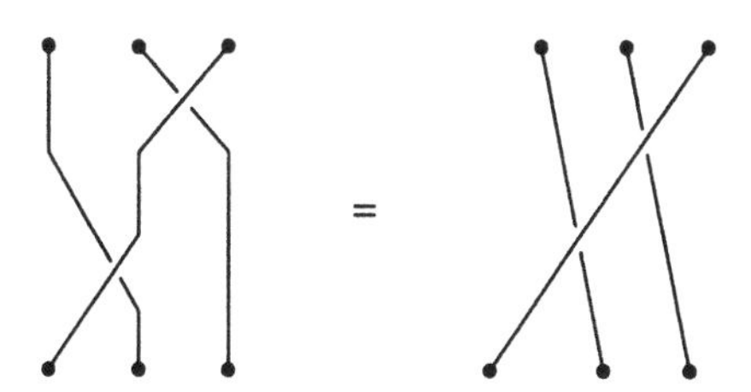

The realization of a braid by twisted strings directly suggests the use of braided categories for string theory in theoretical physics.

5. Braided Coherence

As we have seen in § 1, coherence for a symmetric monoidal category holds; all formal diagrams involving just associativity α and commutativity γ are commutative. This is by no means the case for a braided monoidal category B; given two objects a and b in B, there might be an infinite number of "canonical" automorphisms of $a \square b$, as follows:

$$1, \gamma^2, \gamma^4, \ldots, \gamma^{-2}, \gamma^{-4}, \ldots : a \square b \to a \square b . \tag{1}$$

In this way, a subgroup of the braid group B_2 acts on $a \square b$; as a result, all diagrams in γ do not commute.

The general situation is similar and is described by the following "coherence" theorem of Joyal-Street [1993)]:

Theorem 1. *If B is the braid category and M a braided monoidal category, with M_0 the underlying (ordinary) category, there is an equivalence of categories*

$$\hom_{\mathrm{BMC}}(B, M) \approx M_0 , \tag{2}$$

where $\hom_{\mathrm{BMC}}$ *stands for the category of strong braided monoidal functors $F : B \to M$. The equivalence (2) is given by evaluating each such functor $F : B \to M$ at the object 1 of the braid category B.*

Proof. By Theorem 3.1 the monoidal category M is strongly equivalent to a strict monoidal category S. The braiding γ of M readily translates by this equivalence to a braiding of S, so that the equivalence $M \to S$ is a strong morphism of braided categories. It therefore suffices to prove the theorem with M replaced by a strict monoidal category S. We will then show that there is an isomorphism of categories

$$\hom_{\mathbf{BMS}}(B, S) \cong S_0 , \tag{3}$$

where $\hom_{\mathbf{BMS}}$ stands for strict braided monoidal functors F and the isomorphism is again given by evaluation at $1 \in B$. The correspondence (3) sends each such functor F to the object $F(1)$ of S. Conversely, given an object a of S, we wish to define a strict braided monoidal functor $F = F_a : B \to S$ with $F(1) = a$. Since F is to be strict, it must preserve the product, so we set $F(n) = a^n$. In the braid category B, the maps $n \to n$ must be sent to maps $a^n \to a^n$ in S. These maps $n \to n$ in B are exactly the elements of the n-th braid group B_n, which is generated by the standard maps σ_i, $i = 1, \ldots, n-1$. In particular, $\sigma : 2 \to 2$ must be mapped to

$$F(\sigma) = \gamma_{a,a} : a^2 \to a^2 \quad \text{in } S .$$

In B, the map σ_i (twist string i under string $i+1$) can be written as a sum (i.e., a $\Box$ product)

$$\sigma_i = 1_{i-1} + \sigma_1 + 1_{n-i-1} , \quad a^n \to a^n .$$

Therefore, we must (and do) set

$$F(\sigma_i) = 1_{i-1} + \gamma_{a,a} + 1_{n-i-1} , \quad a^n \to a^n .$$

We must then check that F preserves the defining relations of the braid group B_n. The relations

$$\sigma_i \sigma_j = \sigma_j \sigma_i , \quad |i-j| > 1 ,$$

are immediate, while the relation

$$\sigma_i \sigma_{i+1} \sigma_i = \sigma_{i+1} \sigma_i \sigma_{i+1}$$

follows from the two commutative hexagons for γ, as they are illustrated in the diagram (11) of § 1, for the case when $i = 1$.

To complete the definition of the monoidal functor $F = (F, F_2, F_0)$, we must also produce a suitable map F_0 and a map

$$F_2(m,\ n) : F(m) \Box F(n) \to F(m+n) ,$$

natural in m and n. This must be a map

$$a^m \Box a^n \to a^{m+n}$$

for the chosen $a = F(1)$. But since the sum of a braid $m \to m$ and one $n \to n$ simply lays these braids side by side, we can take this map to be the identity.

We also need a map $F_0 : e' \to F\,e$; here, e' is the unit object of S, while $e = 0$ and so $F(e) = e'$; so we can take F_0 to be the identity; with these choices, F is indeed strict.

Finally, we show that the operation "evaluate at 1" of (3) is an equivalence of categories by showing that it is full and faithful. Indeed, given two strict monoidal functors $F,\ G : B \to S$ and a map $f : F(1) \to G(1)$ in S_0 between their images (by evaluation at 1) in S_0, we wish to have a natural transformation $\theta : F \to G$ for which θ at the object 1 is the given map f. But in B, the object n is the n-fold $\Box$-product of objects 1, while S is a strict monoidal category and so has n-th powers by $\Box$. For a $\Box$-word w with n factors, the strict monoidal functor F yields an isomorphism $F_w : F(1)^n \to F(n)$; also, this word with all n arguments equal to a yields the n-th power of a in B or in S. Hence, the desired natural transformation $\theta : F \to G$ with $\theta(1) = f$ must make the following diagram commute:

$$\begin{array}{ccc} F(1)^n & \xrightarrow{f^n} & G(1)^n \\ \cong\downarrow F_w & & \downarrow G_w \\ F(n) & \xrightarrow[\theta_n]{} & G(n) \end{array};$$

so we must define θ_n by this as $G_w f^n F_w^{-1}$. From the properties of F_w and G_w, it follows that θ so defined is natural.

Note. This proof follows the argument of Joyal-Street in a preprint [1986]; it was not introduced in the subsequent published paper [1993].

The result is a coherence theorem, but not in the usual sense. It does not assert that every diagram in the basic maps α, λ, ρ, and γ commutes, but it does serve to describe all the composites of these maps and then all the endomorphisms they generate for an iterated tensor product. Each of these maps has an underlying braid, as for example in

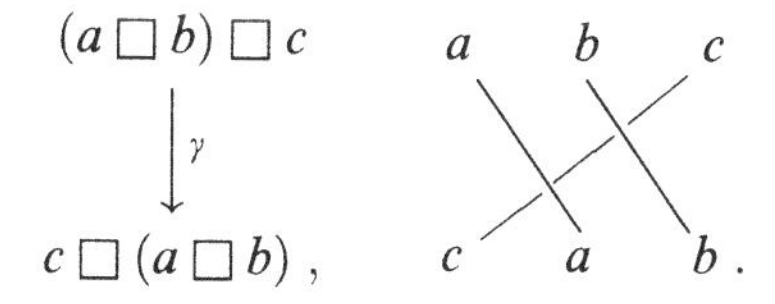

This braid gives the pairing of the variables for this map as a natural transformation. The result can be stated informally (Joyal-Street [1993]) as follows: Two composites of α, λ, ρ, and γ are equal if they have the same underlying braids.

The resulting coherence theorem can be stated as follows:

Theorem 2. *Each composite of the canonical maps acting on an n-fold product in a braided monoidal category M induces a braiding (an element in the braid group). Two such composites are equal for all M if and only if they give the same element in* B_n.

6. Perspectives

The study of braided categories has many connections with other mathematical topics, as well as in the study of parts of quantum field theories in Physics. Some of these connections arise in string theory in Physics. There the paths of elementary particles weaving around each other can form a braid – and then something more, with two strings joining or separating, as in the case of the Feynman diagrams representing the collision and the separation of elementary particles; here, we consider the braid as composed of paths (not by a moving point but by a moving "string" in a different sense, say as an oscillating circular string. Our paths (called strings above) are then replaced by tubes with a (topologically) circular cross section. An array of a finite number of such tubes can then be regarded as the morphisms (the paths) of a braided category. These tubes can be each given a conformal structure and the same applies to the collisions, as in a Feynman diagram (see Mac Lane [1991]. These constructions also play a role in Tannaka duality for compact topological groups (see Doplicher-Roberts [1989]).

In a different direction, the one-dimensional strings in one of our braids may be replaced by ribbons, and these ribbons can given one or more twists, clockwise or counterclockwise (see the ribbon categories of Shum [1994]). There are extensive connections to knot theory (Kauffman [1991, 1993] and again to Physics. The strings of a braid may be replaced by "tangles". In a tangle, a string may start out at the bottom line, twist around various other strings, and then return to a different point on the starting line, as for example in the following diagram:

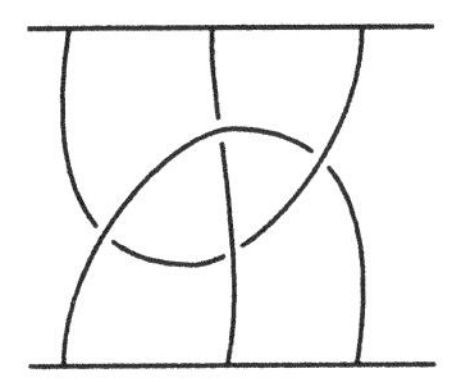

Finally, there are striking connections to Hopf algebras and to quantum groups (which are generalized Hopf algebras). There is an extensive bibliography in the monograph by Schnider and Sternberg [1993].

XII. Structures in Categories

In this chapter, we will examine several conceptual developments. We start with the idea of an "internal" category, described by diagrams within an ambient category. We then go on to study the sequences of composable arrows in a category – they constitute the "nerve" of the category, which turns out to be a simplicial set.

From this point, we turn to consider "higher-dimensional" categories such as a 2-category, which has objects, arrows, and 2-cells between arrows, and so on, to categories with three cells and beyond.

1. Internal Categories

In this section, we will work within an ambient category E which is finitely complete; that is, which has all finite products, pullbacks, and a terminal object. As already observed in our introduction, we can define monoids, groups, graphs, and other types of algebraic objects within E. Following this pattern, we can define a category within E – called a *category object* in E or an *internal category* in E.

Such an internal category $C = (C_0, C_1, i, d_0, d_1, \gamma)$ is to consist of two objects C_0 and C_1 of E, called respectively the "object of objects" and the "object of arrows", together with four maps in E:

$$i : C_0 \to C_1, d_0, d_1 : C_1 \to C_0 , \qquad \gamma : C_1 \times {}_{C_0} C_1 \to C_1 , \tag{1}$$

called identity i, domain d_0, codomain d_1, and composition γ; here, γ is defined on the following pullback $C_1 \times {}_{C_0} C_1$:

$$\begin{array}{ccc} C_1 \times {}_{C_0} C_1 & \xrightarrow{\pi_2} & C_1 \\ {\scriptstyle \pi_1}\big\downarrow & & \big\downarrow{\scriptstyle d_1} \\ C_1 & \xrightarrow{d_0} & C_0 , \end{array} \tag{2}$$

which is thus (for $E =$ **Sets**) just the object of all pairs of composable arrows. These four maps (1) are subject to the following four commutativity conditions, which simply express the usual axioms for a category. Thus,

$$d_0\, i = 1 = d_1\, i : C_0 \to C_0 \tag{3}$$

specifies domain and codomain of the identity arrows and then

$$\begin{array}{ccccc} C_0 \times_{C_0} C_1 & \xrightarrow{i \times 1} & C_1 \times_{C_0} C_1 & \xleftarrow{1 \times i} & C_1 \times_{C_0} C_0 \\ \downarrow{\scriptstyle \pi_2} & & \downarrow{\scriptstyle \gamma} & & \downarrow{\scriptstyle \pi_1} \\ C_1 & = & C_1 & = & C_1 , \end{array} \tag{4}$$

asserts that identity arrows act as such under composition γ, then

$$\begin{array}{ccccc} C_1 & \xleftarrow{\pi_1} & C_1 \times_{C_0} C_1 & \xrightarrow{\pi_2} & C_1 \\ \downarrow{\scriptstyle d_1} & & \downarrow{\scriptstyle \gamma} & & \downarrow{\scriptstyle d_0} \\ C_0 & \xleftarrow{d_1} & C_1 & \xrightarrow{d_0} & C_0 \end{array} \tag{5}$$

which specifies the domain and the codomain of a composite and

$$\begin{array}{ccc} C_1 \times_{C_0} C_1 \times_{C_0} C_1 & \xrightarrow{\gamma \times 1} & C_1 \times_{C_0} C_1 \\ \downarrow{\scriptstyle 1 \times \gamma} & & \downarrow{\scriptstyle \gamma} \\ C_1 \times_{C_0} C_1 & \xrightarrow{\gamma} & C_1 \end{array} \tag{6}$$

which expresses the associative law for composition in terms of the (evidently associative) triple pullback $C_1 \times C_1 \times C_1$. This definition is essentially the same as that previously given in § II.7.(3).

Since these diagrams express the category axioms, a category object in **Sets** is just an ordinary small category. Also, a category object in **Grp** is a category in which both the set C_0 of objects and the set C_1 of arrows are groups, and for which all the structural maps i, d_0, d_1, and γ are homomorphisms of groups. This means for the case of i and d_0 that the diagrams

$$\begin{array}{ccc} C_0 \times C_0 & \xrightarrow{m_0} & C_0 \\ \downarrow{\scriptstyle i \times i} & & \downarrow{\scriptstyle i} \\ C_1 \times C_1 & \xrightarrow{m_1} & C_1 , \end{array} \qquad \begin{array}{ccc} C_1 \times C_1 & \xrightarrow{m_1} & C_1 \\ \downarrow{\scriptstyle d_0 \times d_0} & & \downarrow{\scriptstyle d_0} \\ C_0 \times C_0 & \xrightarrow{m_0} & C_0 \end{array}$$

commute, where m_0 and m_1 are the multiplications in the groups C_0 and C_1. But $C_0 \times C_0$ with $C_1 \times C_1$ give the product category $C \times C$, so these diagrams also mean that the group multiplications m_0 and m_1 together give a morphism of categories; that is, a functor $m : C \times C \to C$ which is associative and has an inverse. Thus, the given category object in **Grp** is the same as a group object in **Cat**.

A similar interchange between algebraic structures holds generally: The category of X objects in that of Y objects is also the category of Y objects in the category of all X objects.

An *internal functor* $f : C \to D$ between two internal categories C and D in the same ambient category E is defined to be a pair of maps $f_0 : C_0 \to D_0$ and $f_1 : C_1 \to D_1$ of E which as the "object" and "arrow" functions make the evident diagrams commute:

$$\begin{array}{ccc} C_1 \times_{C_0} C_1 & \xrightarrow{f_1 \times f_1} & D_1 \times_{D_0} D_1 \\ \downarrow \gamma_C & & \downarrow \gamma_D \\ C_1 & \xrightarrow{f_1} & D_1 , \end{array} \qquad \begin{array}{ccccc} C_1 & \overset{d_0}{\underset{d_1}{\rightrightarrows}} & C_0 & \xrightarrow{i} & C_1 \\ f_1 \downarrow & & f_0 \downarrow & & f_1 \downarrow \\ D_1 & \overset{d_0}{\underset{d_1}{\rightrightarrows}} & D_0 & \xrightarrow{i} & D_1 . \end{array} \tag{7}$$

Similiarly, one may also readily describe an internal natural transformation between two internal functors from C to D.

However, these internal functors $C \to D$ go from C to another internal category D and not from C to the universe E; there is no internal category corresponding to the universe. Thus, internal functors in **Sets** do not include functors $H : C \to$ **Sets** (such as the omnipresent hom-functors). This leads to a reformulation of the concept of such functors H. Since the set C_0 of objects is small and the category of sets is cocomplete we can replace the object function $H_0 : C_0 \to$ **Sets** by a coproduct of sets and its evident projection to C_0,

$$\pi : H_0 = \coprod_{c \in C_0} H_0\, c \to C_0 . \tag{8}$$

The actions of arrows $f : c \to c'$ then combine to yield an "action" map

$$C_1 \times_{C_0} H_0 \to H_0 .$$

Hence, given an internal category C in E, we are led to consider objects in E "over" C_0 such as $\pi : H_0 \to C_0$, $d_0 : C_1 \to C_0$. A left C_0-object in E is thus defined to be an object $\pi : H \to C_0$ over C_0 together with an action map

$$\mu : C_1 \times_{C_0} H \to H , \tag{9}$$

where for this pullback C_1 is an object over C_0 via the domain map $d_0 : C_1 \to C_0$. This action map μ is to be a map "over C_0", in the sense that the following diagram commutes, where p_1 is the projection on the first factor:

$$\begin{array}{ccc} C_1 \times_{C_0} H & \xrightarrow{\mu} & H \\ \downarrow{\scriptstyle p_1} & & \downarrow{\scriptstyle \pi} \\ C_1 & \xrightarrow{d_1} & C_0\,; \end{array} \tag{10}$$

it is also to be just like the action of a group on a set, in that it must satisfy a unit law and an associative law, as follows:

$$\begin{array}{ccc} C_1 \times_{C_0} H & \xrightarrow{i \times 1} & C_1 \times_{C_0} H \\ & {\scriptstyle p_2} \searrow & \downarrow{\scriptstyle \mu} \\ & & H\,, \end{array}$$

$$\begin{array}{ccc} C_1 \times_{C_0} C_1 \times_{C_0} H & \xrightarrow{1 \times \mu} & C_1 \times_{C_0} H \\ \downarrow{\scriptstyle \gamma \times 1} & & \downarrow{\scriptstyle \mu} \\ C_1 \times_{C_0} H & \xrightarrow{\mu} & H\,. \end{array} \tag{11}$$

Such a left C-object is also called an "internal diagram" on the internal category C or (Borceux) an internal base-valued functor. The essential point is to observe that when $E = \mathbf{Set}$, this includes precisely the familiar functors to the ambient category **Set** from a category in **Set**.

A morphism $H \to K$ of such (left) C-objects is then simply a morphism $\phi : H \to K$ in E which preserves the structure involved; that is, which makes both the following diagrams in this morphism ϕ commute:

$$\begin{array}{ccc} H & \xrightarrow{\phi} & K \\ \downarrow{\scriptstyle \pi} & & \downarrow{\scriptstyle \pi} \\ C_0 & = & C_0\,, \end{array} \qquad \begin{array}{ccc} C_1 \times_{C_0} H & \xrightarrow{1 \times \phi} & C_1 \times_{C_0} K \\ \downarrow{\scriptstyle \mu} & & \downarrow{\scriptstyle \mu} \\ H & \xrightarrow[\phi]{} & K\,. \end{array} \tag{12}$$

For ordinary set-valued functors, H and K, this makes ϕ exactly a natural transformation; the first diagram states that ϕ sends each $H(c)$ to $K(c)$, and the second diagram states that this commutes with composition (as required for naturality).

In § 8, we study category objects in groups.

2. The Nerve of a Category

Given a category C (in **Sets**), the pullback $C_2 = C_1 \times_{C_0} C_1$, as used above, consists of the composable pairs of arrows of C. Similarly we consider, as in § II.7.(5), the composable strings

$$\bullet \xrightarrow{f_1} \bullet \xrightarrow{f_2} \cdots \xrightarrow{f_n} \bullet$$

of n arrows, with

$$d_1 f_1 = d_0 f_2, \ldots, d_1 f_{n-1} = d_0 f_n \, . \tag{1}$$

They are the elements of the iterated pullback

$$C_n = C_1 \times {}_{C_0} C_1 \times_{C_0} \cdots \times {}_{C_0} C_1 \quad (n \text{ factors}) \, . \tag{2}$$

With C_0, this sequence of sets

$$C_0, C_1, \ldots, C_n, \ldots$$

actually constitutes a simplicial set (in the same sense as defined in (§ VII.5). For $n = 1$, we already have the "face operators" d_0, $d_1 : C_1 \to C_0$. For $n > 1$, the "face operators" $d_i = C_n \to C_{n-1}$ for $i = 0, \ldots, n$ are defined by deletion or by composition of adjacent arrows as in

$$\begin{aligned} &d_0(f_1, \ldots, f_n) = (f_2, \ldots, f_n) \, , \\ &d_i(f_1, \ldots, f_n) = (f_1, \ldots, f_j f_{j+1}, \ldots, f_n) \, , \quad j = 1, \ldots, n-1 \, , \\ &d_n(f_1, \ldots, f_n) = (f_1, \ldots, f_{n-1}) \, , \end{aligned} \tag{3}$$

while the degeneracies s_j are defined by inserting suitable identity maps $i_{d_0 f}$ at suitable positions, as in

$$\begin{aligned} &s_0(f_1, \ldots, f_n) = (i_{d_0 f_1}, f_1, f_2, \ldots, f_n) \, , \\ &s_j(f_1, \ldots, f_n) = (f_1, \ldots, f_j, i_{d_1 f_j}, f_{j+1}, \ldots, f_n) \, , \end{aligned} \tag{4}$$

for $j = 1, \ldots, n$. The results are again composable strings of arrows. The required identities for face and degeneracy operators, as stated in § VII.5.(11), are readily verified. The geometric meaning may be illustrated by placing the arrows f_i on edges of simplices, so that the compositions are evident, as in

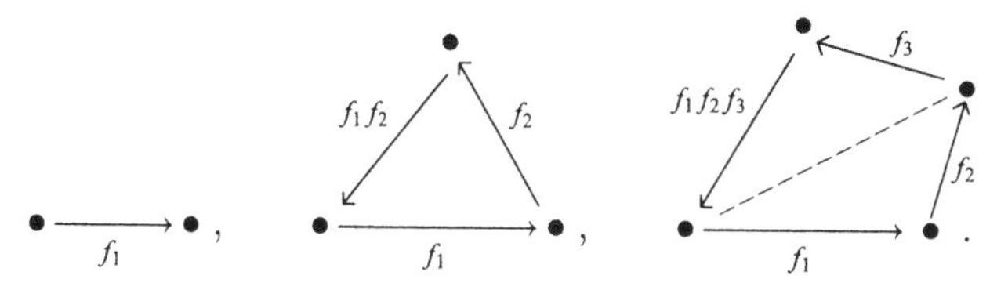

The nerve of an internal category in E is similarly a simplicial object in E.

Exercise

1. Verify the face and degeneracy identities for the operations as defined above for the nerve.

3. 2-Categories

A 2-category is a system of 2-cells or "maps" which can be composed in two different but commuting categorical ways.

A first example (see § II.5) is that in which natural transformations are the "maps". Given three functors

$$R, S, T : C \to B$$

and natural transformations $\sigma : R \xrightarrow{\bullet} S$ and $\tau : S \xrightarrow{\bullet} T$, we have defined in § II.4 a "vertical" composite natural transformation $\tau \bullet \sigma : R \to T$ by $(\tau \bullet \sigma)(c) = \tau c \circ \sigma c$ for each object c of C. This is a first natural transformation σ followed by a second

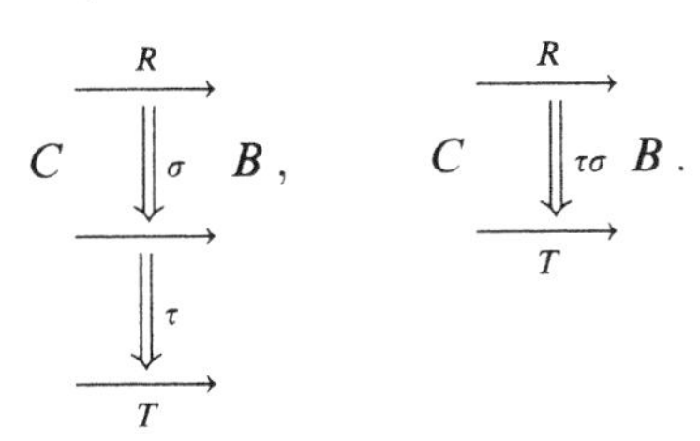

But there is also a horizontal composition of natural transformations, matching the composition of functors (§ II.5) as in the diagram

$$C \underset{S}{\overset{R}{\rightrightarrows}} B \underset{S'}{\overset{R'}{\rightrightarrows}} A, \qquad C \underset{S' \circ S}{\overset{R' \circ R}{\rightrightarrows}} A,$$

(with 2-cells σ, σ' and $\sigma' \circ \sigma$ respectively)

with $(\sigma' \circ \sigma)c = \sigma' S c \circ R' \sigma c$ for any object c in C. Both compositions are associative, and they commute with each other (Theorem II.5.1).

Similarly, there are two commuting ways of composing homotopies between continuous maps. Recall from topology that a homotopy $\theta : f \sim g$ between continuous maps f and g of a space X into a space Y is a continuous deformation

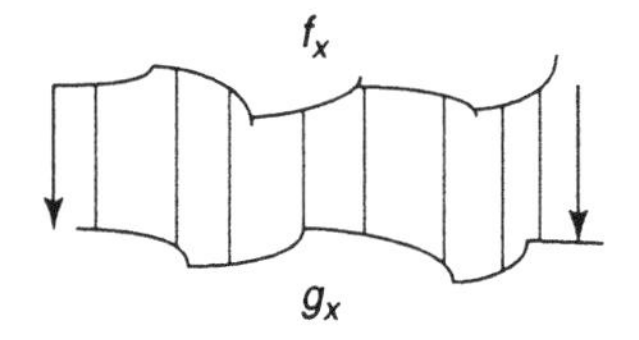

of the map f into g; that is, a continuous map $\theta : X \times I \to Y$, where I is the unit interval and the identities $\theta(x, 0) = f(x)$ and $\theta(x, 1) = g(x)$ hold at the start 0 and the end 1 for each point x in X. A second homotopy $\varphi = g \sim h$ has with θ a vertical composite $\varphi \bullet \theta : f \sim h$ (see below). Also, maps f', $g' : Y \to Z$ and a homotopy $\theta' : f' \sim g'$ between them give with θ a "horizontal" composite $\theta' \circ \theta : f' \circ f \sim g' \circ g$ of the composite maps. However, the expected vertical composite $\varphi \bullet \theta$ of two homotopies, which uses θ for $0 < t < 1/2$ and then φ for $1/2 < t < 1$, is not associative. Hence, to get categoricity we must we must use as 2-cells the homotopy classes of homotopies! Again, the horizontal composition commutes with the vertical one.

These examples (and others) lead to the general notion of a 2-category to be a structure consisting of objects, arrows between the objects, and 2-cells between the arrows, where the 2-cells can be composed in two ways, "horizontal" and "vertical".

Start with an ordinary category C with objects a, $b, \ldots$ and "horizontal" arrows $f : a \to b$. A 2-category Π on C has, additionally, certain 2-cells $\alpha : f \Rightarrow g$ with domain f and codomain g, where f and g are parallel arrows in C, say from a to b, as displayed in

$$\alpha : f \Rightarrow g : a \to b \tag{1}$$

or pictured (vertically) as

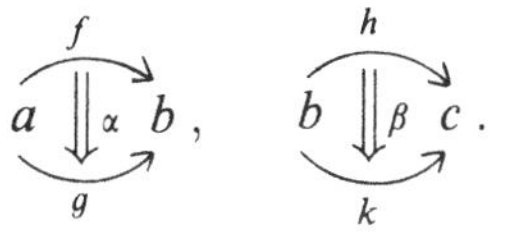

$$a \underset{g}{\overset{f}{\Downarrow \alpha}} b, \qquad b \underset{k}{\overset{h}{\Downarrow \beta}} c. \tag{2}$$

These 2-cells have two different compositions. First, if

$$\alpha' : f' \Rightarrow g' : b \to c \tag{3}$$

is a second 2-cell, there is a horizontal composite 2-cell $\alpha' \circ \alpha$

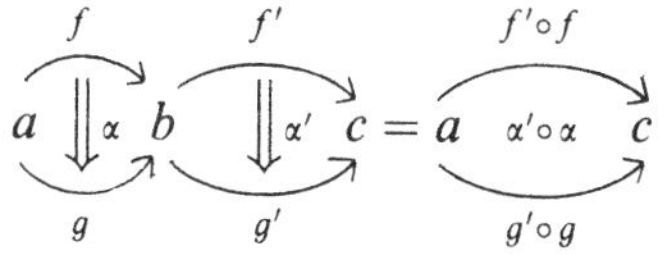

$$a \underset{g}{\overset{f}{\Downarrow \alpha}} b \underset{g'}{\overset{f'}{\Downarrow \alpha'}} c = a \underset{g' \circ g}{\overset{f' \circ f}{\alpha' \circ \alpha}} c \tag{4}$$

which matches (above and below) the given composition of arrows. We require that the 2-cells form a category under this horizontal composition. In particular, this means that there is for each object b an "identity" 2-cell $1 : 1 \Rightarrow 1 : b \to b$, acting as a 2-sided identity for this composition. Also, both "domain" $\alpha \mapsto f$ and "codomain" $\alpha \mapsto g$ are functors from the horizontal category of 2-cells to the horizontal category of arrows.

Moreover, for each pair of objects a, b, the 2-cells from a to b are the arrows of a category under a "vertical" composition, as in

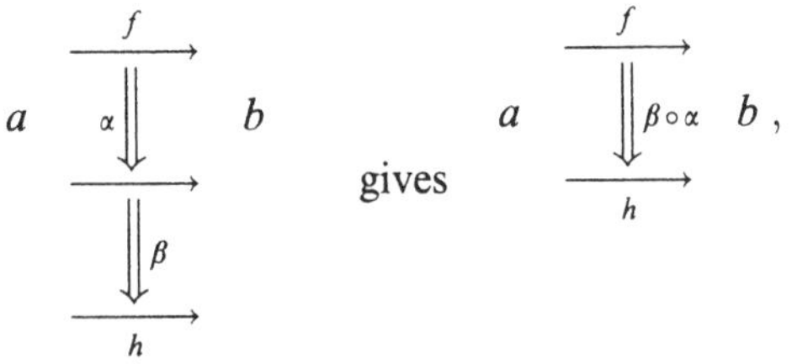

$$a \;\begin{array}{c} \xrightarrow{\;f\;} \\ \Downarrow \alpha \\ \longrightarrow \\ \Downarrow \beta \\ \xrightarrow[\;h\;]{} \end{array}\; b \qquad \text{gives} \qquad a \;\begin{array}{c} \xrightarrow{\;f\;} \\ \Downarrow \beta \circ \alpha \\ \xrightarrow[\;h\;]{} \end{array}\; b\,, \tag{5}$$

with a solid dot denoting this vertical composition. There are also vertical identity 2-cells $1_f : f \Rightarrow f$ for this composition.

Two additional axioms relate the horizontal to the vertical. First, we require that the horizontal composite of two vertical identities is itself a vertical identity, as in the diagram

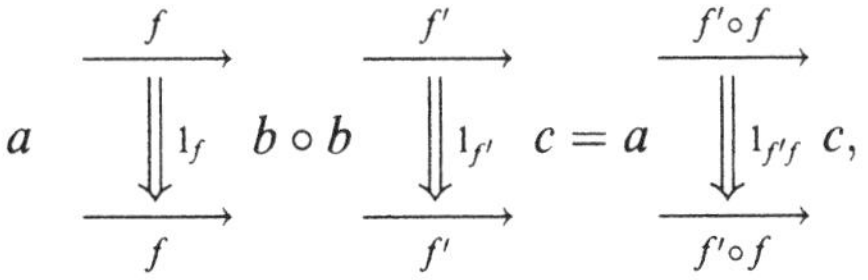

$$a \;\begin{array}{c} \xrightarrow{\;f\;} \\ \Downarrow 1_f \\ \xrightarrow[\;f\;]{} \end{array}\; b \circ b \;\begin{array}{c} \xrightarrow{\;f'\;} \\ \Downarrow 1_{f'} \\ \xrightarrow[\;f'\;]{} \end{array}\; c = a \;\begin{array}{c} \xrightarrow{\;f' \circ f\;} \\ \Downarrow 1_{f'f} \\ \xrightarrow[\;f' \circ f\;]{} \end{array}\; c, \qquad 1_{f'f} = 1_{f'} \circ 1_f\,. \tag{6}$$

Next, given the array of 2-cells

$$\begin{array}{ll} \alpha : f \Rightarrow g : a \to b\,, & \alpha' : f' \Rightarrow g' : b \to c\,, \\ \beta : g \Rightarrow h : a \to b\,, & \beta' : g' \Rightarrow h' : b \to c\,, \end{array} \tag{7}$$

the composites involved in the display

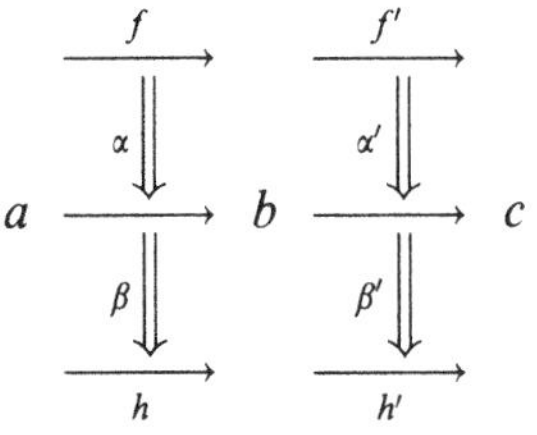

$$a \;\begin{array}{c} \xrightarrow{\;f\;} \\ \Downarrow \alpha \\ \longrightarrow \\ \Downarrow \beta \\ \xrightarrow[\;h\;]{} \end{array}\; b \;\begin{array}{c} \xrightarrow{\;f'\;} \\ \Downarrow \alpha' \\ \longrightarrow \\ \Downarrow \beta' \\ \xrightarrow[\;h'\;]{} \end{array}\; c \tag{8}$$

must satisfy the equation

$$(\beta' \circ \beta) \bullet (\alpha' \circ \alpha) = (\beta' \bullet \alpha') \circ (\beta \bullet \alpha) : f' \circ f \Rightarrow h' \circ h : a \to c\,. \tag{9}$$

Here (as elsewhere), the solid dot $\bullet$ is used for vertical composition and the small circle $\circ$ (or just juxtaposition) for horizontal composition. This axiom (9) is called the "middle four exchange", because it interchanges the middle two arguments in the sequence of the four 2-cells β', β, α', and α.

Note that this structure also provides a horizontal composite of a 2-cell with a 1-cell – just compose the 2-cell with the vertical identity of the 1-cell, on either side, as in $f' \circ \alpha$,

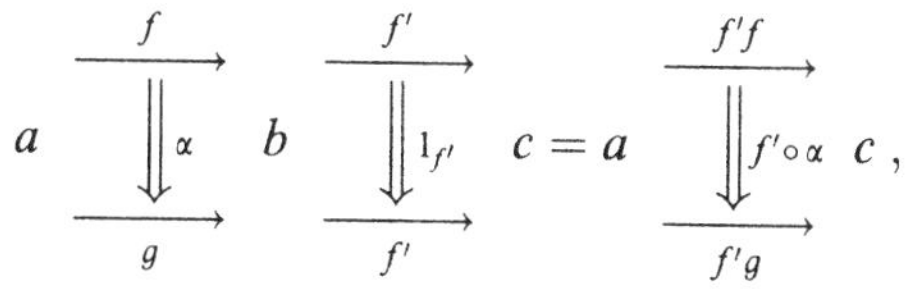

$$a \; \underset{g}{\overset{f}{\Downarrow \alpha}} \; b \; \underset{f'}{\overset{f'}{\Downarrow 1_{f'}}} \; c = a \; \underset{f'g}{\overset{f'f}{\Downarrow f' \circ \alpha}} \; c \,, \tag{10}$$

also written (as a "whisker" f' on the 2-cell α) as follows:

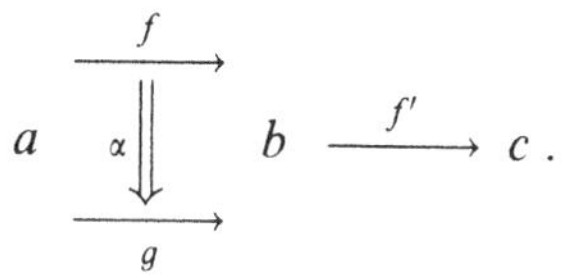

$$a \; \underset{g}{\overset{f}{\alpha \Downarrow}} \; b \xrightarrow{f'} c \,. \tag{11}$$

This definition of a 2-category does include the examples of 2-categories adduced above: the 2-category of topological spaces, continuous maps, and classes of homotopies, and the 2-category **CAT** of small categories, functors, and natural transformations.

It is convenient to write $T(a, b)$ for the vertical category on the objects a and b. Then the middle four interchange (9) and the rule (6) for the vertical identities together mean that horizontal composition is a bifunctor between vertical categories:

$$K_{a,b,c} : T(b, c) \times T(a, b) \to T(a, c) \,. \tag{12}$$

Also, the operation U_a which sends any object a to its identity arrow $1_a : a \to a$ is a functor from the terminal category 1 (with one object, one arrow)

$$U_a : 1 \to T(a, a) \,. \tag{13}$$

These two operations suffice to describe a 2-category in terms of its vertical hom-categories $T(a, b)$ – the description is parallel to the definition of an ordinary category by hom-sets (§I.8). Thus, a 2-category is given by the following data:

(i) A set of objects $a, b, c, \ldots$;
(ii) A function which assigns to each ordered pair of objects (a, b) a category $T(a, b)$;
(iii) For each ordered triple $\langle a, b, c \rangle$ of objects a functor (12), called composition;
(iv) For each object a, a functor U_a as in (13).

These elements of data are required to satisfy the associative law for the composition (iii) and the requirement that U_a provides a left and right identity for this composition.

This set of axioms for a 2-category is equivalent to the previous set. It is exactly like the definition of a category in terms of hom-sets, which have been here replaced by the hom-objects $T(a, b)$. These objects are not just sets but are categories (i.e., objects of the category **CAT**); one says that they are hom-sets "enriched" in **CAT**, the category of all (small) categories. The construction uses the fact that **CAT** has products and a terminal object. More generally, it is often helpful to use monoidal categories V in place of **CAT** and to examine categories "enriched" in V – that is, with hom-objects which are objects of V, with composition and identities as above. (See also §VII.7 and the remark in §I.8 about *Ab*-categories.) The monograph of Kelly [1982] is a systematic examination of such enriched categories; see also Dubuc [1970].

4. Operations in 2-Categories

Many of the properties of functors, as they have been developed in **CAT**, will carry over directly to other 2-categories. Adjunction is an example. Thus, in a 2-category, one says that two 1-cells running in opposite directions between the same two objects, as in the figure

$$a \underset{g}{\overset{f}{\rightleftarrows}} b\,,$$

are *adjoint*, with f a left adjoint to the right adjoint g, when there are 2-cells η and ε ("unit" and "counit")

$$\eta : 1 \Rightarrow gf : a \to a\,, \qquad \varepsilon : f\,g \Rightarrow 1_b : b \to b \tag{1}$$

such that both the following equations hold:

$$(\varepsilon f) \bullet (f\,\eta) = 1_f : f \Rightarrow f\,gf \Rightarrow f : a \to b\,, \tag{2}$$

$$(g\,\varepsilon) \bullet (\eta\,g) = 1_g : g \Rightarrow gf\,g \Rightarrow g : b \to a\,. \tag{3}$$

Indeed, in the 2-category **CAT**, these two equations state exactly the two triangular laws for the unit η and the counit ε of an (ordinary) adjunction between functors (§IV.1.(9)). In the first equation, εf really stands for the horizontal composite $\varepsilon 1_f$, so that Eq. (2) should strictly be pictured as follows:

$$\begin{array}{ccccccc} a & \multicolumn{3}{c}{\xrightarrow{\quad\quad 1 \quad\quad}} & a & \xrightarrow{f} & b \\ \| & & \Downarrow\eta & & \| & & \| \\ a & \xrightarrow{f} & b & \xrightarrow{g} & a & \xrightarrow{f} & b \\ \| & & \| & & \Downarrow\varepsilon & & \| \\ a & \xrightarrow[f]{} & b & \multicolumn{3}{c}{\xrightarrow[\quad\quad 1 \quad\quad]{}} & b \end{array} \;=\; \begin{array}{ccc} a & \xrightarrow{f} & b \\ \| & \Downarrow 1_f & \| \\ a & \xrightarrow[f]{} & b. \end{array} \tag{4}$$

Here, the left-hand side presents two horizontal compositions of 2-cells, followed by a vertical composition of the results. This may be suggestively pictured, omitting the 1_f lower left, as

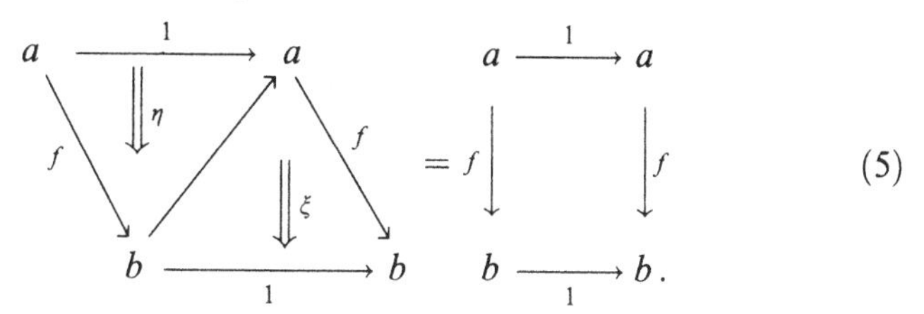

Similarly, the left-hand side of Eq. (3) involves the following vertical composites:

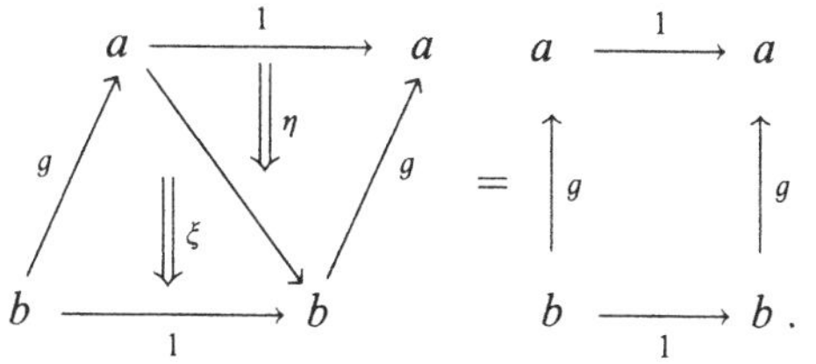

$$\begin{array}{ccccc} a & \xrightarrow{1} & a & & a \xrightarrow{1} a \\ {}^{g}\nearrow & \searrow \Downarrow \eta & \nearrow^{g} & = & {}_{g}\uparrow \qquad \uparrow_{g} \\ b & \xrightarrow[1]{\Downarrow \xi} & b & & b \xrightarrow[1]{} b\,. \end{array}$$

The diagram expresses the fact that the horizontal cmposite $g\varepsilon$ is "pasted" along f to the horizontal composite ηg to get the identity cell from g to g.

In much the same way, we can lift the notion of a (right) Kan extension (§X.3) to 2-categories. Thus, given objects m, c, a and arrows k, t in the following configuration in a 2-category:

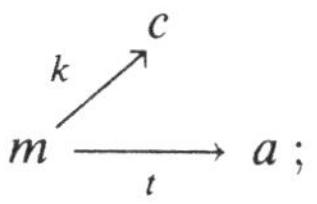

$$\begin{array}{ccc} & c & \\ {}^{k}\nearrow & & \\ m & \xrightarrow[t]{} & a\,; \end{array} \tag{6}$$

a right Kan extension of t along k is an arrow r and a 2-cell $\varepsilon : rk \Rightarrow t$

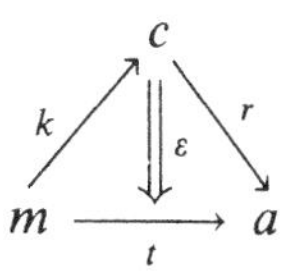

$$\begin{array}{ccc} & c & \\ {}^{k}\nearrow & \Downarrow \varepsilon & \searrow^{r} \\ m & \xrightarrow[t]{} & a \end{array} \tag{7}$$

such that any other such 2-cell α with $\alpha : sk \Rightarrow t$ for some $s : c \Rightarrow a$ is a composite of ε with a unique 2-cell σ, as in

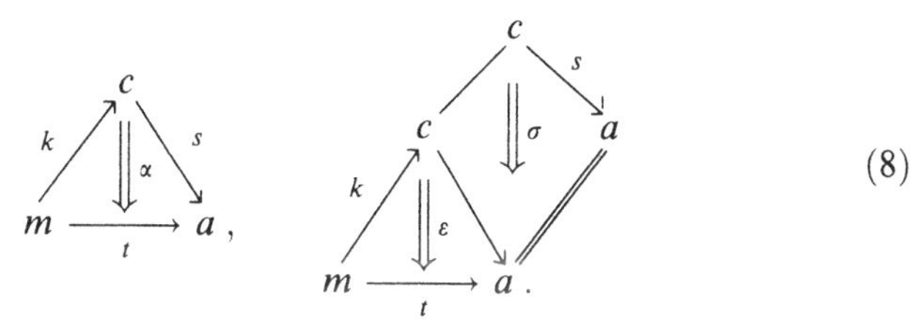

In other words, r and ε form the universal way of filling in the triangle (6), unique up to a 2-cell σ. This configuration is, of course, exactly that already used in §X.3 to describe right Kan extensions of actual functors.

As with all other algebraic objects, we must define the morphisms between 2-categories. They are called 2-functors. A 2-**functor**

$$F : T \to U$$

between two 2-categories T and U is a triple of functions sending objects, arrows, and 2-cells of T to items of the same types in U so as to preserve all the categorical structures (domains, codomains, identities, and composites). If G is a second such functor $G : T \to U$ between the same 2-categories, a 2-natural transformation $\theta : F \stackrel{\bullet}{\to} G$ is a function which sends each object a of T to an arrow $\theta a : Fa \to Ga$ of U in such a way that for each 2-cell $\alpha : f \Rightarrow g$ of T, the equality

$$Fa \underset{Fg}{\overset{Ff}{\Downarrow f\alpha}} Fb \xrightarrow{\theta b} Gb = Fa \xrightarrow{\theta a} Ga \underset{Gg}{\overset{Gf}{\Downarrow f\alpha}} Gb \qquad (9)$$

holds (between the indicated "whiskered" 2-cells). In particular, applied just to the edges (the 1-cells), this means that θ is necessarily an (ordinary) natural transformation between the associated ordinary functors F and G. The reader may check that 2-categories, 2-functors, and 2-natural transformations between them form the objects, arrows, and 2-cells of a 2-category! This category is often called 2-**Cat**.

But there is now a step up to the next dimension; given two 2-natural transformations $\theta : F \to G$ and $\varphi : F \to G$ between the same two 2-functors F and G, there are certain appropriate maps $\mu : \theta \Rightarrow \varphi$, called *modifications*, between transformations. Specifically, such a modification μ is required to assign to each object a of T a 2-cell $\mu_a : \theta_a \Rightarrow \varphi_a$ such that the following 2-composites are equal for every 2-cell $\alpha : f \Rightarrow g$:

$$Fa \underset{Fg}{\overset{Ff}{\Downarrow F\alpha}} Fb \underset{\varphi_b}{\overset{\theta_b}{\Downarrow \mu_b}} Gb = Fa \underset{\varphi_a}{\overset{\theta_a}{\Downarrow \mu_a}} Ga \underset{Gg}{\overset{Gf}{\Downarrow G\alpha}} Gb. \qquad (10)$$

A three-dimensional presentation of this requirement is as follows:

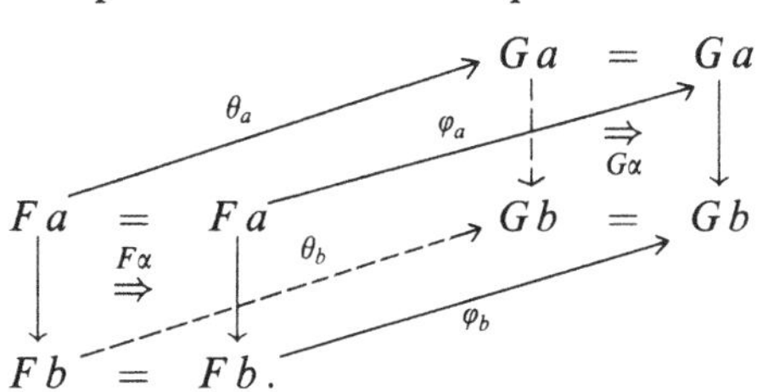

The front face of the cube is $F\alpha$, the back face $G\alpha$. The bottom is μb, the top μ_a, while the right and left side squares commute (because θ and φ are natural). Equation (10) states that the front followed by the bottom equals the top followed by the back.

This suggests that we regard the modification μ as a 3-cell $\mu : \theta \Rightarrow \varphi$. These 3-cells taken with θ's as 2-cells, 2-functors as arrows, and objects as objects together form the data for a 3-category. Just as a 2-category can be defined as a category with hom-sets enriched in **Cat**, so a 3-category can be formally defined to be a category with hom-sets enriched in 2-**Cat**, as we will see in the next section.

5. Single-Set Categories

A category is usually considered as a structure consisting of two sets, a set of objects and a set of arrows. But it is also possible to have a definition which uses only one set, that of arrows, with the objects regarded as special arrows – to wit, as the identity arrows. In § I.1, we have already described such an "arrows-only" definition of a category. Here is a different arrows-only formulation which will be used below to describe n-categories. A category is a set C of arrows with two functions $s, t : C \to C$, called "source" and target", and a partially defined binary operation $\#$, called composition, all subject to the following axioms, for all x, y, and z in C:

The operation $x \# y$ is defined iff $sx = ty$ and then

$$s(x \# y) = sy\,, \qquad t(x \# y) = tx\,; \tag{1}$$

$$x \# sx = x\,, \qquad tx \# x = x\,; \tag{2}$$

$$(x \# y) \# z = x \# (y \# z) \quad \text{if either side is defined}\,; \tag{3}$$

$$ssx = sx = tsx\,;$$

$$ttx = tx = stx\,. \tag{4}$$

Then x is an identity iff $x = sx$ or, equivalently, iff $x = tx$.

In this form, a functor $F : C \to D$ is simply a function from the set C to the set D such that

$$sF = Fs \quad \text{and} \quad tF = Ft : C \to D \tag{5}$$

and also

$$F(x \# y) = Fx \# Fy \tag{6}$$

whenever $x \# y$ is defined (and this, by (5), implies that $Fx \# Fy$ is defined).

This definition of a category or of a functor clearly is equivalent to the standard definition in terms of a set of objects and a set of arrows.

Similarly a 2-category can be considered to be a single set X considered as the set of 2-cells (e.g., of natural transformations). Then the previous 1-cells (the arrows) and the 0-cells (the objects) are just regarded as special "degenerate" 2-cells. On the set X of 2-cells there are then two category structures, the "horizontal" structure $(\#_0, s_0, t_0)$ and the "vertical" structure $(\#_1, s_1, t_1)$. Each satisfies the axioms above for a category structure and in addition

(i) Every identity for the 0-structure is an identity for the 1-structure;
(ii) The two category structures commute with each other.

Here, the condition (ii) means, of course, that

$$s_0\, s_1 = s_1\, s_0\,, \qquad s_0\, t_1 = t_1\, s_0\,, \qquad t_0\, s_1 = s_1\, t_0\,, \qquad t_0\, t_1 = t_1\, t_0 \tag{7}$$

and that, for $\alpha, \beta = 0$, 1 or 1, 0, and for all x, y, u, and v

$$(x \#_\alpha y) \#_\beta (u \#_\alpha v) \#_\alpha (y \#_\beta v)\,, \tag{8}$$

$$t_\alpha(x \#_\beta y) = (t_\alpha x) \#_\beta (t_\alpha y)\,,$$

$$s_\alpha(x \#_\beta y) = (s_\alpha x) \#_\beta (s_\alpha y)\,,$$

whenever both sides are defined.

Since $s_0 x$ and $t_0 x$ are identities for the 0-structure, they are also identities for the 1-structure by condition (i) above. Hence,

$$s_1\, s_0 = s_0\,, \qquad t_1\, s_0 = s_0\,, \qquad s_1\, t_0 = t_0\,, \qquad t_1\, t_0 = t_0\,. \tag{9}$$

With condition (7), this yields also

$$s_0\, s_1 = s_0\,, \qquad s_0\, t_1 = s_0\,, \qquad t_0\, s_1 = t_0\,, \qquad t_0\, t_1 = t_0\,. \tag{10}$$

Together, these rules, with (4), calculate any composite of an s or t with an s or t. The results agree with the intuitive picture of the "edges" of a 2-cell, as follows:

$$s_1 = s_1\, s_1 = t_1\, s_1\,,$$

$$s_0\, s_1 = s_0\, t_1 = s_0\, s_0 = t_1\, s_0 = s_0 \bullet \underset{t_1}{\overset{s_1}{\Downarrow}} \bullet\, t_0 = t_0\, s_1 = t_0\, t_1 = s_1\, t_0 = t_0\, t_0\,,$$

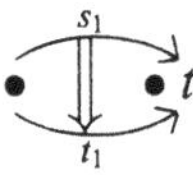

$$t_1 = t_1\, t_1 = s_1\, t_1\,.$$

With this preparation, we can now readily define a 3-category or more generally an n-category for any natural number n. The latter is a set X with n different category structures $(\#_i, s_i, t_i)$, for $i = 0, \ldots, n-1$, which commute with each other and are such that an identity for structure i is also an identity for structures j whenever $j > i$. Put differently, each pair $\#_i$ and $\#_j$ for $j > i$ constitute a 2-category. This readily leads to a definition of the useful notion of an ω-category: $i = 0, 1, 2, \ldots$.

6. Bicategories

Sometimes the composition of arrows in a would-be category is not associative, but only associative "up to" an isomorphism. This suggests the notion of a bicategory, which is a structure like a 2-category, but one in which the composition of arrows is associative only up to an isomorphism given by a suitable 2-cell.

Formally, a bicategory B consists of 0-cells $a, b, \ldots$, 1-cells $f, g, \ldots$, and 2-cells $\rho, \sigma, \ldots$, with sources and targets arranged as suggested in

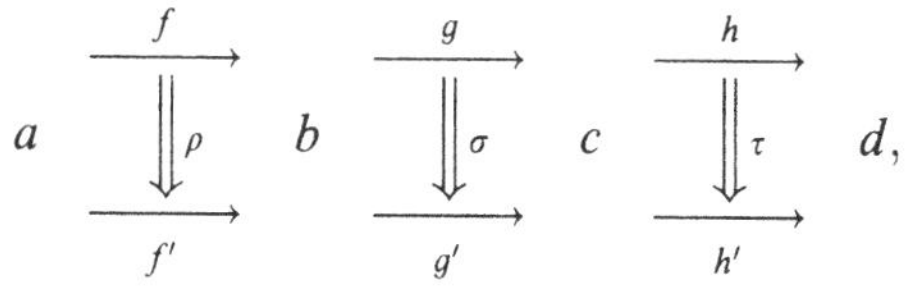

$$\begin{array}{ccccccc} & \xrightarrow{f} & & \xrightarrow{g} & & \xrightarrow{h} & \\ a & \Downarrow \rho & b & \Downarrow \sigma & c & \Downarrow \tau & d, \\ & \xrightarrow[f']{} & & \xrightarrow[g']{} & & \xrightarrow[h']{} & \end{array} \tag{1}$$

where the 1-cells extend horizontally and the 2-cells vertically. Specifically, each 1-cell f has 0-cells a and b as domain and codomain, as in $f : a \to b$, while each 2-cell ρ has coterminal (i.e., parallel) 1-cells f and f' as its domain and codomain. Moreover, to each pair of 0-cells (a, b), there is an (ordinary) category $B(a, b)$ in which the objects are all the 1-cells $f, f', \ldots$ from a to b, while the arrows are the 2-cells between such 1-cells. In this category $B(a, b)$ the "vertical" composition of 2-cells is (of course) associative and has for each object $f : a \to b$ a 2-cell $1_f : f \Rightarrow f$ which acts as an identity for this vertical composition. This composition is denoted by juxtaposition, or by $\circ$.

Next, for each ordered triple of 0-cells a, b, c, there is a bifunctor

$$* : B(b, c) \times B(a, b) \to B(a, c) , \tag{2}$$

called horizontal composition and written as $*$. Thus, given the diagram (1) above, there are composite 2-cells $\sigma * \rho$, $\tau * \sigma$ and composite 1-cells $g * f$ as follows:

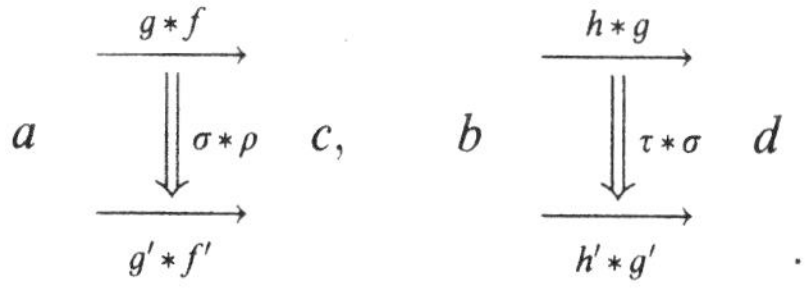

$$\begin{array}{ccc} & \xrightarrow{g*f} & \\ a & \Downarrow \sigma * \rho & c, \\ & \xrightarrow[g'*f']{} & \end{array} \qquad \begin{array}{ccc} & \xrightarrow{h*g} & \\ b & \Downarrow \tau * \sigma & d \\ & \xrightarrow[h'*g']{} & \end{array} \quad . \tag{3}$$

There is also for each 0-cell a an identity 1-cell $1_a : a \to a$ (which is not quite a real identity; see below).

This horizontal composition $*$, although written in the usual order of composition for a category, is not strictly associative, but is associative only "up to" a natural isomorphism α between iterated composite functors, as follows:

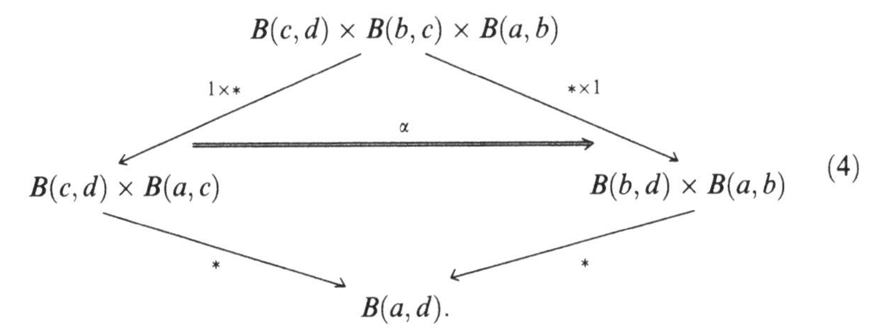

(4)

The requirement that α be a natural transformation of functors amounts to the following commutativity for the 2-cells displayed in (1):

$$\begin{array}{ccc} h*(g*f) & \xrightarrow{\alpha} & (h*g)*f \\ {\scriptstyle \tau*(\sigma*\rho)}\downarrow & & \downarrow{\scriptstyle (\tau*\sigma)*\rho} \\ h'*(g'*f') & \xrightarrow[\alpha]{} & (h'*g')*'f'. \end{array} \tag{5}$$

The purported "identity" arrows 1_a are required to act as identities for the horizontal composition only up to the following isomorphisms λ and ρ, natural in $f \in B(a, b)$:

$$\rho_{a,b} : f * 1_a \Rightarrow f\,, \qquad \lambda_{a,b} : 1_b * f \Rightarrow f\,. \tag{6}$$

These three natural transformations α, λ, and ρ are subject to two "coherence" axioms, as follows. For 1-cells f, g, and h as above and $k : d \to e$, the following pentagon, copied from that required for a monoidal category (§ VII.1.(5)), must commute:

$$\begin{array}{ccccc} k*(h*(g*f)) & \xrightarrow{\alpha} & (k*h)*(g*f) & \xrightarrow{\alpha} & ((k*h)*g)*f \\ {\scriptstyle k*\alpha}\downarrow & & & & \downarrow{\scriptstyle \alpha*f} \\ k*((h*g)*f & & \xrightarrow[\alpha]{\hspace{6em}} & & (k*(h*g))*f. \end{array} \tag{7}$$

For λ and ρ, just as for a monoidal category in § VII.1.(7), the following diagram (left and right identities) must commute:

$$\begin{array}{ccc} g*(1_b*f) & \xrightarrow{\alpha} & (g*1_b)*f) \\ \downarrow{\scriptstyle 1*\lambda} & & \downarrow{\scriptstyle \rho*1} \\ g*f & = & g*f\,. \end{array} \tag{8}$$

For ordinary categories C, there is the familiar transition to the dual or "opposite" category C^{op} (§ II.2), in which the direction of all the arrows is reversed. For a bicategory (or also for a 2-category), there are two such duals; one by reversing the arrows, the other by reversing the 2-cells!

There is also a suitable definition of a functor (i.e., a morphism) between bicategories. We urge the reader to make his own definition or to consult the literature (e.g., Bénabou [1967]).

7. Examples of Bicategories

First, any monoidal category M is also a bicategory B with one 0-cell. Just take the objects $a, b, \ldots$ of the monoidal M to be the 1-cells of B, with the tensor product as composition, and the arrows of M to be the 2-cells. Then the natural isomorphisms α, λ, and ρ given with M are exactly those required for B with the same identities, as expressed in the pentagon and square just above. Reciprocally, a bicategory B with just one 0-cell is just a monoidal category in this sense. For that matter, the coherence theorem for monoidal categories also applies to give a coherence result for the maps ρ, λ, and α in bicategories.

The next example is the category of rings; more exactly, the bicategory in which

0-cells are rings R, S, T,
1-cells are bimodules ${}_SA_R : R \to S$,
2-cells are bimodule homomorphisms $A \to A'$.

As for the composition of 1-cells, if ${}_TB_S$ is a left T, right S bimodule the usual tensor product $B \otimes_S A$ of modules, with elements sums of products $b \otimes a$ for $b \in B$ and $a \in A$ and with $bs \otimes a = b \otimes sa$ for $s \in S$, etc., is evidently a left T and right R bimodule. For a right T and left U-module ${}_UC_T$ the known associativity of the tensor product of modules

$$C \otimes_T (B \otimes_S A') \cong (C \otimes_T B) \otimes_S A' \tag{1}$$

yields the required associativity of this horizontal composition. The vertical composition is the usual composition of bimodule homomorphisms $A \to A' \to A''$.

Next, we describe the bicategory of "spans" – in a base category C which has all pullbacks. We choose to each pair of arrows s, t with a common target object b in C a pullback, to be called the "canonical" pullback of s and t:

$$\begin{array}{ccc} & \longrightarrow & \\ \downarrow & & \downarrow{\scriptstyle t} \\ \bullet & \xrightarrow[s]{} & b. \end{array} \tag{2}$$

A *span* from a to b in the base category C is now defined to be a pair of arrows from some common vertex v to a and to b, as in

$$a \leftarrow v \rightarrow b \,. \tag{3}$$

The bicategory Span(C) is now defined to have as

0-cells the objects of the given category C,
1-cells the spans of C, as above,
2-cells between two such spans from a to b those 1-cells $x : v \to v'$ of C as in

which make the two triangles commute. The vertical composite of two such 2-cells is given by the composite (in C) of the middle arrows x. The horizontal composite of spans from a to b and from b to c is obtained by taking the (already chosen) canonical pullback of the two middle arrows, as in

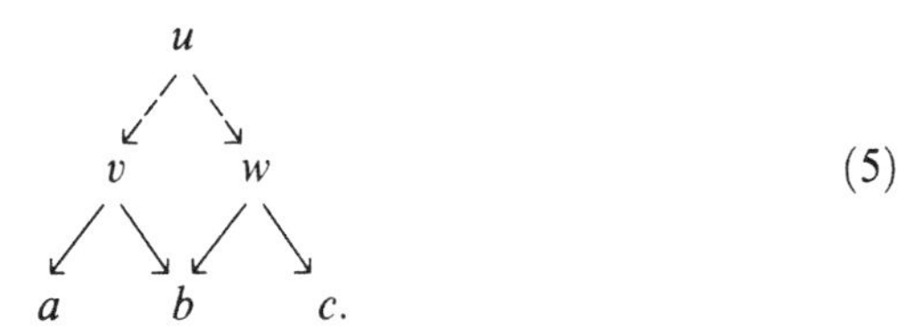

For associativity of this composition, consider three successive spans from a to b to c to d, as in

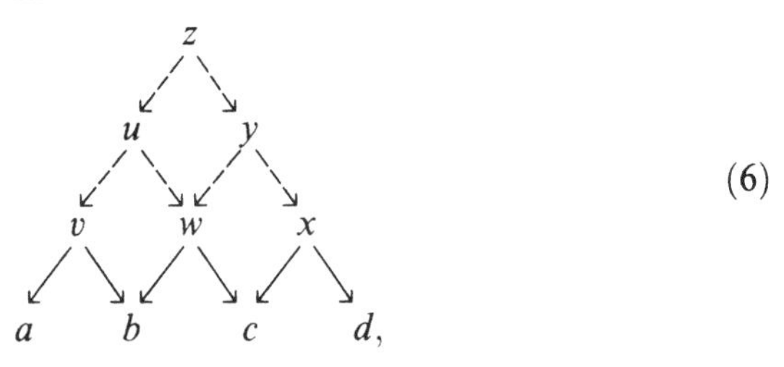

to which there are added above three (canonical) pullback squares, as displayed. Since the squares are pullbacks, the composite rectagles here are also pullbacks. Hence, both associations of the triple composite gives a pullback which might be that with vertex the top 0-cell z of the three given spans. Since the pullback of any such diagram is unique up to isomorphism, this gives the required natural isomorphism α from one triple composite to the other. Thus, it also shows why the pentagonal commutativity condition for α must hold.

The identity span on a 0-cell a is a compound of two identity arrows of C, from a to a. The pullback diagram with a second span from a to b,

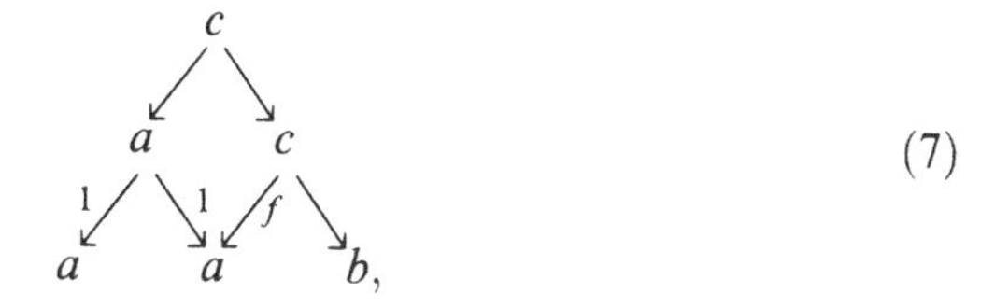

(7)

exhibits the required natural transformation λ on this composite. The remaining verifications of the axioms are left to the reader.

A bicategory is also called a "weak 2-category". This terminology does suggest that there could be a notion of a weak n-category. Such is the case, and it appears with various definitions in aspects of theoretical physics (see Baez et al. [1995, 1996]).

8. Crossed Modules and Categories in Grp

We return to an illustration of category objects. It turns out that a category object in **Grp** is essentially the same thing as one of the "crossed modules" which arise in homotopy theory. We present here a proof, as formulated for me by George Janelidze, of this striking categorical result.

A crossed module is defined to be a pair of groups H and P with an action of P on H and a homomorphism $\propto : H \to P$ respecting this action. If the action of an element p in P on an element h is denoted by h^p, then this means that for h, k in H and p, q in P, $h^1 = h$, and

$$(h^p)^q = h^{pq}, \qquad (hk)^p = h^p k^p, \qquad \propto(h^p) = p(\propto h)p^{-1}.$$

Such crossed modules arise in homotopy theory.

Now recall from § 1 that an internal category in **Grp** consists of two groups C_0 and C_1 and four group homorphisms

$$C_1 \underset{d_0}{\overset{d_1}{\rightrightarrows}} C_0\,, \qquad C_0 \xrightarrow{\;i\;} C_1\,, \qquad \gamma : C_1 \times_{C_0} C_1 \to C_1 \tag{1}$$

satisfying the usual conditions for an internal category. We claim first that in this system we can forget the categorical composition γ; the system is completely determined by the diagram of groups and homorphisms

$$C_1 \underset{d_0}{\overset{d_1}{\rightrightarrows}} C_0\,, \qquad C_0 \xrightarrow{\ i\ } C_1 \tag{2}$$

with

$$d_0 i = 1 = d_1 i \tag{3}$$

and the vanishing of the following commutator subgroup of C_1,

$$[\ker d_0\,, \ker d_1] = 1\,, \tag{4}$$

formed from the kernels of the group homomorphisms d_0 and d_1.

Indeed, to see that the group homomorphisms d_0, d_1, and i together "remember" the composition γ, we write this composition, that of a composable pair of morphisms

$$a \xrightarrow{\ f\ } b \xrightarrow{\ g\ } c\,,$$

as $g \circ f$. Since this is a morphism of the group product, here written as juxtapostion $g_1 g_2$, we have

$$(g_1 g_2) \circ (f_1 f_2) = (g_1 \circ f_1)(g_2 \circ f_2)$$

when both composities on the right are defined (this is again the middle four exchange). In particular, this result applied to the identity arrow 1_b of b gives

$$\begin{aligned} g \circ f &= (1_b 1_b^{-1} g) \circ (f 1_b^{-1} 1_b) \\ &= (1_b \circ f)(1_b^{-1} \circ 1_b^{-1})(g \circ 1_b) \\ &= f(1_b^{-1} \circ 1_b^{-1}) g\,. \end{aligned}$$

Setting $f = g = 1_b$ gives the product formula

$$1_b = 1_b (1_b^{-1} \circ 1_b^{-1}) 1_b$$

and hence

$$(1_b^{-1} \circ 1_b^{-1}) = 1_b^{-1}\,.$$

This also follows from the fact that $(\ )^{-1} : C \to C$ is *a* functor, where C is a group object.

Putting this in (5) gives

$$g \circ f = f 1_b^{-1} g\,.$$

In other words, the categorical composition γ is determined by the group structure of C. Alternatively, we also have

$$\begin{aligned} g \circ f &= (g 1_b^{-1} 1_b) \circ (1_b 1_b^{-1} f) \\ &= (g \circ 1_b)(1_b^{-1} \circ 1_b^{-1})(1_b \circ f) \\ &= g 1_b^{-1} f . \end{aligned}$$

Hence, the group product satisfies

$$f 1_b^{-1} g = g 1_b^{-1} f .$$

In particular, if the object b of C_0 is the unit in the group C_0 – and hence $1_b = i(b)$ is the unit element in C_1, then $fg = gf$. But since b is the unit element of the group C_0, then the category arrow g with domain b must be in $\ker d_0$, while f is similarly in $\ker d_1$. Hence, the result $fg = gf$ means that the commutator (4) above must be the identity.

Conversely, given a diagram (2) in **Grp** with (3) and (4), we can define a categorical composition $\gamma : C_1 \times_{C_0} C_1 \to C_1$ by setting

$$\gamma(g, f) = f 1_b^{-1} g \quad \text{or} \quad g 1_b^{-1} f .$$

A calculation then shows that this γ is the composition for an internal category in **Grp**.

Next, the data (2), (3), and (4) in the category **Grp** determine a crossed module $\alpha : H \to P$ as follows. Take

$$H = \ker d_0 \quad \text{and} \quad P = C_0 .$$

Then the restriction d_1' of d_1 to $\ker d_0$ is a homorphism

$$\alpha : H \to P .$$

Indeed,

$$1 \to \operatorname{Ker} d_0 \to C_1 \xrightarrow{d_0} C_0 \to 1 .$$

is a short exact sequence of groups which determine an action of C_0 on $H = \operatorname{Ker} d_0$ – in the standard way: conjugate each element of $\operatorname{Ker} d_0$ by a chosen representative, in C_1, of each element of C_0. Note that $i : C_0 \to C_1$ has $d_0 i = 1$, so that the short exact sequence above is "split" by i. Conversely, the reader may show that a crossed module $\alpha : H \to P$ determines such a split exact sequence of groups. These reciprocal construction show that the category of crossed modules is categorically equivalent to the category of diagrams (2) satisfying (3) and (4) in **Grp**.

Appendix. Foundations

We have described a category in terms of sets, as a set of objects and a set of arrows. However, categories can be described directly—and they can then be used as a possible foundation for all of mathematics, thus replacing the use in such a foundation of the usual Zermelo-Fraenkel axioms for set theory. Here is the direct description:

1. *Objects and arrows.* A category consists of objects $a, b, c, \ldots$ and arrows f, g, h. Sets form a category with sets as the objects and functions as the arrows.
2. *Domain.* Each arrow f has an object a as its "domain" and an object b as its "codomain"; we then write $f : a \to b$.
3. *Composition.* Given $f : a \to b$ and $g : b \to c$, their composite is an arrow $g \circ f : a \to c$.
4. *Associativity.* If also $h : c \to d$, then the triple composition is associative:

$$h \circ (g \circ f) = (h \circ g) \circ f : a \to d .$$

5. *Identities.* Each object b has an identity arrow $1_b : b \to b$. If also $f : a \to b$, then $1_b \circ f = f$. If also $g : b \to c$, then $g \circ 1_b = g$.

An elementary topos is a category with a certain additional structure: terminal object, pullbacks, truth, a subobject classifier, and power objects (sets of subsets). The axioms for this addtional structure are as follows:

6. *Terminal object.* There is a terminal object 1 such that every object a has exactly one arrow $a \to 1$.
7. *Pullbacks.* Every pair of arrows $f : a \to b \leftarrow c : g$ with a common codomain b has a pullback, defined as in § III.4:

$$\begin{array}{ccc} p & \longrightarrow & c \\ \downarrow & & \downarrow s \\ a & \xrightarrow[f]{} & b . \end{array} \tag{1}$$

   In particular, (take $b = 1$), any two objects a and c have a product $a \times c$.
8. *Truth.* There is an object Ω (the object of truth values) and a

monomorphism $t : 1 \to \Omega$ called truth; to any monomorphism $m : a \to b$, there is a unique arrow $\psi : b \to \Omega$ such that the following square is a pullback:

$$\begin{array}{ccc} a & \longrightarrow & 1 \\ {\scriptstyle m}\downarrow & & \downarrow{\scriptstyle t} \\ b & \xrightarrow[\psi]{} & \Omega . \end{array} \tag{2}$$

9. *Power objects.* To each object b, there is an associated object Pb and an arrow $\varepsilon_b : b \times Pb \to \Omega$ such that for every arrow $f : b \times a \to \Omega$ there is a unique arrow $g : a \to Pb$ for which the following diagram commutes:

$$\begin{array}{ccc} b \times a & \xrightarrow{f} & \Omega \\ {\scriptstyle 1\times g}\downarrow & & \| \\ b \times Pb & \xrightarrow[\varepsilon_b]{} & \Omega . \end{array} \tag{3}$$

To understand these axioms, we observe how they apply to the usual category of all sets. There, any set with just one element can serve as a terminal object 1, because each set a has a unique function $a \to 1$ to 1. For two sets a and b, the pullback of two arrows $a \to 1 \leftarrow b$ is then the usual set-theoretic product, with its projections to the given factors a and b.

For truth values, take the object Ω to be any set 2 consisting of two objects, 1 and 0, while the monomorphism $t : 1 \to 2$ is just the usual inclusion of 1 in 2. Then a monomorphism $m : a \to b$, as in Axiom 8, is a subset a of b. This subset has a well-known characteristic function $\psi : b \to 2$ with $\psi(y) = 1$ or 0 according as the element of y of b is or is not in the subset a. This produces the pullback (2) above.

Axiom 9 describes Pb, the set of all subsets s of b, often called the "power set" Pb. Indeed, one can then set $\varepsilon_b(x, s) = 0$ if the element x of b is in the subset s and equal to 1 otherwise. This does give a pullback, as in (3) above.

These axioms for a topos then hold for the category of sets. They have a number of strong consequences. For example, they give all finite categorical products and pullbacks, as well as all finite coproducts, including an initial object $\emptyset$, the empty set. For example, they provide a right adjoint to the product $a \times b$ as a functor of a; this is the exponential c^b with (see § IV.6)

$$hom(a \times b, c) \cong (a, c^b) .$$

A category of sets can now be described as an elementary topos, defined as above, with three additional properties:

(a) it is well-pointed,
(b) it has the axiom of choice (AC),
(c) it has a natural-numbers object (NNO).

In describing these properties, it is useful to think of the objects as sets and the arrows as functions.

Well-pointed requires that if two arrows $f, g : a \to b$ have $(f \neq g)$ then there must exist an arrow $p : 1 \to a$ for which $fp \neq gp$. The intention is that when the functions f and g differ they must differ at some "point" p, that is, at some element p of the set a.

The *axiom of choice* (AC) requires that every subjection $f : a \to b$ has a right inverse $\tau : b \to a$ for which $f \circ \tau = b$. This right inverse picks out to each point $p : 1 \to b$ of b a point of a, to *wit*, the composite τp which is mapped by f onto p.

The *natural-numbers object* (NNO) can be described as a set N with an initial object $O : 1 \to N$ and a successor function $s : N \to N$ in terms of which functions $f : N \to X$ on N can be defined by recursion, by specifying $f(0)$ and the composite fos. In other words, an NNO N in a category is a diagram

$$1 \xrightarrow{O} N \xrightarrow{s} N$$

consisting of a point O of N and a map s such that, given any arrows $1 \xrightarrow{h} b \xrightarrow{k} b$, there is a unique arrow $f : N \to b$ which makes the following diagram commute:

$$\begin{array}{ccccc} 1 & \xrightarrow{O} & N & \xrightarrow{s} & N \\ \| & & \downarrow{\scriptstyle f} & & \downarrow{\scriptstyle f} \\ 1 & \xrightarrow{h} & b & \xrightarrow{k} & b \end{array}$$

In the usual functional notation, this states that

$$fO = h \, , \qquad fs = kf \; ;$$

that is, f is defined by giving $f(O)$ and then $f(n+1)$ in terms of $f(n)$.

Thus, the category of sets may be described as a well-pointed topos with the AC in which there is an NNO. This set of axioms for set theory is weakly consistent with a version of the Zermelo axioms (the so-called "bounded" Zermelo; see Mac Lane and Moerdijk [1992]). They are originally due to Lawvere [1964], who called them the "elementary theory of the category of sets." Among the other examples of an elementary topos are the sheaves on a topological space; see Mac Lane and Moerdijk [1992].

In these axioms, it is often assumed that the subject classifier Ω has just two elements. This makes it a Boolean algebra.

Table of Standard Categories: Objects and Arrows

Ab	Abelian groups
Adj	Small categories, adjunctions, p. 104
Alg	$\langle\,, E\rangle$-algebras
Bool	Boolean algebras
CAb	Compact topological abelian groups
CAT	Categories, functors
CG Haus	Compactly generated Hausdorff spaces p. 185
Comp Bool	Complete Boolean algebras
Comp Haus	Compact Hausdorff spaces, p. 125
CRng	Commutatative rings, homomorphisms
$\mathbf{Ens}_V$	Sets and functions, within a universe V, p. 11
Euclid	Euclidean vector spaces, orthogonal transformations
Fin	Skeletal category of finite sets
Finord	Finite ordinals, all set functions, p. 12
Grp	Groups and homomorphisms
Grph	Directed graphs and morphisms
Haus	Hausdorff Spaces, continuous maps
L conn	Locally connected topological spaces
K-**Mod**	K-Modules and their morphisms
Mod-R	Right R-modules, R a ring
R-**Mod**	Left R-modules and morphisms
$\mathbf{Matr}_K$	Natural numbers, morphisms rectangular matrices, p. 11
Mon	Monoids and morphisms of monoids, p. 12
Moncat	Monoidal categories and strict morphisms, p. 160
Ord	Ordered sets, order-preserving maps, p. 123
Rng	Rings and homomorphisms
Ses-A	Short exact sequences of A-modules
Set	All small sets and functions
$\mathbf{Set}_*$	Sets with base point
Smgrp	Semigroups and morphisms
Top	Topological spaces, continuous maps, p. 122
Toph	Topological spaces, homotopy classes of maps, p. 12
Vct	Vector spaces, linear transformations
0	Empty category, p. 10
1	One-object category, p. 10
2	Two objects, p. 10
3	Three objects, p. 11
$\langle \Omega, E\rangle$	Universal algebras, type τ, p. 120, 152
$b \downarrow C$	Objects of C under b, p. 46
$C \downarrow a$	Objects of C over a, p. 46
$T \downarrow S$	Comma category, p. 46

Table of Terminology

This Book	*Elsewhere* (for abbreviations, see below)
arrow	map (E & M), morphism (Gr)
domain	source (Ehr)
codomain	target (Ehr)
graph	precategory, diagram scheme (Mit)
natural transformation	morphism of functors (Gr), functorial map (G-Z)
natural isomorphism	natural equivalence (E & M; now obsolete)
monic	monomorphism
epi	epimorphism, epic
idempotent	projector (Gr)
opposite	dual
coproduct	sum
equalizer	kernel, difference kernel
pullback	fibered product (Gr), cartesian square
pushout	cocartesian square, comeet
universal arrow	left liberty map (G-Z)
limit exists	limit is representable (Gr)
limit	projective limit, inverse limit
colimit	inductive limit, direct limit
cone to a functor	projective cone, inverse cone (G-Z)
cone from a functor	inductive cone, co-cone
left adjoint	coadjoint (Mit), adjoint
right adjoint	adjoint (Mit), coadjoint
unit of adjunction	adjunction morphism (G-Z)
triangular identities	ε quasi-inverse to η (G-Z)
monad	triple
biproduct	direct sum (in *Ab*-categories)
Ab-category	preadditive category (old)

Gr = Grothendieck
Ehr = Ehresmann
Mit = Mitchell
E & M = Eilenberg & Mac Lane
G-Z = Gabriel-Zisman

Bibliography

Adamek, J., Herrlich, H., Strecker, G.E. [1990]: Abstract and concrete categories. New York: John Wiley & Sons 1990.

André, Michel [1967]: Méthode simpliciale en algèbre homologique et algèbre commutative IV. Lecture Notes in Mathematics, Vol. 32. Berlin-Heidelberg-New York: Springer 1967.

— [1970]: Homology of simplicial objects. Proceeding of the AMS symposium in Pure Mathematics on Applications of Categorical Algebra, pp. 15–36. Providence: American Mathematical Society 1970.

Baez, J., Dolan, J. [1995]: Higher dimensional algebra and topological quantum field theory. J. Math. Phys. **36**, 60–105 (1995).

Baez, J., Neuchli, M. [1996]: Higher dimensional algebra, braided monoidal 2-algebras. Adv. Math. **121**, 196–244 (1996).

Barr, M., Beck, J. [1966]: Acyclic models and triples. Proceedings of the Conference on Categorical Algebra, La Jolla, 1965, pp. 336–344. New York: Springer 1966.

— [1969]: Homology and standard constructions. Seminar on Triples and Categorical Homology Theory, pp. 245–336. Vol. 80. Lecture Notes in Mathematics. Berlin-Heidelberg-New York: Springer 1969.

Barr, M., Wells, C. [1985]: Toposes, triples, and theories. Vol. 278. Grundlehren der Math. Berlin: Springer 1985.

— [1990]: Category theory for computing science. Englewood Cliffs, NJ: Prentice-Hall 1990.

Bass, Hyman [1968]: Algebraic K-theory. Mathematics Lecture Note Series. New York-Amsterdam: W.A. Benjamin, Inc. 1968.

Bénabou, Jean [1963]: Catégories avec multiplication. C.R. Acad. Sci. Paris **256**, 1887–1890 (1963).

— [1965]: Critères de représentabilité des foncteurs. C.R. Acad. Sci. Paris **260**, 752–755 (1965).

— [1967]: Introduction to bicategories, pp. 1–77. Vol. 40. Springer Lecture Notes in Mathematics. Berlin: Springer 1967.

Birkhoff, Garrett [1967]: Lattice theory. AMS Colloquium Publications, Vol. 25, 3rd Ed. Providence: American Mathematical Society 1967.

Borceux, F. [1904]: Handbook of categorical algebra; vol. 1, Basic category theory, Vol. 2 Categories and structures, Vol. 3 Categories of sheaves. Encyclopedia of mathematics and its applications, Vols. 50, 51, and 52. Cambridge: Cambridge University Press 1904.

Bourbaki, N. [1948]: Éléments de mathématique, Vol. VII., Algèbre, Livre II,

Algèbre multilineaire, ch. 3. Actualités scientifiques et industrielles, 1044. Paris: Hermann 1948.

— [1957]. Éléments de mathématique, Vol. XXII, Thèorie des ensembles, Livre I, Structures, ch. 4. Actualités scientifiques et industrielles, 1258. Paris: Hermann 1957.

Brown, Ronald [1964]. Function spaces and product topologies. Quart. J. Math. **15**, 238–250 (1964).

Buchsbaum, David A. [1955]: Exact categories and duality. Trans. Am. Math. Soc. **80**, 1–34 (1955).

Bucur, I., Deleanu, A. [1968]: Introduction to the theory of categories and functors. London-New York-Sydney: John Wiley and Sons, Ltd. 1968.

Cartan, H., Eilenberg, S. [1956]: Homological algebra. Princeton: Princeton University Press 1956.

Cassels, J.W.S., Fröhlich, A., Edrs. [1967]: Algebraic number theory. London: Academic Press 1967.

Cohen, Paul J. [1966]: Set theory and the continuum hypothesis. New York-Amsterdam: W.A. Benjamin 1966.

Cohn, P.M. [1965]: Universal algebra. New York-Evanston-London: Harper and Row 1965.

Day, B.J., Kelly, G.M. [1969]: Enriched functor categories. Reports of the Midwest Category Seminar, Vol. III, pp. 178–191, Vol. 106, Lecture Notes in Mathematics. Berlin-Heidelberg-New York: Springer 1969.

Doplicher, S., Roberts, J.E. [1989]: A new duality for compact groups. Invent. Math. **98**, 157–221 (1989).

Dubuc, E.J. [1970]: Kan extensions in enriched category theory. Lecture Notes in Mathematics, Vol. 145. Berlin-Heidelberg-New York: Springer 1970.

Dubuc, E.J., Porta, H. [1971]: Convenient categories of topological algebras, and their duality theory. J. Pure Appl. Algebra. **1**, 281–316 (1971).

Dubuc, E.J., Street, R. [1970]: Dinatural transformations. Reports of the Midwest Category Seminar, Vol. IV, pp. 126–138, Vol. 137, Lecture Notes in Mathematics. Berlin-Heidelberg-New York: Springer 1970.

Ehresmann, Ch. [1965]: Catégories et structures. Paris: Dunod 1965.

Eilenberg, S., Elgot, S.E. [1970]: Recursiveness. New York-London: Academic Press 1970.

Eilenberg, S., Kelly, G.M. [1966a]: Closed categories. Proceedings of the Conference on Categorical Algebra, La Jolla, 1965, 421–562. New York: Springer 1966.

— [1966b]: A generalization of the functorial calculus. J. Algebra **3**, 366–375 (1966).

Eilenberg, S., Mac Lane, S. [1942a]: Group extensions and homology. Ann. Math. **43**, 757–831 (1942).

— [1942b]: Natural isomorphisms in group theory. Proc. Nat. Acad. Sci. U.S.A. **28**, 537–543 (1942).

— [1943]: Relations between homology and homotopy groups. Proc. Nat. Acad. Sci. U.S.A. **29**, 155–158 (1943).

— [1945]: General theory of natural equivalences. Trans. Am. Math. Soc. **58**, 231–294 (1945).

— [1947]: Cohomology theory in abstract groups I. Ann. Math. **48**, 51–78 (1947).

Eilenberg, S., Moore, J.C. [1965]: Adjoint functors and triples. Illinois J. Math. **9**, 381–398 (1965).

Eilenberg, S., Steenrod, N.E. [1952]: Foundations of algebraic topology. Princeton: Princeton University Press 1952.

Epstein, D.B.A. [1966]: Functors between tensored categories. Inventiones Math. **1**, 221–228 (1966).

Fox, R.H. [1943]: Natural systems of homomorphisms. Preliminary Report. Bull. Am. Math. Soc. **49**, 373 (1943).

Freyd, P. [1964]: Abelian categories: An introduction to the theory of functors. New York: Harper and Row 1964.

— [1972]: Aspects of topoi. Bull. Austral. Math. Soc. **7**, 1–72, 467–481 (1972).

Freyd, P.J., Scedrov, A. [1993]: Categories, allegories. Amsterdam: North-Holland 1993.

Freyd, P.J., Yetter, D. [1990]: Braided compact monoidal categories with applications to low dimensional topology. Adv. Math. **77**, 1950–1982 (1990).

Gabriel, P. [1962]: Des catégories abéliennes. Bull. Soc. Math. France **90**, 323–448 (1962).

Gabriel, P., Ulmer, F. [1971]: Lokal Praesentierbare Kategorien. Vol. 221. Springer Lecture Notes in Mathematics. Berlin: Springer 1971.

Gabriel, P., Zisman, M. [1967]: Calculus of fractions and homotopy theory. Ergebnisse der Mathematik, Vol. 35. Berlin-Heidelberg-New York: Springer 1967.

Godement, R. [1958]: Théorie des faisceaux. Paris: Hermann 1958.

Gödel, K. [1940]: The consistency of the continuum hypothesis. Studies, Ann. of Math., No. 3. Princeton: Princeton University 1940.

Goguen, J.A. [1971]: Realization is universal. *Math. Syst. Theory **b***, 359–374 (1971).

Grätzer, G. [1968]: Universal algebra. Princeton: Van Nostrand and Co. Inc. 1968.

Grothendieck, A. [1957]: Sur quelques points d'algèbre homologique. Tôhoku Math. J. **9**, 119–221 (1957).

Hatcher, W.S. [1968]: Foundations of mathematics. Philadelphia-London-Toronto: W.B. Saunders Co. 1968.

Herrlich, H. [1968]: Topologische Reflexionen und Coreflexionen. Lecture Notes in Mathematics, Vol. 78. Berlin-Heidelberg-New York: Springer 1968.

Huber, P.J. [1961]: Homotopy theory in general categories. Math. Ann. **144**, 361–385 (1961).

— [1962]: Standard Constructions in abelian categories. Math. Ann. **146**, 321–325 (1962).

Hurewicz, W. [1941]: On duality theorems. Bull. Am. Math. Soc. **47**, 562–563 (1941).

Isbell, J.R. [1960]: Adequate subcategories. Illinois J. Math. **4**, 541–552 (1960).

— [1964]: Subobjects, adequacy completeness and categories of algebras, Rozprawy Mat. **36**, 3–33 (1964).

— [1968]: Small subcategories and completeness. Math. Syst. Theory **2**, 27–50 (1968).

Johnstone, P.T. [1977]: Topos theory. LMS Math. Monographs. New York: Academic Press 1977.

— [1982]: Stone spaces. Vol. 3. Cambridge Studies in Advanced Mathematics. Cambridge: Cambridge University Press 1982.

Johnstone, P.T., Paré, P. [1978]: Indexed categories and their applications, Vol. 661. Lecture Notes in Mathematics. Berlin: Springer 1978.

Joyal, A., Street, R. [1991]: Tortile Yang-Baxter operators in tensor categories. Adv. Math. **71**, 43–51 (1991).

— [1991]: The geometry of tensor calculus I. Adv. Math. **88**, 55–112 (1991).

— [1986]: Braided monoidal categories, revised, Macquarie Math. Reports no. 86081, Macquarie University, Australia.

— [1993]: Braided tensor categories. Adv. Math. **102**, 20–78 (1993).

Joyal, A., Tierney, M. [1984]: An extension of the Galois theory of Grothendieck, Mem. Am. Math. Soc. **390** (1984).

Kan, D.M. [1958]. Adjoint functors. Trans. Am. Math. Soc. **87**, 294–329 (1958).

Kapranov, M., Voevodsky, V. [1994]: Braided monoidal 2-categories and Manin-Schechtmann higher braid groups, Journal of Pure and Applied Algebra **92**, 241–267 (1994).

— [1994]: Braided monoidal 2-categories, 2-vector spaces and Zamolodchikov tetrahedral equations. Symposia in Pure Mathematics, Vol. 56, part 2 pp. 177–260. Providence, RI: American Mathematical Society 1994.

Kauffman, L. [1993]: Knots and physics, 2nd edition. Riveredge, NJ: World Scientific 1993.

Kelly, G.M. [1964]: On Mac Lane's condition for coherence of natural associativities. J. Algebra **1**, 397–402 (1964).

— [1982]: Basic concepts of enriched category theory. LMS Lecture Notes. Cambridge: Cambridge University Press 1982.

Kelly, G.M., Street, R. [1974]: Review of the elements of 2-categories, pp. 75–103. Vol. 420. Springer Lecture Notes in Mathematics. Berlin: Springer 1974.

Kleisli, H. [1962]: Homotopy theory in abelian categories. Canad. J. Math. **14**, 139–169 (1962).

— [1965]: Every standard construction is induced by a pair of adjoint functors. Proc. Am. Math. Soc. **16**, 544–546 (1965).

Lambek, J. [1968]: Deductive systems and categories I. Math. Syst. Theory **2**, 287–318 (1968).

Lambeck, J., Scott, P.J. [1986]: Introduction to higher order categorical logic. Cambridge: Cambridge University Press 1986.

Lamotke, K. [1968]: Semisimpliziale algebraische Topologie. Berlin-Heidelberg-New York: Springer 1968.

Lawvere, F.W. [1963]: Functorial semantics of algebraic theories. Proc. Nat. Acad. Sci. U.S.A. **50**, 869–873 (1963).

— [1964]: An elementary theory of the category of sets. Proc. Nat. Acad. Sci. U.S.A. **52**, 1506–1511 (1964).

— [1966]: The category of categories as a foundation for mathematics. Proceedings of the Conference on Categorical Algebra, La Jolla, 1965, pp. 1–21. New York: Springer 1966.

— [1970]: Quantifiers and sheaves, pp. 329–334. Nice: Actes du congress International des Mathematiciens 1970.

Linton, F.E.J. [1966]: Some aspects of equational categories. Proceedings of the Conference on Categorical Algebra, La Jolla, 1965, pp. 84–95. New York: Springer 1966.

Mac Lane, S. [1948]: Groups, categories, and duality, Proc. Nat. Acad. Sci. U.S.A. **34**, 263–267 (1948).

— [1950]: Duality for groups. Bull. Am. Math. Soc. **56**, 485–516 (1950).

— [1956]: Homologie des anneaux et des modules. Colloque de topologie algébrique, Louvain 1956, pp. 55–80.

— [1963a]: Homology. Berlin-Göttingen-Heidelberg: Springer 1963.

— [1963b]: Natural associativity and commutativity. Rice Univ. Studies **49**, 28–46 (1963).

— [1965]: Categorical algebra. Bull. Am. Math. Soc. **71**, 40–106 (1965).

— [1969]: One universe as a foundation for category theory. Reports of the Midwest Category Seminar, Vol. III, pp. 192–201. Lecture Notes in Mathematics, Vol. 106. Berlin-Heidelberg-New York: Springer 1969.

— [1970]: The Milgram bar construction as a tensor product of functors: 135–152 in the Steenrod Algebra and its Applications, Lecture Notes in Mathematics, Vol. 168. Berlin-Heidelberg-New York: Springer 1970.

— [1991]: Coherence theorems and conformal field theory. pp. 321–328. Category theory. 1991. Proceedings of an International Summer Category Theory Meeting. CMS Conference Proceedings Vol. 13. Providence, RI: American Mathematical Society 1991.

— [1995]: Categories in geometry, Algebra Logic, pp. 169–178 in Mathematica Japonica Vol. 42.

— [1996]: The development and prospects for category theory. Appl. Categor. Struct. **4**, 129–136 (1996).

Mac Lane, S., Moerdijk, I. [1992]: Sheaves in geometry and logic. A first introduction to topos theory. New York: Springer-Verlag 1992.

Mac Lane, S., Birkhoff, G. [1967]: Algebra. New York: Macmillan 1967. 3rd edition. New York: Chelsea 1988.

Makkai, M., Paré, R. [1989]: Accessible categories: The foundations of categorical model theory, Contemporary Math. 104. Providence, RI: American Mathematical Society 1989.

Makkai, M., Reyes, G.E. [1977]: First order categorical logic. Vol. 611. Lecture Notes in Mathematics. Berlin, Springer.

Manes, E. [1969]. A triple-theoretic construction of compact algebras. Seminar on Triples and Categorical Homology Theory, pp. 91–119. Lecture Notes in Mathematics, Vol. 80. Berlin-Heidelberg-New York: Springer 1969.

May, J. P. [1967]: Simplicial objects in algebraic topology. Princeton: Van Nostrand Co., Inc. 1967.

McLarty, C. [1992]: Elementary categories, elementary toposes. Oxford: Clarendon 1992.

Mitchell, B. [1965]: Theory of categories. New York-London: Academic Press 1965.

Negrepontis, J.W. [1971]: Duality in analysis from the point of view of triples. J. Algebra **19**, 228–253 (1971).

— [1971a]: On absolute colimits. J. Algebra **19**, 80–95 (1971).

Paré, R [1971]: On absolute colimits. J. Algebra **19**, 80–95.

Pareigis, B. [1970]: Categories and functors. New York: Academic Press 1970.
Quillen, D.G. [1967]: Homotopical algebra. Lecture Notes in Mathematics, Vol. 43. Berlin-Heidelberg-New York: Springer 1967.
Samuel, P. [1948]: On universal mappings and free topological groups. Bull. Am. Math. Soc. **54**, 591–598 (1948).
Schubert, H. [1970a and b]: Kategorien, Vols. I and II. Berlin-Heidelberg-New York: Springer 1970.
Schnider, S., Sternberg, S. [1993]: Quantum groups: from coalgebras to Drinfeld algebras, Boston: International Press, Inc. 1993.
Schubert, H. [1970a and b]: Kategorien, Vols. I and II. Berlin-Heidelberg-New York: Springer 1970.
Shum, M.C. [1994]: Tortile tensor categories. J. Pure Appl. Algebra **43**, 57–110 (1994).
Solovay, R.M. [1966]: New proof of a theorem of Gaifman and Hales. Bull. Am. Math. Soc. **72**, 282–284 (1966).
Stasheff, J.D. [1963]: Homotopy associativity of H-spaces, I. Trans. Am. Math. Soc. **108**, 275–292 (1963).
Steenrod, N.E. [1940]: Regular cycles of compact metric spaces. Ann. Math. **41**, 833–851 (1940).
— [1967]: A convenient category of topological spaces. Michigan Math. J. **14**, 133–152 (1967).
Swan, R.G. [1968]: Algebraic K-theory. Lecture Notes in Mathematics, Vol. 76. Berlin-Heidelberg-New York: Springer 1968.
Tierney, M. [1972]: Sheaf theory and the continuum hypothesis, pp. 13–42. Vol. 274. Lecture Notes in Mathematics. Berlin: Springer 1972.
Ulmer, F. [1967a]: Properties of dense and relative adjoint functors. J. Algebra **8**, 77–95 (1967).
— [1967b]: Representable functors with values in arbitrary categories. J. Algebra **8**, 96–129 (1967).
Watts, C.E. [1960]: Intrinsic characterizations of some additive functors. Proc. Am. Math. Soc. **11**, 5–8 (1960).
Yoneda, N. [1954]: On the homology theory of modules. J. Fac. Sci. Tokyo, Sec. I. **7**, 193–227 (1954).
— [1960]. On ext and exact sequences. J. Fac. Sci. Tokyo, Sec. I, **8**, 507–526 (1960).

Index

Graduate Texts in Mathematics

(continued from page ii)

129 FULTON/HARRIS. Representation Theory: A First Course.
Readings in Mathematics
130 DODSON/POSTON. Tensor Geometry.
131 LAM. A First Course in Noncommutative Rings.
132 BEARDON. Iteration of Rational Functions.
133 HARRIS. Algebraic Geometry: A First Course.
134 ROMAN. Coding and Information Theory.
135 ROMAN. Advanced Linear Algebra.
136 ADKINS/WEINTRAUB. Algebra: An Approach via Module Theory.
137 AXLER/BOURDON/RAMEY. Harmonic Function Theory. 2nd ed.
138 COHEN. A Course in Computational Algebraic Number Theory.
139 BREDON. Topology and Geometry.
140 AUBIN. Optima and Equilibria. An Introduction to Nonlinear Analysis.
141 BECKER/WEISPFENNING/KREDEL. Gröbner Bases. A Computational Approach to Commutative Algebra.
142 LANG. Real and Functional Analysis. 3rd ed.
143 DOOB. Measure Theory.
144 DENNIS/FARB. Noncommutative Algebra.
145 VICK. Homology Theory. An Introduction to Algebraic Topology. 2nd ed.
146 BRIDGES. Computability: A Mathematical Sketchbook.
147 ROSENBERG. Algebraic K-Theory and Its Applications.
148 ROTMAN. An Introduction to the Theory of Groups. 4th ed.
149 RATCLIFFE. Foundations of Hyperbolic Manifolds.
150 EISENBUD. Commutative Algebra with a View Toward Algebraic Geometry.
151 SILVERMAN. Advanced Topics in the Arithmetic of Elliptic Curves.
152 ZIEGLER. Lectures on Polytopes.
153 FULTON. Algebraic Topology: A First Course.
154 BROWN/PEARCY. An Introduction to Analysis.
155 KASSEL. Quantum Groups.
156 KECHRIS. Classical Descriptive Set Theory.
157 MALLIAVIN. Integration and Probability.
158 ROMAN. Field Theory.
159 CONWAY. Functions of One Complex Variable II.
160 LANG. Differential and Riemannian Manifolds.
161 BORWEIN/ERDÉLYI. Polynomials and Polynomial Inequalities.
162 ALPERIN/BELL. Groups and Representations.
163 DIXON/MORTIMER. Permutation Groups.
164 NATHANSON. Additive Number Theory: The Classical Bases.
165 NATHANSON. Additive Number Theory: Inverse Problems and the Geometry of Sumsets.
166 SHARPE. Differential Geometry: Cartan's Generalization of Klein's Erlangen Program.
167 MORANDI. Field and Galois Theory.
168 EWALD. Combinatorial Convexity and Algebraic Geometry.
169 BHATIA. Matrix Analysis.
170 BREDON. Sheaf Theory. 2nd ed.
171 PETERSEN. Riemannian Geometry.
172 REMMERT. Classical Topics in Complex Function Theory.
173 DIESTEL. Graph Theory. 2nd ed.
174 BRIDGES. Foundations of Real and Abstract Analysis.
175 LICKORISH. An Introduction to Knot Theory.
176 LEE. Riemannian Manifolds.
177 NEWMAN. Analytic Number Theory.
178 CLARKE/LEDYAEV/STERN/WOLENSKI. Nonsmooth Analysis and Control Theory.
179 DOUGLAS. Banach Algebra Techniques in Operator Theory. 2nd ed.
180 SRIVASTAVA. A Course on Borel Sets.
181 KRESS. Numerical Analysis.
182 WALTER. Ordinary Differential Equations.
183 MEGGINSON. An Introduction to Banach Space Theory.
184 BOLLOBAS. Modern Graph Theory.
185 COX/LITTLE/O'SHEA. Using Algebraic Geometry.
186 RAMAKRISHNAN/VALENZA. Fourier Analysis on Number Fields.
187 HARRIS/MORRISON. Moduli of Curves.
188 GOLDBLATT. Lectures on the Hyperreals: An Introduction to Nonstandard Analysis.
189 LAM. Lectures on Modules and Rings.
190 ESMONDE/MURTY. Problems in Algebraic Number Theory.
191 LANG. Fundamentals of Differential Geometry.
192 HIRSCH/LACOMBE. Elements of Functional Analysis.
193 COHEN. Advanced Topics in Computational Number Theory.
194 ENGEL/NAGEL. One-Parameter Semigroups for Linear Evolution Equations.

Printed in Great Britain by
Amazon.co.uk, Ltd.,
Marston Gate.